Pittsburgh-Konstanz Series in the Philosophy and History of Science

Science at Century's End

Philosophical Questions on the Progress and Limits of Science

EDITED BY Martin Carrier,
Gerald J. Massey, and
Laura Ruetsche

University of Pittsburgh Press / Universitätsverlag Konstanz

Published by the University of Pittsburgh Press,
Pittsburgh, Pa. 15261
All material copyright © 2000, University of
Pittsburgh Press, except "Extending Ourselves,"
copyright © 2000, Paul Humphreys.
All rights reserved
Manufactured in the United States of America
Printed on acid-free paper
10 9 8 7 6 5 4 3 2 1

Library of Congress Cataloging-in-Publication Data
Science at century's end : philosophical questions on
    the progress and limits of science / edited by Martin
    Carrier, Gerald J. Massey, and Laura Ruetsche.
         p.   cm.—(Pittsburgh-Konstanz series in the
    philosophy and history of science)
    "Fourth Pittsburgh-Konstanz Colloquium held in
    Pittsburgh, October 3–7, 1997"—Pref.
       Includes bibliographical references and index.
       ISBN 0-8229-4121-X (alk. paper)
       1. Science—Philosophy—Congresses. I. Carrier,
    Martin. II. Massey, Gerald J. III. Ruetsche,
    Laura. IV. Pittsburgh-Konstanz Colloquium in the
    Philosophy of Science (4th : 1997 : University of
    Pittsburgh) V. Series.
    Q175 .S4193 2000
    501—dc21
                                                      99-050941
                                                           CIP

A CIP catalogue record for this book is available from
the British Library.

We dedicate this volume to our friend and colleague,
Gereon Wolters, whose broad shoulders have gracefully carried
much of the weight of the Pittsburgh-Konstanz
collaboration.

Contents

Prospects for the Special Sciences

Preface

The Pittsburgh-Konstanz Colloquium in the Philosophy of Science is the joint undertaking of the Center for Philosophy of Science at the University of Pittsburgh (USA) and the Zentrum Philosophie und Wissenschaftstheorie at the University of Konstanz (Germany). It consists of a series of biennial international conferences that alternate between Konstanz and Pittsburgh. The present volume grew out of the fourth meeting of the Colloquium, which took place in Pittsburgh on October 3–7, 1997, under the rubric "Science at the End of the Century: The Limits of Science." The Colloquium is dedicated to advancing the kind of philosophical understanding of science made possible only by close familiarity with actual and ongoing scientific work. This orientation, which certainly characterized the 1997 conference, is abundantly reflected in the contributions to this volume.

The fourth meeting of the Pittsburgh-Konstanz Colloquium was made possible by financial support from the German-American Academic Council Foundation of Germany and by the Harvey and Leslie Wagner Endowment of the University of Pittsburgh. We gratefully acknowledge their generous assistance.

Our thanks go also to the officers and staff of the Pittsburgh Center for Philosophy of Science who were responsible for the planning and organization of the conference, in particular to Karen Kovalchick, Joyce McDonald, and Jennifer Bushee. Especially warm thanks are extended to Gereon Wolters (Konstanz) and Nicholas Rescher (Pittsburgh), who as co-chairs of the program committee bore the chief burden of getting the conference underway. The program committee itself consisted of Martin Carrier (Bielefeld), James Lennox (Pittsburgh), Peter McLaughlin (Konstanz), Gerald J. Massey (Pittsburgh), Nicholas Rescher (Pittsburgh), and Gereon Wolters (Konstanz).

Martin Carrier
Gerald J. Massey
Laura Ruetsche

Introduction: Science at the End of the Century

Prospects and Limits of Science

Martin Carrier, Gerald J. Massey, and Laura Ruetsche

Since the time of Francis Bacon the progress of science has been part of the agenda of philosophy. As is typical of a philosophical question, the issue has been contentious. Whereas some claim that science will progress continually, others regard science as being constrained by a number of cognitive and moral restrictions. Famous among the latter is Emil Du Bois-Reymond, who suggested in 1880 that there are *Welträtsel* or riddles of the universe, some of which science will never be able to resolve. Du Bois-Reymond included "the essence of matter and force" and "the emergence of sensory experience or consciousness" in this class of supposedly incessantly mysterious riddles. *Ignorabimus*, Du Bois-Reymond felt he had to acknowledge: we will never know.

This position met with strong opposition. In the Vienna Circle, for instance, it was held that no meaningful question is inaccessible in principle by science. Either seeming riddles are shown to arise from conceptual confusion—and thus are debunked as pseudo-problems—or they are transformed into well-posed empirical issues that for this reason will be subject to scientific treatment. Consequently, there can be no unanswerable questions. Either they are meaningless or they can be addressed by science.

To be sure, Du Bois-Reymond's thesis, as well as its Viennese retort, attest to their age; the landscapes of science and philosophy of science alike have profoundly changed in the course of the twentieth century.

1

Still, the issue driving the debate is as pressing as ever. It concerns the question whether fundamental science is subject to in-principle constraints. Distinguished points in time, such as the turn of the century, provide an appropriate opportunity to ask anew if science can be expected to continue to produce novel insights into the workings of nature, or, alternatively, if epistemic limitations on basic science are yet in the offing.

It goes without saying that science is subject to a multitude of constraints. For instance, physical theories themselves pose limits on epistemic goals. Scientific progress discloses features of nature that make other epistemic objectives unattainable: Special relativity restricts signal transmission to the velocity of light. Indeterminacy relations of quantum mechanics thwart the simultaneous prediction of position and momentum. General relativity prohibits gaining information from the interior of a black hole. Likewise, complexity is a major source of scientific limitation. The intricate entanglement of a host of influences frequently blocks cognitive penetration. The complex interaction of different factors prevents humans from accounting for and predicting the future course of the relevant systems. Finally, science is constrained by moral considerations. This applies in particular to biological research, such as experimentation on human subjects or their cloning. But issues of this kind, and the applied sciences in general, are not the focus of the present volume. We are not concerned with what nature allows us to do or what we are allowed to do to nature—or to our neighbor for that matter. Rather, we are concerned with prospects and limitations of what we can know about nature. The issue is whether there are epistemic limits to fundamental science.

The volume comprises two major topical blocks. In the first one, constraints are addressed that are possibly relevant for the whole of science; in the second one, possible limitations of specific branches of science are considered.

The volume opens with Paul Humphreys distinguishing between limits and limitations of science. A limit is an "in-principle" epistemological constraint, whereas a limitation is an "in-practice" epistemological or pragmatic constraint. Humphreys contends that unreasonable expectations about limits should not be misconstrued as limitations, and that science is not as limited as some current philosophical views would have it. In particular, he argues that "generative atomism"—the view that all objects are constituted by elementary particles and arise from the latter by a determinate construction process—is an inappropriate ideal for science. Humphreys contends that generative atomism fails to take account of objects that are not mere composites of fundamental entities. Consider the hydrogen molecule. The two atomic nuclei "share" the electrons of the hydrogen atoms that form the molecule. In the molecule, the electrons have lost their identity and constitute an integral, essentially molecular entity. As a result, the molecule does not consist of the two atoms. The atoms have "fused" and thereby have brought forth the new molecular entity. Humphreys's moral is that science is in no way limited by the failure to trace back the properties of complex entities to the properties and relations of their constituents.

The questions of the scope and limit of scientific progress have occupied Nicholas Rescher for decades. The volume's next three essays address Rescher's work on the limits of science; they are accompanied by Rescher's own omnibus commentary. In the first chapter of the Rescher trio, Robert Almeder enumerates seven main pillars of Rescher's general thesis on the nature and limits of natural science. The pillars include Rescher's rejection of scientism, his defense of fallibilism, his defense of inductive methodology in the presence of the problem of induction, the eternity of science and scientific progress, the constraints laid upon scientific progress by economic forces (leading inevitably to a logarithmic retardation of the progress of science), the ways in which science only approximates truth at any given time (and why these ways pose no deep skepticism for scientific realism), and the ultimate failure of science to deal with the increasing complexity of social problems in the future. Almeder further observes how some of Rescher's earlier views on scientific realism and the approximative view of scientific theories have evolved to what one may regard as a more satisfying form of realism.

In "Toward the Pale," Laura Ruetsche confronts two claims Rescher offers in *The Limits of Science* with a recent, and bold, cosmological proposal. Rescher's first claim is that the plasticity of scientific explanation ensures that no scientific question is in principle unanswerable. His second is that a mechanism of erotetic propagation ensures that there will always be unanswered scientific questions. The cosmological proposal is Lee Smolin's: that black holes give birth to entirely new universes, and that these universes differ in fecundity. On this basis, Smolin claims to achieve a statistical explanation of the values of a host of constants. The salient point with respect to Rescher's argument is that Smolin's theory casts old arguments in a new explanatory role. That is, well-known statistical arguments furnish explanations of the values of the pertinent constants. And in changing the explanatory principles in this way, Smolin hopes to block the emergence of new questions in a methodologically acceptable way. Ruetsche intends Smolin's approach to illustrate a tension between Rescher's commitment to scientific plasticity and his mechanism of erotetic propagation.

While Rescher assumes the unending proliferation of new issues of research in the course of scientific progress, Jürgen Mittelstrass envisages the additional continual emergence of new purposes for research. According to Mittelstrass, Aristotelian and Newtonian physics strove for completely different goals, and it is to be expected that future science will proceed toward yet different ends. On the other hand, Mittelstrass follows Rescher in maintaining that the success of science in its cognitive and physical mastery over nature continues to be science's ultimate measure of validation.

In his chapter "How to Pile Up Fundamental Truths Incessantly," Martin Carrier explores the options for reconciling two apparently conflicting but equally widely shared views. First, epistemic progress is unlimited and goes on forever. Second, science manages increasingly to reveal the innermost secrets of nature. The conflict arises from the fact that if you get more and more things right, sooner or later you should be done. Carrier argues that the assumption of emergent laws would be sufficient for bringing both views into harmony with one another. In addition, he suggests that there are prima facie candidates for such laws, namely laws connecting functional properties. His conclusion is, first, that there are possible worlds in which the conflict between unending fundamental progress and scientific realism is resolved, and, second, that the actual world may be of the requisite kind.

Nicholas Rescher asks the question: "Are there any limits to the problem-solving capacity of computers?" That is, are there any in-principle constraints on the capacity of computers to provide answers to cognitive problems in an appropriate way? In order for a computer to solve problems, it must issue not only statements that happen to be correct but also responses with a rationale suitable for establishing their credibility. The most significant type of limitations computers might face is of the kind that other problem solvers (such as humans) could overcome. Rescher argues that he has identified such limitations. His claim is that owing to the intricacies of self-reference no computer could answer the question as to the limitations of the general capacity of computers to solve questions. But humans could answer this question because no self-reference is involved in this case.

In his chapter on the "Limits to Self-Observation," Thomas Breuer points out restrictions for testing a universal theory. Such a theory is supposed to cover the whole of nature—including the measuring apparatus used for checking the accuracy of the theory's predictions. Consequently, each empirical test of such an overarching account represents a "measurement from inside": the observer is part of the observed system. Intuitively speaking, Breuer's argument says that in the case of an internal observer the enclosing system has more degrees of freedom than the registering device. This fact entails limitations to the accuracy of the results an internal observer is able to obtain: she cannot distinguish empirically among all the distinct states. It follows that there are in-principle constraints for obtaining the database relevant for a universal theory. No observer is able to determine precisely all the states of the universe, and this fact might constitute a serious impediment for examining the correctness of such a theory. A theory of the universe as a whole might be true, to be sure, but we might be at a loss to confirm it.

In his chapter "Blinded to History?" Alfred Nordmann explores a tension between the claim of science to produce universal truths and its commitment to acknowledge the contingency or "historicity" of nature. Although modern science deals with temporal developments on a vast scale, it refers these to an ultimately unchanging and ahistorical nature. Indeed, Nordmann presents as a sort of transcendental requirement of science that change is conceptualized within an invariant framework of immutable law. This condition constitutes a limit of science and is constitutive of the possibility of objective scientific knowledge. Correspondingly, the supposed fundamental invariance of nature or the assumption that all natural processes involve a mere rearrangement of unchanging elements functions as a productive "founding myth" of science. Case studies from the development of evolutionary biology and the establishment of modern chemistry show how scientists struggle against this limit while having to reintroduce it at levels of greater abstraction. Since the ontological postulate of an unchanging lawful nature coincides with the epistemological demand for a separable context of justification, science must blind itself also to its own historicity and is therefore prevented from knowing itself.

Hans Julius Schneider addresses "Metaphors and Theoretical Terms" and advocates a hermeneutic understanding of some important concepts of psychology. This understanding, if adequate, would place severe limitations on any scientific investigation of the "mental events" for which these concepts allegedly stand. Schneider argues within the framework of Von Wright's interventionist theory of causation and Wittgenstein's language-games semantics. His contention is that there is a purely psychological use of mental concepts. That is, such concepts do not refer to any neurophysiological processes—or any natural objects in the first place. Rather, they derive their meaning from cultural language games and contribute to generating cultural objects. Such objects are accessible only to a hermeneutic approach that takes the use of metaphors into account.

Margaret Morrison takes up the issue of "Unity and the Limits of Science," that is, the possible limitations on science's capacity to give a unified account of phenomena. She stresses, first, that the issue of scientific unification must be dealt with in a context-specific fashion. Unity is not a metaphysical problem that could sensibly be addressed for science as a whole; rather, science's success and failure in achieving unity should be analyzed for particular cases. Examining such instances of unification in science, Morrison points out that covering a class of phenomena in a unified fashion in no way entails giving a substantial theoretical explanation, let alone a causal account. Unification is not tantamount to explanatory power.

The second major topic addressed by Morrison is the relationship between overarching theoretical principles and more specific models. She emphasizes the comparative autonomy of models and the use of disparate models. Whereas some models clearly derive from an underlying theory, others do not. Nuclear structure is largely accounted for by drawing on phenomenological models that owe very little to the underlying theory of quantum chromodynamics. In such cases, models provide a conceptualization of the relevant phenomena different from the one provided by the theory. In addition, a variety of models may be needed to cope with different aspects of the same set of phenomena. These features induce a significant amount of disunity into science. This disunity should be accepted as an integral part of present-day science; it is as much part of science as the considerable unification produced by its theories.

The remaining contributions consider limits of a less general nature —limits encountered in physics, in biology, or in modeling.

Gordon Fleming surveys the disorienting landscape of competing interpretations of and approaches to quantum theory more than seventy years after the formulation of the basic ideas. He presents two conjectures on possible resolutions of the present malaise. First, the future discovery of the breakdown of superposition may emerge from dynamical state reduction theories, or from quantum gravity research, or from a hypothetical "energy" contribution from the quantum state itself, or, finally, as an emergent phenomenon in the domain of complex systems. The second conjecture assumes quantum theory, in essentially its present form, to never be falsified. In that scenario, Fleming suggests, various interpretations will come to be seen as similar to "choosing a gauge" in electrodynamics—helpful for solving certain problems but devoid of physical significance. One then seeks a "gauge-invariant" formulation to identify the physically deep aspects of the theory.

The chief claim of Alex Rosenberg's essay is that the structure of biological theory, unlike the structure of the physical sciences, is due to cognitive limitations of humans. Whereas the physical sciences reflect the blueprint of nature, biological theory gives testimony to human epistemic weakness. If our capacities of perception, calculation, and memory were far grander than they actually are, biology would employ generalizations of a structure widely different from the ones it uses now. The salient feature is the recourse to functions. Natural selection operates on the effects of the relevant structures. But the same effects can be produced by quite different structures, that is, selection is blind to structures. Consequently, biological regularities address functions; they connect functionally described states with one another. But these functional states are realized physically in heterogeneous ways, and this is why they do not form natural kinds. They are of a different nature.

But turning to structural kinds, that is, waiving functional individuation of states, is too high a price to pay for laws. Functionally uniform states group together what is significant for us; recourse to them makes intelligible explanations in biology possible in the first place. This situation brings severe restrictions of biological knowledge in its train. Biological explanations fall short of reaching the bottom of the phenomena. Nonbiological explanations are more fundamental in accounting for life than biological ones. Biology, just like other "special sciences" but in contrast to the physical sciences, is relativized to human interests and abilities.

Giora Hon and Rosenberg agree that the epistemic achievements of the physical sciences are unattainable in biology. However, whereas Rosenberg focuses on the structure of biological theory, Hon claims that the bearing of the experimental method is subject to serious restrictions in the realm of the living, so that the evidential basis of biological theories is inferior to the one underlying physical theories. The central thesis of Hon's chapter on "The Limits of Experimental Method" is that experiments fail to grasp the nature of organisms. Experiments involve two basic tasks, namely the preparation of the system and the acquisition of information about it through testing. The essential feature of the method of experimentation is testing by parameter variation. Some properties of the prepared system are altered and it is observed which other aspects of the system change as well or remain invariant. Hon's claim is that living systems do not possess states and respond in too indeterminate a fashion to external interventions. For this reason there is no experimental access to life. A state description fixes all relevant properties of the system and thus determines its future development. Hon argues that as a result of the intricate interaction of the parts of a cell, there is no such state description to be given. But if the states of an object are not specifiable unambiguously, no preparation of the object can be made and the experimental method finds no point to latch on to.

Deborah Mayo's chapter on "Models of Error and the Limits of Experimental Testing" explains her error-statistical version of the position of new experimentalism. New experimentalism contends that the performance and analysis of experiments is less dependent on higher-level theories than traditionally assumed. Rather, experiments are typically directed at scrutinizing more specific or restricted models, and the assessment of their epistemic bearing requires statistical methods. One of Mayo's objectives is to develop methods for coping effectively with epistemic restrictions. As she argues, limitations and error are best overcome by using error-statistical methods. Her charge against traditional ways of assessing evidential relevance relations, such as the Bayesian measure of confirmation, is that they afford merely a "white-glove" logical analysis: the truth and the significance of the evidence are accepted as given. The error-statistical approach, by contrast, stipulates criteria for the reliability and bearing of data. Mayo's criteria include considerations of how probable it is that a hypothesis would pass a given test even if it were false. These criteria apply to the entire procedure that is supposed to confer support on the hypothesis in question. It is the empirical study or experiment that is subjected to an analysis of its probative force. The contention is that error-statistical methods provide a powerful toolkit for producing trustworthy results. These methods serve to push the limits of science further away.

Whereas Mayo considers prima facie limits placed on science by data that are noisy, messy, or riddled with error, Manfred Stöckler addresses prima facie limits imposed by complexity. In his essay "On Modeling and Simulations as Instruments for the Study of Complex Systems," Stöckler defends the claim that simulations constitute a revolutionary new tool for science in that they enable one to follow out the consequences of assumptions. Without the help of computer simulations it was simply inscrutable what certain hypotheses or equations actually implied. Numerical models, along with powerful computational devices, make it possible to understand—and even visualize—what a theory maintains; they sometimes serve to endow the theory with empirical content in the first place. However, as Stöckler argues, simulation is not an activity intermediate between theory and experiment. It is and remains theoretical in kind. Simulation is derivation writ fast. From the methodological point of view, nothing new accrues from using computers. The virtues of simulations are purely pragmatic.

Most of the main chapters in the volume are accompanied by commentaries. At a conference, commentaries serve to stimulate discussion. They are included here for two different reasons. First, a commentary often provides an improved access to the content of its main text and to the intentions of its author; second, commentaries frequently make substantial claims worthy of being considered. Well-written commentaries constitute contributions in their own right. In particular, some of the commentaries in this volume develop a sort of counterpoint to the claims entertained in the main chapters and thus provide a kind of living proof of the assertion that philosophy is essentially characterized by debate and controversy. Good philosophy is contentious philosophy. Unanimous agreement testifies only that the relevant field has lost its energy and may as well be pronounced dead. We trust that the present volume gives sufficient evidence that philosophy of science is very much alive.

Frontiers of Knowledge

1

Extending Ourselves

Paul Humphreys
Department of Philosophy, University of Virginia

"Limit science" is a usefully ambiguous term. Limit science is in one sense the best that we can do. But in the other, punning, sense it suggests that the best we can do is, well, the best that we can do and thus falls short of what is desired. So it will help if we use different words for the two senses—"limit" for the ideal achievement and "limitation" for falling short of that or some other desired goal. We must then be careful that unreasonable expectations about achieving limits are not automatically misconstrued as limitations. There is a limit to how fast a Ferrari F50 can go, but it is only in special circumstances, and perhaps under unrealistic expectations, that the limit should be construed as a limitation. Paying attention to limitations is not a concession to pragmatics; it is rather a needed concession to reality, to the fact that we have limited faculties and must do our best to work around them.

This distinction between limits and limitations will help us when we look at what we expect science to do, for very frequently what are perceived as limitations on current scientific achievements are perceived as such only relative to some inappropriate ideal drawn from a limit science that is characterized in an a priori fashion. Often, such an ideal results from the adoption of some ontology that happens to be particularly well suited to a convenient representational technique. We can call this first sort of limitation a *limitation from inappropriate ideals,* and I shall illustrate such a limitation later when discussing generative atomism.

Conversely, one of the most important potential sources of limitations on science comes from the fact that science is, for us, science done by humans.[1] This is inescapable, but it must not be misconstrued. Results from some otherwise scientific enterprise that were completely inaccessible to us as epistemic agents would be, for us, no science at all. Yet there is a tendency to go too far in the other direction and to make contingent features of humans too constraining. One of the limitations that certain sorts of epistemologies tend to impose on science is forcing anthropocentric constraints on the output of various scientific methods. The best known of these constraints, of course, involves observability, but I shall make a case that certain conceptions of mathematics impose related constraints on science. We can call this second kind of limitation an *anthropocentric limitation*.

Limitations from inappropriate ideals tend to result from goals that are too ambitious, whereas anthropocentric limitations tend to result from goals that are not ambitious enough. I shall here make a case for two ways in which we can avoid anthropocentric limitations and one way in which we can avoid limitations from inappropriate ideals. These will not determine exactly where we should aim, but they will, I think, suggest that science is not as limited as some contemporary views have indicated.

"In Principle" versus "In Practice"

A great deal of what philosophers do is concerned with what can be done in principle. The reasons for this focus are entirely understandable, and some of them involve the following:

1. The fact that we want to abstract from the contingencies and the limitations of our human and historical situations.

2. The fact that a considerable amount of attention has been paid to negative results: if some goal—such as providing models that are unique up to isomorphism for a first-order theory—cannot be achieved even in principle because the theory is noncategorical, then it is pointless pursuing that end.

3. The fact that there remains a trace of the view that philosophy is and should be concerned with necessary truths, and that in consequence dealing with the merely contingent is irrelevant. Some of this interest in the "in principle" is legitimate, but much of it is not (or, to put it less contentiously, there is room for both views). Talk of what

can be deduced in principle from a set of laws is too remote from the way that theories are actually applied, in somewhat the way that an existence proof for a solution to a differential equation form is often of little help in producing a solution to an equation of that form.

In contrast, I believe that there is a legitimate role in philosophy of science for examining how science is applied within the contingent circumstances that hold in present times. The issue of how science is applied surely concerns primarily how it is applied in practice and not how it is applied in principle. None of this is intended to devalue those activities that look at idealized science. It is rather that there should be a place for some general principles about how scientific theories are applied in practice.

The Limits of Detectability

It is worth asking this question about limit science: Supposing our sensory abilities or our mathematical abilities had been different than they are. Would we arrive at the same limit of scientific inquiry? That is, is limit science invariant under changes in the epistemic community that produces it, or do the evolutionarily contingent senses we possess affect the form even of limit science? This is not an easy question to answer, not the least because we have no idea what our limit science would be like, but if the answer were to be that limit science is *not* invariant under such changes, then science would lose a great deal of the objectivity that one wants from it. One of the principal reasons for dealing with the limit is that it allows us to accommodate false starts, partial access to the truth, historically contingent technological factors, and so on, the idea being that in the long run the effects of these will become arbitrarily small. Now as I have mentioned, traditional empiricism does place serious anthropocentric limitations on current science. Nanoscientists who grew up naturally observing elementary particles and not much else would be faced with the difficult task of synthesizing the ontology of macroscopic objects from particle physics, perhaps from within a conceptual framework that was inherently quantum mechanical. As we know, even with direct access to the macroscopic realm this is a formidable task, and it might never be accomplished by the nanoscientists who lacked that access. Nevertheless, one purpose of science is to allow us to transcend the contingent limitations imposed on us by evolution, and the issue is the extent to which this is possible.

What exactly is the attraction of anthropocentric empiricism?[2] For those who are drawn to it, much of the attraction lies in the security that empirical data give us. When in doubt about some conclusion, one can always resort to the data to justify it. A common justification for relying on such empirical evidence is that it is direct rather than inferential, the tacit assumption being that the more direct the evidence the more veridical it is. That assumption is not generally true, for it also requires the detector to be content preserving. A contact print from a photographic negative will be inaccurate if the photographic paper has deteriorated; direct translations are often inaccurate because the concepts in the receptor language do not match up to those in the original language; and direct audition of an orchestra in an acoustically muddy auditorium provides an inferior experience of the music compared to an engineered recording. (Traditionally this kind of objection would have been dealt with by an appeal to ideal observers, but what is ideal is exactly the issue we are investigating here.)

What undoubtedly motivates many empiricists, then, is the need for a sense of security in their evidence. I do not say that this is the only reason for wanting to be an empiricist, but it is a very common one. There are two aspects to this, which, following an old scientific tradition, we may call accuracy and precision. A quantitative datum is accurate if it closely approximates the true value of the quantity from which it is derived; data have a high degree of precision if the empirical variance of the data is small. (One may subdivide this definition into individual precision and collective precision, with the former being concordance of data from a single observer and the latter being the reproducibility of data from different observers.) A sequence of accurate data drawn from the same source will be precise, but precision does not entail accuracy of the individual data points. Perhaps accuracy is the more important virtue for most empiricists—they do not want to be mistaken—but precision without accuracy could well be important to an antirealist empiricist.[3] However, empiricists generally want their data to be both accurate and precise, and the idea is that the smaller the gap between the data and their origin, the likelier it is that both accuracy and precision will be achieved. I shall thus call any detector, including any of the five traditional human senses, *reliable* if it is both accurate and precise.

Now, we know that particular detectors have different domains of reliability and that those domains are often quite limited in scope. So let us start with the idea that human sensory capacities are merely particular kinds of instruments, sometimes better but often worse than other detectors. Human observers have fair reliability within certain restricted domains. For example, if a flying ant and a Southern termite are compared in good light, then the correct classification of each can be made accurately by many different observers with minimal training, thus making unaided vision a reliable comparative detector of flying ants and Southern termites. Yet despite the empiricists' elevation of human detectors as the ultimate standard, the domains of reliability for human senses are highly restricted, and even within those domains humans are often not the best we can do.

We can begin to approach the issue by considering the ways in which scientific instruments allow us to transcend the limitations of our native senses. Much of the debate about observables has focused on examples of devices through which we enhance our existing sensory inputs—instruments such as microscopes, telescopes, and so on, which amplify our visual sense. This is undeniably interesting, but one of the most important features of scientific instruments is their ability to put us into contact with properties of objects that are natively almost or completely inaccessible to us. Indeed, one of the misleading conclusions that tends to be drawn from the microscope and telescope discussions is that it is objects that are observed or not observed and that problems of observation are scaling problems. In fact it is *properties* of objects that are observed, and occasionally what property of an object is observed will make a difference to how it is classified. For example, the galaxy NGC 309 is usually classified as a classic spiral galaxy on the basis of its image in visible light, thus putting it into the same category as the Milky Way. But when viewed in the near infrared at a wavelength of 2.1 micrometers, the disk has only two arms instead of three, and the central disk is close to an ellipsoid. This would put it into the class of barred spirals, which are in some ways quite different from the classic spirals (see Wolff and Yeager 1993, 20).

So let us look at instrumentation that detects things by means that are quite different from those that humans use. To start with a simple case, consider the problem of identifying another human being. Visual reidentification will often do under reasonably good conditions of original identification, but it breaks down when the identification has to take place after a long passage of time, or when the individual has radically altered his or her appearance. Fingerprints can be allowed as good old-fashioned observables, relying on a visual comparison between an existing set of prints and a fresh set,[4] but fingerprint patterns can be altered by plastic surgery. In contrast, DNA profiles are far more reliable than any of these processes—a probability of, say, one in forty million that this individual's DNA pattern came from another human is a level of reliability that is hard to match. Suppose that we came naturally equipped with DNA analyzers—tiny samples of blood were automatically taken from anyone we encountered and the analysis was instantly compared with a vast database of named DNA profiles from those whom we had encountered previously. Wouldn't you prefer that over old-fashioned visual recognition? No more embarrassing incidents of forgetting someone's name or, as I did not long ago, of talking animatedly to a colleague about his research on ancient Greek texts only to realize some time later that I had been talking to the chairman of the department of anthropology. (Virginians are awfully polite.)

Now there is something special about this case, because the reliability of the DNA sequencing can be assessed by means of a direct comparison between the output of the matching process and a sample taken from the individual whose identity can be determined on old-fashioned empiricist grounds. (The comparison is made, of course, in uncontroversial cases in which the identity of the individual is beyond question and the identification takes place within the domain of human reliability. The increased reliability and the wider domain of the instrument come into play in the hard cases, in which inductive extrapolation is required from the successes in the uncontroversial cases.)

So consider a somewhat different case, that of temperature. Is temperature an observable? Not according to most empiricist criteria. Certainly, there is no sense in which we can get up close and inspect the temperature, to use one of van Fraassen's criteria. At best, humans are highly unreliable observers of temperature even within a narrow range, which we may generously put at between $-100°F$ and $+150°F$. Outside that range it is inaccessible to human observation; within that range we may be accurate to within a few degrees in the central region of $0°-100°F$. (Internal body temperatures are a different matter: our range of accessibility is much smaller, but we are correspondingly more sensitive to minor differences.) So we have used thermometers for over

three hundred years to extend our sensory abilities for temperature.[5] How do we do this? By a process that is universal in all
instrumentation—by converting input from a domain in which the
instrument is an accurate or precise detector into an output that lies
within the domain of human perceptual accuracy and precision.

For the purposes of this chapter, I am going to grant the traditional
empiricist that his criteria (whatever they may be) are the grounds for
assessing accuracy. Thus, for example, who this person really is will
ultimately be decided by direct observation (identification by next of
kin under epistemologically favorable conditions, for example), and
the identification of when the fixed points on a temperature scale (such
as the triple point or the boiling point of water) have been reached can
also be decided, at least approximately, by direct observation. Then we
can make the following claim: the calibration of the instrument that
fixes the accuracy can be made on traditional empiricist grounds,[6] but
in both the DNA case and the temperature case it is the superior
precision of the instrument and its wider domain of applicability that
override any appeal to mere observables. Recall that in DNA testing
one compares a reference sample of blood or other bodily fluid from an
individual whose identity is known by empiricist criteria with a new
sample for a match or a failure to match. It is the precision of the DNA
tests that allows the correct matching to the reference sample because
of the sharp reproducibility of outcomes on samples from the same
source in a wide variety of contexts.

In the temperature case, it is the precision of a platinum resistance
thermometer that, having been calibrated in a vapor/liquid/ice-I mixture, allows the reidentification of that triple point (= 0°C) temperature
in a different substance—a sample of alcohol, for example. This reidentification relies primarily on the zeroth law of thermodynamics: if
one body (the vapor/water/ice-I mix) is in thermal equilibrium with a
second body (the platinum resistance thermometer) via a diathermal
wall, and the second body is isolated and found to be in thermal
equilibrium with a third body (the alcohol) via a diathermal wall, then
the first is in thermal equilibrium with the third, and hence they have
the same temperature. To be clear what is happening here: calibration
points in thermometry are almost always phase transition points,
which have the virtue of being sharply discontinuous and very stable.
In the case of the freezing and boiling points of water, they happen also
to be directly observable—you can just *see* (to within reasonable

limits) when water boils.[7] But the same temperatures (0°C and 100°C) in alcohol are not observable in that sense. There are no phase discontinuities to mark them at these temperatures in alcohol. So we need the thermometers to extend the range of application of the fixed points to other materials. The wider the domain of precise measurement, the greater can be the extent of reliable measurement.

I can now draw the main philosophical conclusion from these considerations. Whereas direct observation may well be the initial reference point in grounding claims to knowledge, it is often abandoned and replaced with instrumental justification. That is, once the reliability of a given instrument has been established, we do not, in the last resort, turn to human observations to ground claims to knowledge, even when the domains of human perception and of the instrument overlap. Rather, the observational basis is abandoned and the instrumental data become the foundation.

Of course the kinds of cases in which we can directly calibrate the instruments are very limited. But they do illuminate at least one debate about the nature of observations. Hacking's well-known discussions of how we see through microscopes were intended at least in part as a response to Dudley Shapere's account of direct observation (see Hacking 1983, 182–85). To recap their positions very briefly, but in a way that still captures an important difference in outlook, Shapere analyzed direct observation in this way: "x is directly observed if (1) information is received by an appropriate receptor and (2) that information is transmitted without interference to the receptor from the entity x (which is the source of the information)" (Shapere 1982, 492).

A more important aspect of Shapere's analysis was the claim that even direct observation is a function of current theory, and it was this aspect that was addressed by Hacking's claim "One needs theory to make a microscope. You do not need theory to use one" (Hacking 1983, 191), substituting for direct observability an experimental criterion of existence. As I have remarked elsewhere, the kind of bench microscopes that Hacking was discussing have been deliberately constructed to be idiot- (or student-) proof (Humphreys 2000). Many instruments are not like that. To use a magnetic resonance imager, for example, or even to read X rays correctly, one needs to know a considerable amount about how the machine operates and how it interacts with its target. So I believe that Shapere was correct about the dependence of observations on background knowledge, but—and here is where the present account differs from his—that he failed to recognize the important role that sheer reliability plays in much instrumental observation.[8]

What about the important cases in which we extrapolate beyond the range of empirically accessible base cases? Here *there are no general answers*—you have to know how the instruments can go wrong, and that requires a good deal of subject-specific knowledge. This lack of generality is a widely applicable feature of instruments, however. You need to know how they work in order to identify and correct errors. Of course this requires the use of theory to some extent. But since we are concerned with reliability, this is not a reason to retreat to a minimalist empiricism. The reason is that the theories used to correct practical temperature scales (such as the expansion characteristics of glass, the annealing characteristics of platinum, the use of virial coefficients in gas thermometers) are so well confirmed in areas outside thermometry that it would require one to reject an enormous amount of established theory to deny that these error-correction techniques are well founded and themselves reliable. There is a direct parallel here with a better-known case: of course it is possible to be a 4004 B.C. creationist, but if you are, you had better be willing to give up most of what we know about radioactive carbon dating, paleontology, sedimentary rock formation, cosmological time scales, planetary formation, and so on.

And Also for Mathematics . . .

There are some parallels between what I have said about observation and the rise of computational science in the past fifty years. Just as instrumentation extends the domain of our senses, so computer-assisted science extends the range of our natural mathematical abilities. Because I have written at length on this topic elsewhere, I shall restrict myself here to a few general remarks (see, e.g., Humphreys 1991, 1994, 1995a, 1995b). If one has in mind the ideal that scientific theories should be formulated in mathematical terms and that explicit predictions should be made from them in the form of analytic solutions to the relevant equations, then one is likely to be disappointed at the relatively small class of cases in which this is possible in practice.

These analytically solvable cases are extremely valuable as calibration points against which the accuracy of (usually discrete) approximative methods in numerical mathematics can be assessed, just as the directly observable fixed points are used in thermometry.[9] Moreover, one sees very much the same extension of domains in computational science as we saw for scientific instruments, and the same point holds about needing to know how the instrumental extension works. Discrete approximations to continuous functions, random number generators for Monte Carlo methods, truncations in double precision arithmetic, time step size, and so on can all go disastrously wrong if inappropriately chosen, and although automatic error-correction algorithms can sometimes be used, they are of severely limited applicability.

Again, similar points hold regarding abandoning humans as the ultimate justification for a claim to having a correct computation. The cumulative probability of error when multiple stages in a computation are required makes humans too unreliable a foundation. If your tax return is as complicated as mine, then you no longer trust your own arithmetical abilities to justify the amount owed. You use the computational resources of tax software instead.

This said, the use of computational methods vastly increases the range of applicability of many theories. What are often considered to be laws within the ideal picture sketched earlier are instead simply computational templates. Well-known examples are Newton's second law, Schrödinger's equation, expected utility maximization, and conservation of energy principles. In each of these cases the relevant mathematical framework imposes constraints on the development of the system's states, but the framework cannot be used until some specific additional material—such as the force function, the Hamiltonian, the utility function, the exact form of the potential energy, and so on—has been inserted into the template.[10] As should be well known by now, most specific applications of these very general frameworks require the use of computational mathematics in order to bring the abstract theory to bear on a concrete case. Far from viewing the need for numerical computations as a limitation, it should be seen as a method for vastly increasing the range of applications of a given theory. This is not to deny that there are serious and philosophically interesting issues concerned with the use of numerical approximations in science, but I have discussed these issues elsewhere (see, e.g., Humphreys 1991, 1994, 1995a, 1995b).

Generative Atomism

Where does the idea of limit science come from? The optimistic concept involves a sense of completeness: the idea that everything that is true, or everything that is knowable, has at the limit been subsumed under the final theories of science. "Limit science" also conveys the sense of being the end point of a process, and the additional sense of that process converging on that limit. Although it is consistent with these ideas to have as limit science simply a vast collection of individual and independent facts knowable in totality only by a superhuman intelligence, this is obviously not what is usually meant by limit science. Now there is no point in speculating in detail about what the limit will look like, but there are certain broad structural features that are often imputed to limit science. For many years, the image many had of limit science had these components:

1. Limit science consisted of a collection of fundamental facts—usually regarded as relatively small in number, and almost always conceived of as being about fundamental physics—that describe the properties of the basic constituents of nature. These fundamental facts would be of four kinds:
  a. A list of the fundamental kinds of entities.
  b. Facts about the permanent properties of these fundamental constituents, such as their charge and mass, and specific quantitative values for these properties. (If these values, such as the value of the charge, were defining features of the fundamental objects, then (a) and (b) could be collapsed into a single category.)
  c. Facts about what transitory properties the fundamental entities could have, regarding such features as spin, position, and momentum. In contrast to (b), some of these would not require specific values, whereas others would. (For example, the value of spin for fermions would be required in (b), but not whether it was up or down for a particular particle.)
  d. Facts about the laws that govern the distribution and dynamics of the fundamental entities and their properties.
2. Limit science contained the view that all objects in the universe are composed of the fundamental entities: what we would now call mereological composition (that complex systems and their constitu-

ents stand in the whole-part relationship) or, weaker but still perfectly general, mereological supervenience (that the parts determine the whole).

3. Limit science suggested that all the facts about the universe follow from, are determined by, the fundamental facts together with specifications of state. The ideal of a recursively axiomatizable theory is simply one extreme of this view, and what was meant by "are determined by" ranged from "are deducible from," through "semantically entail" or "can be embedded in," to "supervene upon." The idea then was that in the limit of scientific progress—that optimistic Peircean limit of knowledge—an optimal version of limit science would result, with a fixed ontology of objects and a final set of laws and theories. There would be a pragmatic element to the choice of this final set, and it was an open question as to how realistically such a final theory could be interpreted.

It would be superfluous to go into detail about all the ways in which this ideal picture has fallen under criticism in the past three decades. The view that various irreducible special sciences have autonomous domains of inquiry has undermined the belief that limit science can be identified with limit physics; the belief in incommensurable theories has undermined the idea of continuous progress toward a limit; the essential underdetermination of theories by evidence and the existence of theories whose axioms do not even form a recursively enumerable set has lead some to question the intelligibility of encapsulating all the truths in a final axiomatic theory.[11] Yet this ideal picture retains its attractions. It is, for example, very close to the position that David Lewis currently holds, with his adherence to perfectly natural properties discoverable by physics as part of the fundamental structure of the world, the Humean supervenience of most features on patterns of local properties at space-time points, the Peirce/Ramsey treatment of laws, the advocation of a wide domain for mereology, and so on.

I have run through this familiar picture for one reason only. When we ask what are the limits on science, we tend to assess limitations in terms of how far short of this ideal of limit science we have fallen. And there are more than a few philosophers, not to mention social studies scholars, who hold that we have fallen so far short that science's claims to knowledge are irrevocably undermined. I suggest that some of these perceived limits on science are the result of chasing the wrong ideal and that developments in science itself in the last two decades or so ought to result in a revision of this generative conception of science.

Within this optimistic picture of limit science, one version has had a particular attraction. I shall call this version *generative atomism*. Generative atomism has an upward and a downward component. The upward component says that there is a collection of elementary objects from which all other legitimate objects in the domain are to be constructed and there is a fixed set of rules (usually required to be computable) that govern the construction process. Examples of this that are the stock in trade of philosophers are the objects of propositional logic (in which both syntax and semantics are generative), primitive recursive functions, mereological wholes and their parts, formalized proofs, mosaics, computer-generated embroidery, and many functionally characterized systems, such an automobiles.[12] All of this is powerful and lovely stuff, and it fits admirably with the combinatorial syntax of standard languages. It is no slight to suggest that not everything in our world is amenable to the methods of generative atomism. Since the subject matter should determine the method, to insist upon generative atomism is to impose a limitation from an inappropriate ideal.

The downward component of generative atomism is of equal interest. It requires three things of the atoms, and I take each of these three conditions to be individually necessary for something to be an atom of a generative system. The atoms must be (1) functionally indivisible, (2) individually distinguishable, and (3) immutable. For functional indivisibility to be satisfied, no proper part of an atom can carry properties appropriate to the domain when separated from the other parts, if indeed such separation is even possible. So, for example, no part of a noncompound word carries meaning or has a referent, no subsentential unit of language carries a truth value, and no part of a basic functional unit in a car (such as a broken nut) has an autonomous automotive function. Regarding individual distinguishability, at the type level each primitive sign in syntax must be distinguishable from every other primitive type, and at the token level each letter in a text must be unambiguously distinguishable from the neighboring letters. To maintain immutability, the constraint is that even when embedded in larger units the atoms must retain their core properties (and, for (2), their distinguishability). Thus a brick does not change its core properties when mortared into a wall, spark plugs retain their basic functional features when screwed into an engine block, and sentences retain their truth value when made part of larger logical complexes (at least within truth-functional semantics).

An Aside on Supervenience

In asserting that basic physical phenomena fix all other features of this world, a common way of construing this is to claim that nonphysical features supervene upon physical features. There is, as is well known, a variety of ways of making this claim precise, but the core ideas are contained in the standard definition of strong supervenience: "When A and B are families of properties closed under Boolean operations, then A *strongly supervenes* on B just in case, necessarily, for each x and each property F in A, if x has F, then there is a property G in B such that x has G, and *necessarily* if any y has G it has F" (Kim 1993, 65).

I want to draw attention to a feature of this definition that is easily overlooked. Although the supervenience relation itself holds between properties, the definition is formulated in terms of property instances or realizations, and more importantly, the definition requires that it is *the same object* x that has both the subvenient and the supervenient properties. That is to say, if F and G occur at different levels in the layered hierarchy of properties, and all the higher-level properties arise from supervenience relations to lower-level properties, then supervenience simply does not address the possibility that new objects have the higher-level properties.

This commitment to the same object having both the subvenient and the supervenient properties is, I believe, an integral part of the supervenience approach. Within its origins in moral philosophy, it has to be the very same thing that possesses both the physical attributes and the moral qualities that supervene upon them. It is specific brain processes and structures that possess both neurological properties and intentional states (assuming narrow supervenience). To abandon this dual aspect approach would be to abandon the idea that it is the very same physical object that has both kinds of properties.

We could, however, generalize the definition in this way: Say that an individual x of type X mereologically supervenes upon a class of objects $Y = \{y_1, \ldots, y_m\}$ just in case, necessarily, for any object x of type X there are objects $y_1, \ldots, y_n$ belonging to Y of types $Y_1, \ldots, Y_n$ and some relations $R_1, \ldots, R_m$ such that $R_1(y_1, \ldots, y_n), \ldots, R_m(y_1, \ldots, y_n)$ hold and necessarily, whenever $y_1, \ldots, y_n$ exist and $R_1(y_1, \ldots, y_n), \ldots, R_m(y_1, \ldots, y_n)$ hold, x exists.[13]

Then let F be a single property and $A = \{G_i\}$ be a set of properties that is closed under Boolean operations. Then F strongly supervenes upon A just in case, necessarily, for each x such that $F(x)$, there are individuals $y_1, \ldots, y_n$ such that x strongly supervenes upon $y_1, \ldots, y_n$ and

$$G_1(S(y_1, \ldots, y_n)), \ldots, G_m(S(y_1, \ldots, y_n)) \qquad (1)$$

(where S is a selection operator that picks out the appropriate number of arguments) and necessarily, if any $y_1, \ldots, y_n$ satisfy (1), then $F(x)$.

Atomism Again

I now suggest that generative atomism is flawed as an account of the world in two ways. It fails to give a proper place to objects that are not mere composites of atomic entities and it fails to allow for certain kinds of emergent properties. To see this we do not need to refer to exotic features of physics such as nonlocal EPR correlations, or to extremely complex biochemical or neurophysiological phenomena. The evidence already exists in widely accepted and basic features of particle physics and molecular chemistry.

The basic motivation for the idea that "molecules are nothing but collections of atoms" is a compositionality view: molecules are simply mereological sums of atoms. At the token level, this mereological view involves a two-way determination: given the atoms, one necessarily has the molecule, and given the molecule, one must have the atoms. This view requires that (1) the atoms are separately individuated and (2) they retain their identity when in the mereological sum, that is, they satisfy the individual distinguishability and immutability conditions described earlier. (We have of course abandoned the idea that the atoms are physically indivisible, but we retain the idea of functional indivisibility to the extent that no proper part of an atom has the properties required to generate a molecule of the right kind.) This mereological view would be correct were the simple solar system picture of atoms true, but it is false not only for molecules but also for atoms and for larger entities, such as lattices.

Perhaps the simplest example of this involves the hydrogen molecule. When two hydrogen atoms are widely separated, each can be considered to be composed of a single proton and an associated electron, and all four entities in the separated-atoms system are distinguishable from one another, the electrons from the protons in terms of charge value and mass, and the individual electrons and individual protons from one another by virtue of their wide spatial separation. In contrast, when the atoms are spatially proximate to one another, because of the well-known indistinguishability of electrons the electron wavefunctions overlap to such an extent that it is impossible to individuate them. Although the standard textbook treatments talk of two indistinguishable electrons, what one really has here is a joint probability distribution within which there is no sense to be made of separate, coexisting particles. As one account puts it: "Because of the complete space overlap of the wave functions of the indistinguishable electrons in H_2, it is definitely not possible to associate a particular electron with a particular atom of the molecule. Instead, the two electrons . . . are shared by the molecule, or shared by the bond itself" (Eisberg and Resnick 1985, 421).

This lack of distinguishability would be bad enough for atomism, but I believe that asserting something stronger is warranted. It is not that there are two electrons and one cannot say which belongs to which atom. It is that it no longer makes sense, given the complete spatial overlap of the wavefunctions, to consider the electrons as anything but a fused entity, a single "object" that is shared by the two protons (which do retain their separate identities).[14] If this is so, then generative atomism fails for this case. At the very least, we do not have the original two electrons, and a fortiori we do not have the original two atoms because objectively speaking it is no longer possible to say that we have, for example, the original "left-hand" electron.[15]

Standardly in setting up the standard formalism for indistinguishable particles, the notation labels them in a way that simply indicates numerical difference, rather than qualitative difference. But as Paul Teller (1995) has suggested, there is a danger of interpreting these labels as indicating a kind of individual essence, or haecceitism, or (to use Robert Adam's felicitous term) "primitive thisness" about the electrons. As an alternative, Teller suggests that electrons and similarly indistinguishable particles can be aggregated but not counted. In that spirit I can put the idea of fusion of objects this way: the fused entity has a parameter attached to it that we can call a "number," but it does not designate the result of a counting operation applied to the number of constituents in the fusion. To provide a partial analogy, consider poker chips in a casino. The basic units are red chips, and as soon as you have accumulated two red chips you can trade them in for a blue chip that is worth two units. The blue chip is not composed of two red chips and you cannot count its two components because it does not have any, but it behaves exactly as if there were two such units present. Moreover, under certain conditions you can decompose it by trading it in for two red chips. We recognize that counting systems are like that—that even though, for example, the higher-order units are not literally composed of ten units, they act as if they were. This is an imperfect analogy, of course, because the examples involve interpreted representational systems. Yet I believe the analogy does give us some idea of what the fusion of objects might be like.

Compare this with a traditional example of something that looks like fusion. We take two lumps of putty, we knead them together, and we have a lump that has twice the mass of the originals. There is a sense in which we have lost the original lumps of putty qua lumps, but we know that if we had colored the original lumps the mixture could now be seen to be composed of the original two masses. The belief here is that at the atomic level, the atoms of each lump would be preserved and the original "left-hand lump" would be present as a scattered object within the whole. If we believe that scattered objects exist, then the lumps are still there, not qua lumps but scattered around the jointly occupied volume.

Lumps of putty are of course not atoms of anything—they are neither immutable nor indivisible. Yet these examples indicate something important—it looks like generative atomistic physicalism must frequently appeal to some finer-grained ontology in order to retain its mereological orientation. And that in turn seems to mean that if you want to be a physicalist of the type I have described, then you must make your physicalist commitments at a very fundamental level indeed. This would be a useful argument in favor of a certain kind of fundamental physicalism, were it not to fail when emergence happens.

What does this argument establish? At least this: that at levels above, but not far removed from, what we currently consider to be the most fundamental, mereology cannot account for the relationship between atoms, nucleons, and electrons on the one hand, or between covalent bonded molecules and atoms on the other. This leaves open the possibility that there is a mereological relation between quarks and these larger objects, perhaps one of mereological supervenience, but that relation would have to skip at least one level, and it would fail to provide us with the kind of explanation of exchange forces that contemporary accounts do.

Summary

Sweeping conclusions about the presence or absence of limits on science are not likely to be rewarding at the present time, whether those conclusions are optimistic, as in unifying accounts, or pessimistic, as in pluralistic approaches. I have examined two ways in which we can be more optimistic that some of the epistemological limitations on science are being relaxed, and one way in which an overly optimistic unifying method will likely have to be more modest. None of this should be construed as supporting the idea either that science is so massively fragmented that nothing general can be said about it or that science cannot provide us with objective knowledge about the world.

NOTES

I am grateful to Fritz Rohrlich and to Bill Wimsatt for comments on a preliminary version of this chapter and to Charles Klein for numerous illuminating conversations.

1. I say "for us" because science is not an essentially human activity. Enterprises such as mathematical modeling and randomized field experiments could easily be done by cognitive agents that are very different from us.

2. Perhaps there are nowadays few adherents to a position that gives a privileged place to traditional observables. Maybe old-fashioned empiricism is a moribund position, but the philosophy of science literature is still replete with references to it; see, for example, Kitcher (1993, 222–33).

3. The term "antirealist empiricist" is not pleonastic. It is true that most empiricists tend to be antirealists, but one does not want to trivialize the debates between empiricists and realists by ruling out by definition the possibility of reconciling the two. I, for one, am an empirical realist.

4. Much of the work of fingerprint comparison is now automated, but for the purposes of the example we can take the traditional unaided process as the focus.

5. John Locke's diaries contain quite extensive records of daily temperature readings.

6. The importance of calibration to reliable detection has been discussed in detail in Franklin (1986).

7. Of course, this directly observable aspect is far too crude for sophisticated thermometric purposes, but it illustrates again how scientific procedures sharpen up ordinary practice. Historically we begin with everyday practice and gradually refine it.

8. A very important issue (but one that cannot be addressed here owing to lack of space) is what exactly are the differences between purpose-built detectors, such as the neutrino detectors Shapere considered in his article, and the widely available instruments that are used across a huge array of laboratories. Human senses are, in their versatility, much more similar to the latter than the former.

9. I do not mean here to identify computational science with the use of numerical mathematics. Algebraic and geometric computational methods are likely to play an increasingly important role. I mention numerical methods only because they are at present the dominant aspect of simulations.

10. When a distinction is made between a law and initial conditions or boundary conditions, the latter are ambiguous. A specific value of initial conditions is generally meant, but the specification of initial condition types is an important intermediate stage in the application of a theory.

11. It is worth pointing out that the current emphasis on the disunity of science is by no means new. Patrick Suppes has been urging what he calls "scientific pluralism" throughout his career (see, e.g., Suppes 1978), and Bill Wimsatt has held a piecemeal approach in the biological and physical sciences since the early 1960s (see the introduction and epilogue to his collection of papers, Wimsatt forthcoming).

12. Equally well known is the suggestion that real (rather than simulated) connectionist systems are not generative in this sense, and perhaps not even atomistic.

13. Charles Klein has pointed out to me that this requires a commitment to transworld identity of the objects involved. If that bothers you, then the quantificational phrases should be interpreted in terms of counterpart theory.

14. I discuss the fusion of property instances in Humphreys (1997). The present discussion of fused objects extends the treatment in that paper.

15. In his *Parts of Classes,* David Lewis (1991, 75) asserts of a somewhat similar case: "Suppose it turned out that the three quarks of a proton are exactly superimposed, each one just where the others are and just where the proton is. (And suppose that the three quarks last just as long as the proton.) Still the quarks are parts of the proton, but the proton is not part of the quarks and the quarks are not parts of each other." Uncharacteristically, Lewis does not give an argument for this being so. In the absence of such an argument, I can take it as no more an objection to the view proposed here than his preceding example of angels dancing on the head of a pin. Although Quine's quip "No entity without identity" is not a good argument against possibilia, it is a sound doctrine to apply to actualia, and unless angels and quarks have primitive thisness, the doctrine would seem to eliminate Lewis's examples.

REFERENCES

Eisberg, R., and R. Resnick. 1985. *Quantum Physics of Atoms, Molecules, Solids, Nuclei, and Particles.* 2nd ed. New York: Wiley.
Franklin, A. 1986. *The Neglect of Experiment.* Cambridge: Cambridge University Press.

Hacking, I. 1983. *Representing and Intervening.* Cambridge: Cambridge University Press.

Humphreys, P. 1991. "Computer Simulations." In A. Fine, M. Forbes, and L.Wessels, eds., *PSA 1990,* vol. 2. East Lansing, Mich.: Philosophy of Science Association, 497–506.

———. 1994. "Numerical Experimentation." In P. Humphreys, ed., *Patrick Suppes: Scientific Philosopher,* vol. 2. Dordrecht: Kluwer, 103–21.

———. 1995a. "Computational Science and Scientific Method." *Minds and Machines 5*: 499–512.

———. 1995b. "Computational Empiricism." *Foundations of Science* 1: 119–30.

———. 1997. "How Properties Emerge." *Philosophy of Science* 64: 1–17.

———. 2000. "Observation and Reliable Detection." In M. Dalla Chiara, R. Guintini, and F. Laudisa, eds., *Language, Quantum, Music.* Dordrecht: Kluwer, 19–24.

Kim, J. 1993. *Supervenience and Mind.* Cambridge: Cambridge University Press.

Kitcher, P. 1993. *The Advancement of Science.* Oxford: Oxford University Press.

Lewis, D. 1991. *Parts of Classes.* Oxford: Basil Blackwell.

Shapere, D. 1982. "The Concept of Observation in Science and Philosophy." *Philosophy of Science* 49: 485–525.

Suppes, P. 1978. "The Plurality of Science." In P. Asquith and I. Hacking, eds., *PSA 1978,* vol. 2. East Lansing, Mich.: Philosophy of Science Association, 3–16.

Teller, P. 1995. *An Interpretative Introduction to Quantum Field Theory.* Princeton, N.J.: Princeton University Press.

Wimsatt, W. Forthcoming. *Piecewise Approximations to Reality.* Cambridge, Mass.: Harvard University Press.

Wolff, R. S., and L. Yeager. 1993. *Visualization of Natural Phenomena.* New York: Springer-Verlag.

2

The Limits of Generative Atomism

Comment on Paul Humphreys's "Extending Ourselves"

Peter McLaughlin

Department of Philosophy, University of Konstanz

Science, Kant tells us in the *Prolegomena,* knows "limits" but no "bounds" or "bounds" but no "limits"—the translation is to a certain extent arbitrary and Kant's use of terms is not entirely consistent. What he means is that some things lie forever beyond the reach of science because they are not the kinds of things that science can tell us about: primarily questions of metaphysics or morals. However, within its own proprietary realm (that is, the empirical knowledge of nature) science has no bounds or limits: its progress will never be completed.[1] As plausible as this position is as an abstract ideal or a regulative principle, it does not seem to have guided the proclamations of the practitioners and publicists of modern science.

The dream of a final theory—a *Weltformel*—is endemic to modern science. The goal or ideal of a science that correctly and completely describes how the world really is, is, I think, what most scientists strive to attain. On occasion some actually believe they are on the brink of reaching this limit of science. Physicists of just about every other generation since the mid-nineteenth century seem to have expected the near-completion of their science and thanked their lucky stars that they were born a generation before all the interesting stuff had already been discovered. The oft-quoted paradigm of exaggerated confidence in our impending arrival at limit science is Denis Diderot's proclamation in 1754 that mathematics was basically complete; he expected it to stay put forever where it had been left by Euler, d'Alembert, and the Bernoulli family: "They have erected the Pillars of Hercules," he wrote. "We shall not pass beyond them."[2] And it would not surprise me if every generation of physicists had produced a suitably quotable expression of similar hubris. If you view science through the metaphor of the scientist sailing out beyond the Pillars of Hercules like Columbus to discover new lands, it is quite reasonable to conclude that the big continents will be discovered first and are perhaps now all already taken, and thus that (for obvious reasons) it is only the smaller islands that are still undiscovered. In an effectively finite world, geographical exploration must be subject to diminishing returns. Thus future Nobel Prizes for navigation of this kind can only be awarded for discovering ever smaller sandbars.

The notion that the interesting and exciting discoveries in science will run out soon, that the next generation of scientists will only add decimal places to today's discoveries, is probably harmless. At the latest it is exposed for what it is in the next generation. Such notions do not necessarily even claim to specify the content of future ideal science: they sometimes just announce the imminent arrival of this still-unspecified content. However, some more sedate notions may actually be more problematical, for instance, the notion that although one cannot indeed specify the content of future ideal science or even predict its arrival date, one can nonetheless determine its *form*. Emil Du Bois-Reymond, for instance, though he did not predict what the world formula would be or when it would be discovered, nonetheless did purport to know what its mathematical and ontological form would be: a set of differential equations governing the forces and motions of elementary particles in space. The limits of science for him were the limits to what Laplace's Demon could know: the "mechanics of atoms" (Du Bois-Reymond 1916, 16, 23).

Paul Humphreys has taken up two somewhat differing problems connected with the conceptualization of the form or structure of this ideal or "limit" science, each of which might mislead us into seeing deficiencies in real science because of a false or ungrounded view of ideal science. The first problem is caused by an explicit philosophical theory of what ideal science should look like (logical empiricism). The second is based on an implicit metaphysics that is most often not even recognized as such (generative atomism). Humphreys characterizes the first problem, because of its specific content (observability by the unaided senses), as an "anthropocentric constraint," but he clearly means to describe not an idol of the tribe but an idol of the theater. The second problem is of course an idol of the marketplace.[3]

I shall have little to say about his critique of logical empiricism, which I basically support, except to point out—and peremptorily to reject—the poison pill of scientific realism insinuated within this sugar coating. If we stipulatively define the pragmatist's fundamental concept of *reliability* in terms of the realist's fundamental notion of *accuracy*, as Humphreys does, then we have settled the issue before we get started—but this leads us to sectarian quarrels that are not directly at issue here. I shall concentrate my remarks on Humphreys's second point, which is dealt with in the second half of his chapter.

Not only have scientists long dreamed of a final theory, often they have dreamed the *same* dream: what Humphreys calls "generative atomism." In recent discussions this view also passes by the name of "microdetermination" or "mereological determination"; it sometimes used to be called "ontological reductionism"; Immanuel Kant called it simply "mechanism."[4] Generative atomism may be conceived of as a methodological prescription, such as "a system *is to be explained* in terms of the intrinsic properties and lawlike interactions of its parts," or as a metaphysical postulate, such as "every system *consists of and is determined by* the system-independent properties and the interactions of its parts." Both versions preclude so-called downward causality (macrodeterminism or holism), but they leave open the question of how often the level reduction should be iterated: whether every part is in turn (to be taken as) a system or whether there are ultimate parts.

I think it is unquestionable that some form of generative atomism has been the dominant self-understanding of modern science and that it still exerts considerable appeal. It is one of the notions that distinguished seventeenth-century science from that of Aristotle's medieval followers. For Thomas Aquinas, a natural body was distinguished from an artificial one precisely by the fact that its component parts lost their individuality in the compound (*miscibilia enim in mixto non salvantur*). Only in an *artificial* body did the parts remain intact like bricks in a wall (1961–67, Bk. 4, §35). For the heroes of the Scientific Revolution, on the contrary, neither the primary or essential properties of bodies nor the substances themselves came and went depending on context. Isaac Newton, let us remember, was looking for the most general properties of the ultimate particles—properties these particles would have independent of any external system, for instance, even if they were alone in empty absolute space. He shared the common belief of mechanical philosophers that the analysis of a phenomenon was both an ascent from the particular to the general *and* the dissection of a system into the parts that could be used to recompound it. The smaller the part, the more universal the property. Abstraction from the particular to the general and dissection of the whole into its parts could be done with the same scalpel or lens; they were basically the same activity. As Newton put it, "By this way of Analysis we may proceed from Compounds to Ingredients, and from Motions to the Forces producing them; and in general, from Effects to their Causes; and from particular causes to more general ones, till the Argument end in the most general. This is the Method of Analysis" (1952 [1730], 404, Qu. 31). This identity of dissection and theory formation, of decomposition and idealization, is also well illustrated in the traditional empiricist metaphor of the clock that the scientist opens up in order to observe the intrinsic properties of its internal parts.[5]

There have of course been doubters along the way, but they have been comparatively few in number. Leibniz, for instance, was not entirely in agreement with this position; Kant, though he subscribed to it, found it somewhat less than evident and therefore wrote half a book trying to justify the position; and Ernst Mach urged against atomism that perhaps "Nature does not begin with elements as we are obliged to begin with them."[6] But the empirical success of the scientific program linked with generative atomism swept aside all dissent as long as this dissent was merely philosophical criticism of its metaphysical presuppositions. Humphreys points, however, to some recent inner-scientific problems for this view. Electrons in a hydrogen molecule, like protons in the nucleus of a larger atom, seem to lose their individuality. For Aquinas this is the most natural thing in the world. But for most later thinkers, from Descartes to Du Bois-Reymond, this is absurd: there are no properties without a substance to anchor them; if you lose your individuality, you have lost all you have to lose. Thus, we may now also have empirical reasons for concrete doubt as to whether ideal or limit science will accord with this metaphysical program. Generative atomism may not be merely metaphysically suspect; it may also be incompatible with the best science available.

But should something like generative atomism be viewed solely from the perspective of the unacceptable limits it *now* places on science? After all, it did determine the self-understanding of scientists through more than three hundred years of scientific progress. Can we drop it without a replacement? Although the philosopher may deny that anything distinctive can be said about perfected or limit science and the historian may present one case after another in which one generation of scientists transcends the limits envisioned by a previous generation, still the scientists themselves apparently just will not stop talking about the form and content of future ideal science—and using metaphysical assumptions to do this. The reason for this may lie in the fact that they view themselves as actively pursuing this ideal science (that gets it right) and thus have to make some conjectures about the nature of what they are pursuing. Before you set off in pursuit of an ideal, you must make a (one hopes informed) decision about which direction to follow—at least whether to veer to the left or veer to the right; you must have some notion of what such a science is like and where it might lie. An abstract ideal can only actually be pursued by treating it as something *like* some end to which we already possess the means, that is, as something that can be acquired by techniques similar to those available. The ineffable cannot even be pursued. The scientists seen by their own generation as pursuing the ultimate truth may be seen by a later generation as merely exploring the limits of their conceptual scheme or exhausting the potential of their conceptual tools. But as long as you want to characterize the actions of scientists by their intended ends, it would seem that some anticipation of the results of science is demanded. If metaphysical assumptions are necessarily part of such anticipations, we may be stuck with them or some equally metaphysical successor.

Let's take a somewhat more mundane goal—flying. I assume that humans have wanted to fly since our ancestors first saw a bird. But before the abstract end of flying can become a deliberate decision to take some concrete actions to make some particular kind of flying machine in order to achieve this end, you have to survey your real options. What it actually means to *want* to fly at some particular time will change with the options actually realizable. Whether the pursuit of the goal of flying leads you to join feathers and wax (Daedalus and son) or to sew a giant balloon (Montgolfier brothers) or to put a propeller on an internal combustion engine (Wright brothers) depends on the historically given means available, which thus determine what actual *operative* goal notions we can have. So, too, the decision regarding which way to go in pursuit of the ideal science will depend on the scientific means available at the time—perhaps with metaphysics to fill in the gaps. The notion that the ideal science consists of ever more elegant equations applied to ever smaller particles is on a par with the notion that the ideal flying machine consists of ever sturdier balloons applied to ever lighter baskets—it's the best projection or anticipation available at a certain historical constellation. We must always leave open the possibility that some newly developed means could enable and thus generate new and unforeseen ends that we might still want to call "flying"—and that newly developed scientific means might produce ends that we would still want to call scientific, though they are not dreamed of yet in our philosophies.

NOTES

1. Kant (1987 [1790]), §57. The *limits* that science is supposed to have are called "Schranken"; the *bounds* that it is not supposed to have are called "Grenzen."
2. "Within the next century we shall not count three great mathematicians. This science will stand quite still where it was left by the Bernoullis, Eulers, Maupertuis', Clairauts, Fontaines, and d'Alemberts. They have erected the Pillars of Hercules. We shall not pass beyond them" (Diderot 1964 [1754], 180–81). See Rescher (1978, 9) for this and many other examples.
3. Leibniz once characterized the notion of absolute space, which he saw connected to generative atomism, as a *culturally* based idol of the tribe, as the *idolon tribus* "of some modern Englishmen"; third letter to Clarke §2, GP 7, 363, and a letter to Remond (3/27/1716) GP 3, 673. For discussion see Freudenthal (1986, 86).
4. See Klee (1984), Kim (1992), and Hoyningen-Huene (1994). Kant writes in the *Critique of Judgment* (1987 [1790], §77, B351): "If we consider a material whole, in terms of its form, as being a product of its parts and of their forces and powers for joining together on their own accord, then we imagine a mechanical kind of production of the same."
5. The classic source for this metaphor is Roger Cotes's preface to Newton's *Mathematical Principles of Natural Philosophy* (1934 [1726], xxvii–xxviii).
6. Mach (1976, 229). On Leibniz see Freudenthal (1986); on Kant see McLaughlin (1990).

REFERENCES

Aquinas, T. 1961–67. *Liber de veritate catholicae fidei contra errores infidelium, seu summa contra gentiles*. Torino and Rome: Marietti.

Diderot, D. 1964 [1754]. "De l'interprétation de la nature." In *Oeuvres philosophiques*. Paris: Garnier.

Du Bois-Reymond, E. 1916. *Über die Grenzen des Naturerkennens, Die Sieben Welträtsel: Zwei Vorträge*. Leipzig: Veit.

Freudenthal, G. 1986. *Atom and Individual in the Age of Newton*. Dordrecht: Reidel.

Hoyningen-Huene, P. 1994. "Zu Emergenz, Mikro- und Makrodetermination." In W. Lübbe, ed., *Kausalität und Zurechnung: Über Verantwortung in komplexen kulturellen Prozessen*. Berlin: De Gruyter, 165–85.

Kant, I. 1987 [1790]. *Critique of Judgment*. W. Pluhar, transl. Indianapolis: Hackett.

———. 1994 [1783]. *Prolegomena to Any Future Metaphysics that Can Qualify as Science*. P. Carus, transl. Chicago: Open Court.

Kim, J. 1992 "'Downward Causation' in Emergentism and Nonreductive Physicalism." In A. Beckermann, H. Flohr, and J. Kim, eds., *Emergence or Reduction? Essays on the Prospects of Nonreductive Physicalism*. Berlin: De Gruyter, 119–38.

Klee, R. L. 1984. "Micro-Determinism and Concepts of Emergence." *Philosophy of Science* 51: 44–63.

Leibniz, G. W. 1875–90. *Die philosophischen Schriften*. 7 vols. C. I. Gerhardt, ed. Berlin (reprint: Hildesheim: Olms, 1978; abbreviated as GP).

Mach, E. 1976. *Die Mechanik*. G. Wolters, ed. Darmstadt: Wissenschaftliche Buchgesellschaft.

McLaughlin, P. 1990. *Kant's Critique of Teleological Judgment*. Lewiston, N.Y.: Mellen.

Newton, I. 1934 [1726]. *Mathematical Principles of Natural Philosophy*. F. Cajori, transl. Berkeley: University of California Press.

———. 1952 [1730]. *Opticks*. New York: Dover.

Rescher, N. 1978. *Scientific Progress: A Philosophical Essay on the Economics of Research in Natural Science*. Oxford: Blackwell.

3

The Limits of Natural Science

Rescher's View

Robert Almeder
Department of Philosophy, Georgia State University

Nicholas Rescher's view on the nature and limits of natural science includes the following basic items[1]:

1. There are some nontrivial, empirically answerable questions that we cannot answer by appeal to the methods of testing and confirmation proper to the natural sciences. Answers to questions about the validity of the inductive method itself, or about the basic beliefs upon which natural science rests at any time, for example, we cannot establish directly and noncircularly by appeal to inductive methods.

2. The methods of induction are also limited in that even when one's particular beliefs about the world turn out to be fully and warrantedly assertable under the usual methods of testing and confirmation in natural science, such warranted assertability is evidence neither of their certainty nor of any strong guarantee of their truth. In fact, for this reason, all beliefs in natural science are fallible, and hence, however well confirmed or confirmable, they are defeasible.

3. Moreover, although the products of inductive methodology may be limited in these ways, the methodology is more than justified as a methodology reliably providing knowledge of the external world because in the long run, and on the whole, it leads often enough to truth. Thus, for example, the limits of natural science do not extend to a purely instrumentalist interpretation of theories.

4. The number of nontrivial, empirically answerable questions that science can in principle answer at any given time is inexhaustible, or indefinitely large; and so natural science will never, even unto eternity, answer all the questions that it can in principle answer.

5. Whatever progress natural science will make unto eternity by way of answering more and more nontrivial questions in time, the progress will be increasingly limited by economic forces and an expanding of natural resources, so that, under the best of circumstances, there will be a logarithmic retardation in scientific progress as time advances. This retardation will result from the ever-increasing cost of the technology necessary for doing science in conjunction with ever-increasing global demand for social services outside science.

6. Scientific theories are limited in that while seeking to describe accurately and completely the nature, theoretical entities, and structure of the world as it is, they can never quite achieve as much— although, to be sure, they approximate or come close to doing as much.

7. As natural science progresses, it is inevitable that the technology required for solving social problems will become so complex (as will the social problems themselves and the answers offered) that science will not be able to deliver the technology or answers necessary to solve the social problems that will inevitably emerge. Natural science, then, in its applicative function, is limited in that it will not indefinitely be able to provide adequately for important social progress and welfare because the answers provided by technology become ever more complex, and the technology becomes ever more costly and difficult to control in the interest of promoting social welfare.

After offering an exposition of the reasoning Rescher offers for the first six of these items, I shall examine briefly the picture presented. In so doing, we will see that although one might conceivably question some of the finer implications of Rescher's thesis on the eternity of scientific progress, we can nonetheless applaud the impressive answers offered to the general question of the limits and scope of natural science. Let us turn then to the reasoning Rescher offers in support of the first six claims about the limits.[2]

Scientism and Legitimately Answerable Questions outside Science

For Rescher, it is a mistake to believe that the only legitimately answerable questions about the world are those that can be answered by appeal solely to the methods of testing and confirmation in natural science. Nor has he argued that whether anybody knows anything at all is a scientific question. In fact, he finds such positions strikingly self-defeating because, as he says, "To engage in rational argumentation designed to establish the impossibility of philosophizing is in fact to engage in a bit of it."[3] Like Carnap, for example, he accepts the view that whether one's methods of testing and confirmation in natural science are themselves epistemologically acceptable is a function of whether they ultimately guarantee us precise sensory predictions and control (or applicative success), which is the primary end of cognitive inquiry. Trying to determine inductively whether one's methods are acceptable seems viciously circular if one's purpose is to establish the validity of the inductive methods of natural science (Rescher 1984, 12ff.). Along with Aristotle, Carnap, Sellars, and many others, Rescher often asserts that the first principles and the methodology of natural science cannot be justified or established in any noncircular way by explicit and direct appeal to those very principles in question.[4] And yet, for Rescher, there must be a solution to the problem of induction because, contrary to what the skeptic suggests, we do have knowledge of the physical world, thanks to the use of inductive methods. We know, for example, that atoms exist, and this latter bit of knowledge depends on a good deal of inductive inference. Nor can we establish the validity of inductive methods a priori, as some (e.g., Bonjour 1985, 1998) have suggested. Rather, their validity can, and should, be established *pragmatically,* that is, by directly seeing whether, when adopted, they lead ultimately to the observed satisfaction of the primary goals of science. If they do, although this pragmatic form of justification may indeed be empirical in a general way, it is not *scientific* in the sense of proceeding directly from standard testing and confirming whether the primary cognitive goal has been satisfied (Rescher 1984, 12ff.).

The skeptic, incidentally, may demand more justification for inductive reasoning and first principles. But Rescher's response is that, for various reasons, such a demand roots in a sterile Cartesianism feeding upon a faulty argument whose conclusion is that all beliefs about the world are truly doubtful and therefore in need of justification.[5] Once this skeptical conclusion is eliminated, we need not worry about the charge of there being no noncircular defense of induction or of basic principles, because these basic beliefs will not be the product of beliefs directly justified by appeal to the methods of induction. We can simply start the system with beliefs free of all real or honest doubt, and out of that set of initial credibilities build noncircularly the edifice of natural science (Rescher 1977b, 12, 66ff., 99ff., 189, 197–200).

Other philosophers have thought that pragmatic justification "obviously" does not work as a noncircular way of answering legitimate questions about the adequacy of induction or of first principles. After all, they say, claims about the pragmatic adequacy of any system of beliefs are themselves empirical claims, which can be justified only on inductive grounds. Rescher has replied, however, that there is nothing at all *viciously* circular about observing whether one's methods of testing and confirmation, and the assumptions upon which they rest, lead generally to reliable predictions of sensory experience.[6] We will return to this discussion later. For the moment it will suffice to note that, if Rescher is correct regarding this way of justifying basic beliefs and inductive methodology, there will be a way to defend the epistemological foundations of science without falling into rank subjectivism or skepticism based upon an infinite regress, while simultaneously avoiding vicious circularity and any appeal to a purely a priori defense of foundational beliefs in natural science. Accordingly, although natural science has certain limits in that its foundational beliefs and methods cannot be established noncircularly by direct appeal to propositions established in natural science, nevertheless it will not follow from all that that science is limited in that its beliefs at any level are based upon beliefs that ultimately cannot be objectively vindicated.

At any rate, Rescher is certainly no naturalized epistemologist, singing unreservedly the praises of natural science and announcing the death of philosophy or traditional epistemology.[7] His saying that there are some answerable questions about the world that science cannot answer differs, of course, from saying that there will always be unanswered questions *in* natural science, or questions that science *can* answer but will not, for various reasons, succeed at doing unto eternity.

Naturally, too, as is evident in his adoption of fallibilism, natural science is limited in that it cannot provide us with logically certain beliefs about the physical world. But this only means that all knowledge about the physical world is defeasible; and there is a world of difference between defeasible knowledge and no knowledge at all (Rescher 1977b, 201ff.).

The Limits of Scientific Methodology

In *Methodological Pragmatism* and elsewhere, Rescher takes a cue from C. S. Peirce and argues against another skeptical argument about the possibility of scientific knowledge. In so doing, he points to other limits of natural science without abandoning the claim that we have knowledge in science about the world.[8] The skeptical thesis in question consists in avowing that the methodology of natural science can fail in any given case to provide us with knowledge of the world. Some robustly confirmed theories or beliefs established under inductive methodology have turned out to be false, and so there is inductive evidence that they will continue to do so. For this reason, the real *probability* of error attending any claim offered by inductive methodology more than suggests that there is no knowledge of the world to be had under the method of the natural sciences. We need only admit that often in the past some of our most cherished and robustly confirmed beliefs have turned out to be disturbingly false. This argument itself, incidentally, is an inductive argument based upon the frequency of past failures among robustly confirmed beliefs in natural science.

Rescher, in response, argues that whereas *thesis pragmatism* is unacceptable for such reasons and, if followed, would support the skeptic's claims, nevertheless *methodological pragmatism* is sound. The latter amounts to asserting that the general methods of the natural sciences provide knowledge. They are selected out by nature because they tend to be generally reliable in producing true beliefs in the long run, although to be sure in any given case the methods may fail us. In short, along with Peirce, Rescher claims that the validity of inductive inference rests on the fact that if we follow inductive methods carefully, we are more likely in the long run to find the truth than if we were to follow any other method, and that is the reason why evolution and natural selection favor the use of the scientific method.[9] This means that while highly confirmed beliefs in the history of science may subsequently be found to be false, the inductive method (or the hypothetico-deductive method) is, with suitable provisos, generally reliable most of the time. So, to the skeptic who argues that truth-value revision in natural science shows that the inductive method fails to provide knowledge in any given case, Rescher's reply is that there is no justification for such a strong skepticism about the failure of scientific methodology. It is still the methodology that turns up beliefs most likely to be true, and, more often than not in the long run, what is most likely to be true is true.[10] So, even though science may be limited in that not *all* robustly confirmed beliefs in science are true, neither can it be the case that none, or very few, are true. And if the skeptic urges that knowledge is a function of certainty (of the Cartesian sort), Rescher's response is that such a criterion is too high relative to what our deepest intuitions reveal about what we all know.

Scientific Progress in General

Rescher has also argued that the number of nontrivial, empirically answerable questions in natural science is limitless. No matter how many questions we can (and do) answer in natural science, there has always been a large number of unanswered questions and indeed, typically, the answering of questions produces more questions. Hence, if one believes in induction, the future should be like the past in that there will always be interesting empirical questions to answer. In short, for Rescher, scientific progress will continue unto eternity by way of answering more and more questions in time. Hence, natural science is limited in not being able to answer all *theoretically* answerable questions. By implication, there is no substance to the claim that natural science could come to some end in answering all legitimately answerable questions. Those questions are indefinitely many.

Interestingly enough, however, although this thesis implies that scientific progress will continue for an indefinitely long period (because there will always be nontrivial, scientifically answerable questions), the progress that science makes will be drastically limited as time goes on, such that, *for all practical purposes,* science will come to an end, although not literally so. Let me explain.

Progress and Economic Constraints

In *Scientific Progress* and elsewhere, Rescher has urged that even though natural science will progress for an indefinitely long period, there will be a logarithmic retardation in scientific progress. Such retardation will occur because there will be an ever-increasing economic cost of scientific research combined with the necessity to finance other human needs competing with scientific progress in a world of finite and decreasing resources and productivity (even in the presence of advances in technology). So, for Rescher, although scientific progress will enter into a stage of progressive deceleration, still, given an indefinite future, scientific progress will never stop in some final theory of the empirical world. In defending this thesis, Rescher claims some

dependence by way of inspiration on the important work done by
Peirce on the economics of research and, of course, adds a few points
Peirce left out. Here is his basic argument that natural science is limited
in crucial ways by economic forces.

Given Planck's principle that every advance in science makes the
next advance more difficult and requires a corresponding increase in
effort, ever-larger demands are made of the researchers in science as it
progresses. If we combine this conclusion with the thesis that real
qualitative progress in science is parasitic upon advances in technology, and that these advances in technology are in turn ever more costly,
then we shall need to conclude that scientific progress will become
progressively slower because experiments will become increasingly
more difficult to conduct for being ever more expensive in a world of
finite and decreasing resources. So, if we factor into our understanding
of scientific progress the cold, hard economic facts of increasing cost
for technology and decreasing resources and productivity, then a
logarithmic retardation in scientific progress is inevitable. Thus, as
Rescher sees it, although major scientific discoveries will never come to
a complete stop (because new questions will always emerge with each
discovery), the time between significant discoveries grows larger with
the passage of time. In providing an analogy for this, Rescher asks us
to consider such progress as similar to a person's entering a Borgesian
library, but with a card that entitles the bearer to take out one book the
first week, one for the second two weeks, one for the following three
weeks, and so on, with increasingly long periods of delay unto eternity.
The library has an unlimited, or inexhaustible, number of books in it;
every time somebody takes out a book somebody adds, we may suppose, other books to it. In this model, our ever-expanding knowledge
grows at an ever-decreasing rate. Scientific revolutions will go on forever, but the time gap between them moves toward infinity and carries
with it a substantial slowdown in science as an activity (Rescher
1977a, 230).

In discussing the economics of deceleration in scientific progress,
Rescher considers the possibility that, as technology develops, the cost
of progress will become less expensive rather than more expensive. He
argues that whereas this prospect is certainly feasible, it is not realized
in practice, and hence a nonlinear cost increase is required to maintain
a constant pace of progress in science, at least into the indefinite future.
Furthermore, that there could not be a finite number of empirically

answerable questions follows, Rescher argues in the same place, from what he terms the "Kant-proliferation effect" (1977a, 245). This is to say, along with Kant, that unending progressiveness follows from the very real phenomenon that in the course of answering old questions we constantly come to pose new ones. Thus, no matter how much science progresses, there will still be questions that need answers. By implication, there will be questions that we will never, in fact rather than in principle, be able to answer, no matter how sophisticated we become in the practice of natural science. Moreover, even in the infinitely long run, these questions will remain unanswered because answering them depends upon a greater concurrent commitment of resources than will ever be marshaled at any time in a *zero*-growth world: "They involve interaction with nature on a scale so vast that the resources needed for their realization remain outside our economic reach in a world of finite resource-availability" (Rescher 1977a, 250).

Scientific Realism and Approximationism

As instrumentalists would have it, scientific theories provide no knowledge (rather than purported knowledge) of the real world. Given usual instrumentalist arguments, science is therefore crucially limited, or fails completely, in its claims to describe the world correctly. Rescher's most recent response to this line of reasoning is straightforward: no such limit exists. Science provides a correct picture of the world even if its correct descriptions are partial and always in some important sense incomplete and more in the nature of *approximative estimations* rather than correct descriptions. As Rescher now sees it, his position falls somewhere between antirealism and realism.

More specifically, in *The Limits of Science*, Rescher describes *scientific instrumentalism* as claiming that we need not suppose that theoretical entities exist at all, and that the belief that they do is merely a useful thought-fiction helping to provide explanations for observable phenomena. Scientific realism, on the other hand, is the view that the theoretical entities asserted to exist in natural science "do actually exist as scientific theorizing characterizes them. They are real items of the world's furniture and do indeed possess substantially the descriptive constitution ascribed to them by science" (Rescher 1987, xii). Rescher objects to scientific realism, as just defined, on the grounds that philosophers typically offer it as the only plausible and best explanation of the predictive success of the scientific theories asserting the existence of theoretical entities, when in fact we know that past outmoded theories with wonderful predictive powers turned out to be not even approximately true in their main claims (Rescher 1987, 70ff.). He also objects to the antirealism of the purely instrumentalist variety for the reason that there must be some truth, or "some kernel of truth" (1987, 71), to theories, if they are to have the systematic and far-reaching instrumental success they sometimes do. Predictively successful theories *as a whole* are not demonstrably true for being predictively successful or empirically adequate; nor are they purely instrumental devices simply because, as approximative estimates, the picture they present of the world always falls short of descriptive accuracy even as approximations. In the end, for Rescher, scientific theories, *as a whole*, do not so much actually describe reality as estimate its character in ways that are epistemically privileged and worthy of being accepted as human knowledge, although, to be sure, the net of fallibilism falls everywhere.

Rescher calls the position he ultimately adopts "approximationism" and has characterized it as follows:

While the theoretical entities of natural science do not actually exist in the way current science describes them, science does (increasingly) have "the right general idea." Something roughly like those putative theoretical entities does exist—something which our scientific conception only enables us to "see" inaccurately and roughly. Our scientific conceptions aim at what exists in the world and only hit it imperfectly and "well off the mark." The fit between our scientific ideas and reality is loose and well short of accurate representation. But there indeed is some sort of rough consonance. (1987, xii)

In adopting a position that falls under *approximationism,* Rescher sometimes characterizes his position in *The Limits of Science* as one of adopting *realism of intent* rather than achievement—"a realism that views science not as actually describing reality but as merely *estimating* its character" (1987, 71). Some philosophers will naturally characterize Rescher's position as a form of scientific realism simply because it asserts the essential point that some of the claims made in natural science (including the claims about the existence of theoretical entities) are approximately correct descriptions of an independently given world not of our making. That is all some philosophers require for scientific realism. Alternatively, however, some philosophers will call it a straightforward form of scientific instrumentalism simply because Rescher argues later in *Scientific Realism* that we do not know which propositions are approximately true and thus which are doing the work of the theory.

In the end, whether we call his position ultimately realist or instrumentalist seems incidental to Rescher's basic claim that natural science is limited in that, even under the best of circumstances, the best of theories can be construed as only rough approximations, and the predictive success of them is no high road to the literal truth of the claims made in them.[11] As a matter of fact, for Rescher, the approximate truth of the theory, or the approximate estimations of the theory's truth, turn out to be the reason why the theory is predictively successful. But what exactly does he mean by the approximative truth of a theory?

On reflection, Rescher says in *Scientific Realism* that the success of scientific theories as predictive devices rests on some true propositions in the theories; but we do not know, and cannot say, exactly which ones are true.[12] He then claims that the basic lesson in all this is that we not couple the pragmatic success of theories with the claim that the *theories* get it right, either fully or approximately (1987, 73). This last quote seems to be a bit at odds with the earlier reasons given for the success of theories, but this latter explanation of the success of scientific theories, rather than the earlier explanation offered previously, is demonstrably more compelling, for reasons we shall see shortly.

In sum, Rescher's explanation of the success of scientific theories seems more in line with adopting the view that science in some way, even if only approximately and roughly, succeeds at any given time in accurately describing the external world. That sounds more like realism than antirealism—unless we agree with Rescher that realism amounts to the view that not only does science deliver up true statements about the world, but also we know just which statements are true. It certainly seems, however, that Rescher adopts realism if realism is defined solely in terms of science providing some beliefs that are accurate (even if incomplete or approximative) descriptions of independently given reality—which explain the long-term predictive success of the theory—even if we cannot say which sentences in the theory (or which sentences assumed or implied by the theory) are in fact the ones that are more or less true and doing the work of the theory.

Appraisal

By way of assessing briefly Rescher's views on the limits of natural science, it is difficult to disagree with his firm rejection of scientism.

Nor does this rejection of what has come to be known as "the replace-ment thesis" imply an abandonment of a rich naturalism, unless one happens to define stipulatively a rich naturalism in terms of the re-placement thesis. Science can still be the most privileged of methods for understanding the nature of physical objects and the regularities governing their activities in this world, even if there are some legiti-mate questions about this world that are not scientific questions. Any-thing else, as Rescher has argued, would be a self-defeating and in-coherent form of naturalism.[13]

Second, with regard to Rescher's distinctive views on the economic limits on scientific progress owing to the ever-increasing cost of tech-nology, it is again equally difficult to disagree with the impressive contours of the thesis. Admittedly, in a world of finite and decreasing resources, where global population tends to increase exponentially, there will inevitably be a time when the technological cost of scientific progress will become a real impediment to robust scientific progress, and in all likelihood there will be a logarithmic retardation of scientific progress for that reason. On this score, however, there may well be a question of whether global population will continue to increase non-arithmetically into the indefinite future. Some demographers, for ex-ample, now predict that the world population will peak soon and then enter into a period of indefinite decline (Eberstadt 1997, 3–22). That the world's population has in the past tended to increase exponentially may well be the result of certain factors no longer present. This is an empirical question, of course, and one best left to demographers. Even so, it is worth raising the issue because if contemporary demographers are correct in their estimations, then the cost of the new technology necessary for robust and continuing progress may not become so high as to induce the logarithmic retardation Rescher predicts. The only question then is whether the cost of technology for doing science would continue to escalate for the simple reason that, owing to in-creasing scarcity of natural resources, global productivity would drop dramatically enough in the presence of escalating costs for other hu-man needs, even in a world of decreasing population.

This seems plausible enough. But it is nonetheless within the realm of possibility that a cost-effective technology could arise for the safe use of atomic or solar energy. If so, then even though such energy would not be an unlimited resource, it might take a very long time for the logarithmic retardation in scientific progress to emerge. But presumably Rescher's point is that the logic of logarithmic retardation in scientific progress ties tightly to the fact that, in this world, natural resources for economic productivity are finite and decreasing, and the demand on them is at least constant if not escalating. In such a world the only way to stop the progressive deceleration in scientific progress would be to guarantee that demand for natural resources extend only to those resources that are basically renewable in a way that establishes an equilibrium between the demand and the supply without threatening the capacity for renewal itself to meet existing demand.

Third, on the question of scientific realism, Rescher's earlier views seem different from his later views. In the earlier works (including *The Limits of Science*) his position seems to be one of claiming that scientific theories aspire or intend to succeed in correctly describing the external world, but that, as Laudan and others have argued, the history of science justifies the view that science never succeeds, and never will succeed, in correctly describing the external world. That is why, for example, in *The Limits of Science* Rescher's position seems to end up idealistic in endorsing the view that natural science does not in fact ever succeed in correctly describing the external world. But later in *Scientific Realism* (and in *Methodological Pragmatism*) we hear him saying that the success of natural science under the methods of the natural sciences indicates that at some level or other there must be some truth in the system. He says, for example:

While action on false belief . . . can on occasion succeed—due to chance or good luck or kindly fate or whatever—it cannot do so systematically, the ways of the world being as they are. . . . it is effectively impossible that success should crown the products of systematically error-producing cognitive procedures. . . .

Inquiry procedures which systematically underwrite success-conducive theses thus deserve to be credited with a significant measure of rational warrant. (1977b, 89–90)

Put somewhat differently, Rescher ultimately adopts the argument that the overall and systematic success of scientific theories as predictive devices cannot plausibly be explained in the absence of believing that there has to be some "kernel of truth" to them. As we saw earlier, this is a point Rescher emphasizes in *Scientific Realism,* and he thus comes to refine his "approximationist realism" as a position asserting that we cannot account for the systematic or overall predictive success of some of our theories without assuming that some of the claims made in the theories are accurate descriptions of the external world, even if those descriptions are approximate or incomplete, and even if some of the claims made in the theories are complete failures as successful descriptions. Moreover, in *Scientific Realism* he argues (as we also saw) that just as when a theory fails us we are not in a position (being holists of sorts) to say which sentences failed at describing the world, so too when a theory works well we are equally at a loss to point out which of the beliefs are the successful describers of fact. Naturally, the success of Rescher's position depends on showing that all proposed alternative explanations for the general predictive (or "applicative") success of the theories or hypotheses in question are not as plausible as the assertion that some of the sentences in any successful theory must succeed in correctly describing the world even though we may not be able to pick out the ones that do from the ones that do not. So scientific *theories* may all be false and ultimately go the way of the geocentric hypothesis, but even then some of the claims made in them must succeed in correctly describing the external world.

Notice then that Rescher's view here is more realistic than the "realism of intent" or the "realism of aspiration" he defended in *The Limits of Science* and elsewhere. Notice further that the position, as so construed, is consistent with perpetual revolutions in natural science. It is also consistent with the claim that "all theories are born dead" because, as theories, they will be replaced by successor theories. Finally, notice that in this final and refined picture, according to Rescher, we do not have an effective decision procedure for determining which sentences of our theories are doing the successful describing and hence carrying the success of the theories. Of course, not all of the propositions in a successful theory can be true. If they were all true, then scientific theories would never get replaced. And if none of them were true, then we would have no way to explain the predictive success of such theories. Successful theories require true propositions, but eternal revolution in science requires a good deal of false but robustly confirmed beliefs.

As so stated, then, Rescher's scientific realism is (pending refutation of instrumentalists' alternative explanations of success) quite defensible as a statement on the theoretical limits of scientific theory in providing for an understanding of an independently given world. It implies that whatever we take as criteria for truth (presumably robust confirmability or warranted assertibility, suitably explicated in terms of coherence with initially given data) are sometimes satisfied; but it does not imply, even with a high degree of probability, that any particular proposition one might name as being completely justified is in fact satisfying the correspondence definition of truth, although we know that some of them must be doing so. In this it differs, of course, from classical scientific realism, because the latter typically asserts that we know, and can say, *which* of our beliefs about the external world are in fact the ones describing the external world. But if the *core* of classical realism is the belief that some of our robustly confirmed beliefs about the external world are more or less correct descriptions of that world, and thus that science succeeds at any given time in telling us in some important way how the world really is, then Rescher's position is surely consistent with the *core* of classical scientific realism. As so construed, with the suggested provisos, Rescher's position is an attractive, compelling, and preeminently defensible answer among all other proposed answers to the question of the extent to which natural science can and does succeed in some important measure in accurately describing the external world.

Fourth, in discussing Rescher's views about the foundations of empirical knowledge, we saw earlier that for Rescher there can be no noncircular inductive justification of foundational beliefs that serve as the basic evidence for other beliefs in the structure of science. Rescher's claim is essentially that when it comes to justifying basic beliefs, we start with certain *presuppositions* that serve as evidence for other beliefs; and if what is demonstrated either inductively or deductively under these presuppositions satisfies the purpose for which our methodologies are developed, namely to predict successfully and precisely our sensory experience, then that would be pragmatic justification (or perhaps vindication) of the presuppositions as items of knowledge, even though they themselves are not directly inferred by conscious appeal to other known propositions as their evidence. The central claim, then, is that if the presuppositions were not items of knowledge, they could not in the long run, or systematically, lead to pragmatically successful results. Here again, we would have no way of explaining our success in predicting our sensory experience if the presuppositions upon which we rested the whole system were not as warranted as any result producing human knowledge anywhere else in the system. Is this way of justifying or vindicating basic presuppositions as items of knowledge still an instance of viciously circular reasoning because it appeals to inductive reasoning to establish that certain beliefs have been successful?

Some philosophers have claimed as much.[14] But Rescher's basic point seems to be that we cannot plausibly justify first premises about the world a priori, and we simply must avoid both the infinite regress and directly viciously circular justifications. All this we do by stopping the regress at basic propositions functioning as presuppositions in starting the system, and then indirectly and retroactively justifying the presuppositions as items of knowledge as a result of the empirical adequacy of the system taken as a whole over a long period of time. In this way basic propositions are items of knowledge without being conclusions from premises known and directly supporting them. They are, then, directly known without justification, but only seen retrospectively as items of knowledge because that would be the best explanation of the success of the system that they initiate.

Whether we call this "vindicating" or "justifying" first premises is immaterial to the claim that the activity is not an example of vicious circularity because no *particular* proposition established by simple induction actually enters into the justification of the claim that the presuppositions must be true. Rescher's procedure seems similar to Goodman's and Carnap's mode of *vindicating* certain answers that cannot be *justified* because justification is defined as an activity for authorizing moves within the object language, whereas the question of the epistemic status of the presuppositions of the method adopted in the language can only be a matter of vindication in terms of the whole system and its general empirical adequacy. In this way, first premises can be known to be items of knowledge even though they are not directly inductively justified, although, to be sure, the judgment that the system as a whole is adequate is an inference based upon observation. Doubtless, this issue deserves a deeper discussion and a finer distinction as to kinds of inductive inference. But if we are willing to draw a distinction between valid inductive inference *within* the language of science and valid inductive inferences *about* the language of science, we need not see vicious circularity in Rescher's way of providing for the epistemological foundations of scientific knowledge.

Finally, in spite of all this sweetness and light, one might plausibly register a contrary, and certainly a minority, belief—even if only to raise the issue for further discussion. It has to do with Rescher's thesis that scientific progress will continue forever and will never be completed, even in the presence of a logarithmic retardation in progress, because the number of nontrivial, empirically answerable questions that can in principle be answered in natural science is inexhaustible. Recall that Rescher's argument for this thesis is essentially that since there have always been questions generated by past successes in answering questions in science, there is good inductive reason to believe that as science continues there will continue to be questions emerging from all the questions answered in the advances of science. Even if there happen to be a finite number of nontrivial questions at any given time, there will always be more questions to answer at any time in the indefinite future. In fact, it looks as if this is a straightforward and uncomplicated inductive generalization from the past history of science. But this inductive argument for the thesis may be problematic. Let me explain briefly.

Rescher's thesis that the number of nontrivial, empirically answerable questions is indefinitely many would be undermined if the number of nontrivial questions that science and technology could answer were finitely many in a world existing for an indefinitely long period, and in which more and more nontrivial questions get answered under the methods of the natural sciences and advancing technology. Suppose, for example, that Bertrand Russell's celebrated chicken had strong evidence that his master-farmer had only enough food to feed him for 750 days. In that case, the chicken should not conclude that because the farmer had fed him for 750 days in the past the farmer would feed him tomorrow, on day 751. Simple induction from the past is a valid indicator of the future only under certain clear proviso clauses to the effect that there is no good reason to suppose that the conditions warranting the inductive inference will change.

So it seems that the basic inductive inference supporting the belief in the eternal progress of science *assumes* that the total number of nontrivial questions is not finitely many in a world existing for an indefinitely long period, and in which progress is defined in terms of answering more and more nontrivial questions in time. If the *total* number of such questions were finitely many, then the fact that past progress in natural science always generated more questions than existed before the progress would by no means entail that there always be such questions to answer in the future. In other words, from the fact that there have always been questions generated in the past by answers generated from science and technology, it does not follow that there

56   Robert Almeder

will always be such questions in the future *unless one already knows or assumes that the total number of nontrivial questions in the indefinite future is not finitely many.* Similarly, if Russell's chicken had not assumed that there would be enough food for the farmer to feed him for an indefinitely long period, then he might have been better prepared for his eventual demise. At this point, one might suggest that the belief that the total number of nontrivial questions at any given time is indefinitely large is no better supported by our experience than is its denial. Indeed, what is the inductive evidence that there are indefinitely many nontrivial, empirically answerable questions? Again, it cannot derive simply from the fact that in the past there have always been questions generated by answers that were provided.

Conclusion

Suitably qualified in small ways, Rescher's central views on the limits of natural science are impressive and quite defensible. Whether we are talking about the limits placed on natural science by the progressive and inevitable lack of economic resources to conduct science in a vigorous fashion, or whether we are talking about the limits of science in delivering up in any noncontroversial ways the propositions that in fact succeed in correctly describing the world in some fashion, or whether we are talking about the limits of science in providing a justification for its own basic beliefs, or whether we are talking about the limits accruing to the fact that there must be some interesting questions about the world that are not scientific questions, or whether we are talking about the limits science will inevitably face in seeking to solve pressing social issues because of the increasing complexity of technology and scientific answers, we see a persuasive panoramic picture of the limits of science along with unbounded enthusiasm for what science has done and continues to do in providing specific predictions of our sensory experience.

NOTES

1. The principal sources of Rescher's views on the limits of natural science are: *The Limits of Science* (1984), *Methodological Pragmatism* (1977b), *The Coherence Theory of Truth* (1973), *Scientific Progress* (1977a), *Empirical Inquiry* (1981), *Scientific Realism* (1987), *Cognitive Systematization: A Systems-*

Theoretic Approach to a Coherentist Theory of Knowledge (1979), *Scepticism* (1978), and *Complexity* (forthcoming).

2. I agree, incidentally, with the arguments offered by Rescher in his forthcoming book *Complexity* for the sixth stated limit; but limits of space prevent me from delving further into the arguments here.

3. See, for example, Rescher (1984, 3–4, 15, 108, and 210). For his argument on the self-defeating nature of scientism, see Rescher (1992).

4. See Sellars (1968, 101–15) and Carnap (1980). On this item, see also Rescher's response to Laurence Bonjour's piece in Sosa (1979) and Rescher (1982, 25ff.). See also Bonjour (1985, 222ff.).

5. See Rescher (1984, 12ff.) as well as Rescher (1978, 175–84; 1977b, 12ff.). Actually Rescher's position is a bit more subtle then meets the eye. In the earlier works (1973, 1977b, 1981, 1984) scientific claims, when fully confirmed, are often characterized merely as estimates of truth; the truth is in fact never reached in science, and therefore realism is a matter of intent rather than accomplishment—there is no difference between what we think to be so and what is so (1981, 223). In *The Limits of Science,* for example, he says: "Our science, as it stands here and now, does not present the real truth. The best it can do is provide us with a tentative and provisional estimate of it. . . . The standards of scientific acceptability do not and cannot assure actual, or indeed, even probable or approximate truth" (Rescher 1984, 77–79). And, in the same place, he goes on to claim that the history of science teaches us (like the preface paradox) that "we know, or must presume, that (at the synoptic level) there are errors though we certainly cannot say where and how they arise" (Rescher 1984, 83).

But, as we shall see, this earlier view, relative to the position staked out in *Scientific Realism,* turns out to be true of scientific theories in general, but in a way that allows some sentences in successful scientific theories to be true (or have some kernel of truth) even though we may not be able to pick out the sentences that are true, and even though they are always only partial descriptions (Rescher 1984, 84). Later in *Scientific Realism,* as we shall see, he makes it more explicit that some of the sentences in successful theories are correct (true) descriptions even if we cannot say which sentences they are. But even in *Empirical Inquiry* (Rescher 1981, 236–37) he tells us that the long-term general (or systematic) predictive success of theories across the board is evidence of truth somewhere in the theory. More on this later.

6. See, for example, Rescher (1977b, 99ff.). The same point is made by Goodman (1965, 66–67).

7. See notes 2 and 3.

8. In other places Rescher has argued that we do not have any scientific knowledge of the world other than purported knowledge. But I am urging that if knowledge is robustly confirmed belief, for reasons already noted, then Rescher's views amount to the view that we indeed have such knowledge in science, and that there is nothing inconsistent between that and endorsing fallibilism.

9. See Rescher (1977b, Chapters 3 and 4, 12ff.). Also see Almeder (1982, 11ff.).

10. See Rescher (1977b, 12, 66ff., 99ff., 189, 197–200). Rescher, incidentally, adopts a correspondence definition of truth and a coherence criterion of truth, defending a carefully crafted coherence theory of justification and, in general, a coherence theory of knowledge because it is the demonstrated coherence of our beliefs that establishes the presence of knowledge. In this he is willing to settle for the generally acceptable criteria for the determination of truth in natural science; see Rescher (1973, Chapter 2). Although any one of the most robustly confirmed of scientific beliefs may turn out to be false, it is more likely that, in the long run, many of the beliefs so confirmed will be true. One cannot succeed, he thinks, at having under this method so many beliefs that succeed in predicting precisely one's sensory experience unless some of those beliefs are true (see Rescher 1977b, 89–90).

11. He says:

The best explanation of the applicative efficacy of a belief is not its actual *truth,* but simply that its departures from the truth do not make themselves felt so strongly at the prevailing level of operation as to invalidate our predictions and interactions *at this level.*

Its predictive and applicative success similarly does not show the *truth* of our theoretical picture of nature, but only its *local empirical indistinguishability* from the truth (whatever that may turn out to be) at the present level of observational sophistication. Our praxis within the parametric region in which we currently operate presumably works well in point of prediction and application simply because it operates at a level of crudity that avoids *detectable* errors. The most we can claim is that the inadequacies of our theories—whatever they may ultimately prove to be—are not such as to manifest themselves within the inevitably limited range of phenomena that currently lie within our observational horizons. This circumstance certainly has cognitive utility. But this is a far cry from justifying the claim that the theories that work well "as best we can tell" are thereby *correct.*

In general, the most plausible inference from the successful implementation of our factual beliefs is not that they are right, but merely that they (fortunately) do not go wrong in ways that preclude the realization of success within the applicative range of the particular contexts in which we are able to operate at the time at which we adopt them. . . . In going for the *best* explanation of a family of successful implementations, the destination we reach is not "right" but merely "good enough for present purposes." (Rescher 1987, 71–73)

12. He says:

The lesson is again straightforward: when things go right with a prediction to which various theories contribute, we cannot tell specifically where to attribute responsibility for this success. All that we know is that some truth or other entailed by the T-conjunction is in a position to assure P that something or other both true and P-yielding is assured by the T-conjunction. When we meet with predictive success in applying one of our scientific theories, the conclusion we can draw is not that the theory itself gets the matter right, but that there is something other about it that is true . . . some consequence of it regarding whose nature we have no further information whatsoever.

The Duhem case of a failed prediction and this contrast case of a successful one run altogether parallel. When all we know is that a prediction underwritten by a group of theses fails, we have no idea of just where to lay the blame. Analogously, when all we know is that a prediction underwritten by a group of theses succeeds, we have no idea of just where to give the credit. . . .

Thus, when a theory is "confirmed" through the success of its applications, we cannot appropriately conclude that it is (likely to be) true. We do not know that the credit belongs to the theory at all—all we can conclude is that an otherwise unspecifiable aspect of the overall situation can take credit for the achievement of success—and emphatically not the whole theory as it stands. Columbus found land on the far side of the Atlantic, as he expected to— but certainly not because his theories were correct that China and the Indies lay some 3,000 to 4,000 miles across from Europe. . . .

The successful theory has some otherwise unidentifiable kernel of truth about it, and the superior theory manages to enlarge this kernel. But this says nothing whatsoever about the truth or probable truth or approximate truth of those theories themselves. (Rescher 1987, 73)

13. See note 3. For a full discussion and defense of the view that the replacement thesis, as well as the transformational thesis, is indefensible, and that therefore any naturalism based upon them must fail, see Almeder (1998).

14. In criticizing Rescher's pragmatic defense of basic beliefs, Laurence Bonjour, for example, has accused Rescher of vicious circularity:

There can be no doubt, of course, that the actual pragmatic success which results from acting upon a set of beliefs is indeed directly caused by the external world and beyond our control. *But can pragmatic success itself play any role in the cognitive system?* I do not see how it can. What seems to play such a role in the system is not pragmatic success itself, but rather *beliefs* or *judgments* to the effect that such success had been obtained. It is only such beliefs that can cohere, or fail to cohere, with the other conceptual elements of the system. And these beliefs or judgments, far from being directly caused by the impact of the world, are in Rescher's system highly indirect products of the cognitive machinery and thus dependent upon the operation of precisely the other elements over which they were supposed to provide an independent control. We do not somehow have direct unproblematic access to the fact of pragmatic success but must determine it, if at all, via some complicated process of observation and assessment. And this means that pragmatic success, contrary to Rescher's claim, does not provide the genuine input from the world which might answer the objections considered above. . . . Thus it appears that Rescher's solution to the input problem is not satisfactory, which leaves him with no response at all to the alternative coherent system objection and only a viciously circular solution to the problem of truth. (Bonjour 1985, 228)

In reply to this objection Rescher has said:

Judgments about pragmatic efficacy . . . are judgments of a very peculiar sort in falling within a range where our beliefs and "the harsh rulings of belief-external reality" stand in particularly close position. If I misjudge that the putative food will nourish me, no redesigning of my belief system will eliminate my pangs from my midriff. If I misjudge that the plank will bear my weight, no realignment of my ideas will wipe away the misery of that cold drenching . . . it is hard—if indeed possible—to reorient or recast our thought so as to view a failure in the pragmatic/affective sector as anything but what it is. . . .

To say all this is not, of course, to gainsay the plain truth that pragmatic success or failure do not operate directly in our belief system, but only operate there via *judgments* of success or failure. It is simply to stress that the coupling between actual and judged success and— above all, between actual and judged failure—is so strong that no very serious objection can be supported, by pressing hard upon a distinction which makes so little difference. (Rescher, "Reply to Bonjour" in Sosa 1979, 174)

Bonjour thought this reply an *ignoratio elenchi* (Bonjour 1985, 250).

REFERENCES

Almeder, R. 1982. "Rescherian Epistemology." In R. Almeder, ed., *Praxis and Reason*. Washington, D.C.: University Press of America, 11–36.
———. 1998. *Harmless Naturalism*. Chicago, Ill.: Open Court Publishers.
Bonjour, L. 1985. *The Structure of Empirical Knowledge*. Cambridge, Mass.: Harvard University Press.
———. 1998. *In Defense of Pure Reason*. Cambridge: Cambridge University Press.
Carnap, Rudolf. 1980. "Empiricism, Semantics, and Ontology." *Revue Internationale de Philosophie* 4: 20–41.
Eberstadt, N. 1997. "World Population Implosion?" *The Public Interest* 129: 3–22.
Goodman, N. 1965. *Fact, Fiction and Forecast*. Indianapolis, Ind.: Bobbs-Merrill.
Rescher, N. 1973. *The Coherence Theory of Truth*. Oxford: Clarendon Press.
———. 1977a. *Scientific Progress*. Oxford: Basil Blackwell
———. 1977b. *Methodological Pragmatism*. Oxford: Basil Blackwell.
———. 1978. *Skepticism*. Oxford: Basil Blackwell.
———. 1979. *Cognitive Systematization: A Systems-Theoretic Approach to a Coherentist Theory of Knowledge*. Oxford: Basil Blackwell.
———. 1981. *Empirical Inquiry*. Totowa, N.J.: Rowman and Littlefield.
———. 1982. "Response to Robert Almeder." In R. Almeder, ed., *Praxis and Reason*. Washington, D.C.: University Press of America, 36–38.
———. 1984. *The Limits of Science*. Berkeley: University of California Press.
———. 1987. *Scientific Realism*. Dordrecht: Reidel.
———. 1992. "Philosophobia." *American Philosophical Quarterly* 39(3): 301–2.
———. Forthcoming. *Complexity*.
Sellars, W. 1968. *Science and Metaphysics*. New York: Humanities Press.
Sosa, E., ed. 1979. *The Philosophy of Nicholas Rescher*. Dordrecht: Reidel.

4

Toward the Pale

Cosmological Darwinism and *The Limits of Science*

Laura Ruetsche
Department of Philosophy, University of Pittsburgh

Philosophy, some have said, is the history of future science. A metaphysician is, after all, the sort who looks at a pink ice cube and sees— or rather, hopes for—a successor to particle physics unparticulate in nature (Sellars 1962, §VI). Metaphysics, so practiced, is not the inquiry that comes after physics, but the one that goes before. Should science reach a state beyond which it *cannot* progress, a state of science completed or science incompletable, science would cease to have a future substantially distinct from its present. Science at its limit would mark the end of metaphysics, so conceived.

The central theses of Nicholas Rescher's 1984 study *The Limits of Science* should comfort those who have a stake in the perpetuity of metaphysics, so conceived. Does science confront limits in the form of insolubilia, questions it can entertain but cannot answer? No, says Rescher, invoking the *plasticity* of science. Will science reach a limit in the form of a terminus, a state of completion? No, says Rescher, invoking the erotetic fecundity of science, its capacity to beget open questions without end. I propose here to review one argument Rescher offers *against* insolubilia and another he offers *for* the perpetual incompleteness of science.[1] The juxtaposition of these arguments indicates a loophole in the argument for perpetual incompleteness, which indicates in turn a strategy for completing science.

Lee Smolin, a physicist at Penn State, has been promoting what he calls the theory of cosmological natural selection, and what I will call *cosmological Darwinism* (Smolin 1997). Smolin is too responsible to claim that this openly speculative and diversely promissory theory is the ultimate (as in final) theory of physics. But he does claim that cosmological Darwinism illustrates the form an ultimate theory might take. *And* he seems to me to exploit the loophole in Rescher's perpetual incompleteness argument to make this claim. As those who lift with their backs instead of their legs are prone to discover, one way to gain an appreciation of rules is to break them. I propose here to gain an appreciation of Rescher's principled commitment to the perpetual provisionality of science by examining how Smolin's cosmological Darwinism would break that commitment.

Against Insolubilia

Rescher can resist insolubilia because he refuses to "legislate *a priori* . . . just what explanatory principles [science] may and may not use" (1984, 112). Philosophers are typically concerned with the normative. One way a philosopher of explanation might express this concern is the way of high-handed legislation, of determining standards of explanation *for* science, and judging scientific episodes against those standards. Emphasizing as always the primacy of practice, Rescher rejects this way. "Science," he writes, "is not a fixed object but a temporal and transient one" (1984, 130). A concern with explanatory norms properly enlightened by an appreciation of science's plasticity would (presumably) be driven by attention to the ground-level work of actual scientists, as they lodge and dislodge explanations—cosmologists confronting the horizon problem, say, or molecular biologists dismantling the central dogma. Absent such attention, philosophers cannot discern the shapes of explanation in the sciences and cannot assume the congruence of those shapes to some ordained essence of explanation.

Eschewing high-handed legislation, Rescher reasons as follows: if there is no essence that legitimate scientific explanations must instantiate, then neither is there anything essentially inexplicable. And if nothing is essentially inexplicable, no scientific question is *in principle* unanswerable. To be sure, a given explanatory regime, governing a particular state of science, may oversee apparent insolubilia. Consider "the riddle of existence": *Why is there something rather than nothing?* Under an explanatory regime wherein nothing counts as an explanation for the existence of a thing but the existence of another thing, connected to the first by a relation of (causal) necessitation, the riddle stumps us. For it delimits that which is to be explained in such a way that nothing is left to do the explaining! Does it follow that *no* explanation is possible? Only, Rescher contends, if we are committed to "a principle of hypostatization to the effect that the reason for anything must ultimately inhere in the operation of things" (1984, 120).

But owing to the plasticity of science, we needn't be so committed. For a clear shot at explaining why there is something rather than nothing, we need only reconfigure the explanatory terrain, by adopting alternative principles of explanation. Rescher takes "perhaps the most promising prospect" to be a teleological principle "to the effect that things exist because 'that's for the best'" (1984, 124).[2] For the sake of his argument against insolubilia, Rescher need not maintain that this maneuver answers the riddle of existence. He need only maintain that science's plasticity enables the rearrangement of the explanatory terrain that the maneuver exemplifies, and that such rearrangement can render apparent insolubilia explicable. The moral is that no scientific question is in principle unresolvable. "Since ultimacy is never absolute but framework internal, we can defeat it by shifting to a variant explanatory framework" (Rescher 1984, 127).

This raises yet another way to express a concern with the normative. Suppose we allow that legitimate scientific practice can include restructuring of the explanatory terrain. Surely we would wish to distinguish responsible from irresponsible restructuring! A cheap way to render everything explicable would be to enumerate our scientific commitments, and adopt with respect to that enumeration the novel explanatory principle that the *n*th commitment explains the (*n* + 1)st (we would also have to stipulate that the first commitment is self-explanatory, but why not?). Worse than cheap, such an explanatory principle is worthless. What distinguishes the explanatory restructuring it effects from the worthwhile explanatory restructuring effected by, say, Darwin's idea that the character and diversity of organic life may be explained by appeal to the operation of a statistical mechanism? Rescher contends that what distinguishes true-born explanatory principles from windeggs are pragmatic criteria of efficacy relative to scientific aims of prediction and control: "in the final analysis, praxis is the arbiter of theory" (1984, 46). Principles underlying the cheap explanations do not frame predictions; Darwin's principles do. For Rescher, the latter, but not the former, represent responsible explanatory expansion.

For Incompleteness

Can science reach state of *erotetic completeness,* that is, a state in which no legitimate question lacks an answer? William James opines (as only William James could) that "it seems *a priori* impossible that the truth should be so nicely adjusted to our needs and powers [that science completes itself] . . . in the great boarding house of nature, the cakes and the butter and the syrup seldom come out so even and leave the plates so clean" (as quoted in Rescher 1984, 33). Rescher agrees, and he reinforces his agreement by describing how scientific answers beget fresh scientific questions.

The mechanism of question propagation Rescher envisions is not the crude one, mastered by most toddlers, of using the answer to each "Why?" question as the object of the next "Why?" question. Indeed, Rescher disables this infantile mechanism of question propagation during his discussion of the riddle of existence. The teleological answer to the riddle—that there is something rather than nothing because it is fitting that there be—leaves the "residual issue: 'But why should what is fitting exist?'" Rescher writes, "We shall have to answer the question simply in its own terms: 'Because that's fitting.' Fitness is the end of the line" (1984, 124). So much for the infantile mechanism.

That questions have presuppositions (cf. Belnap and Steel 1976) suggests to Rescher a more sophisticated mechanism of erotetic propagation. A *presupposition* of a question is any statement entailed by every adequate response to that question. To ask what is the ratio of sodium to chlorine in table salt is to presuppose that salt contains both elements. The replies "1:1," "3:2," and "22:35" all imply that both elements are present. The exclamation "But salt's composed of *calcium* and chlorine!" does not answer but instead dismisses the question. To entertain a question is to accept its presuppositions: taking salt to be composed of nothing but calcium and chlorine, we cannot sanely ask about its ratio of sodium to chlorine. Mindful of the phenomenon of presupposition, Rescher introduces the notion of the *question agenda* of a "state of [purported] scientific knowledge." Call such a state *S,* and model it as some set of commitments. For example, let *S* = {there are quarks, tectonic plates, and mitochondria; human beings are not descended from neanderthals; *C. elegans* has 302 cells; it is fitting that there is something, . . . }. The question agenda of *S* consists of all questions all of whose presuppositions lie in *S*. These are just the questions posable relative to *S*. "How many types of quarks are there?" would belong to the question agenda of the set of commitments I just began to enumerate. "Does the line of descent from *H. neanderthalis* to *H. sapiens* pass through intermediate species?" does not.

Now Rescher maintains that "the motive force of inquiry is the existence of questions that are posable relative to the body of knowledge but not answerable within it" (1984, 30). The mechanism of erotetic propagation he has in mind runs as follows:

1. The question agenda of a state of science *S* contains at least one *open question*, a question whose presuppositions lie in *S* and whose answer does not.

2. To close this question is (at least) to add its answer to *S*. A set of commitments *S'*, not identical to *S*, results.

Comment: *Closure is not the only option. The question could remain open, or it could be dismissed. If it remains open, the incompleteness of science follows immediately. If it is dismissed, S must be revised so that it no longer contains the question's presuppositions. If the dismissal proceeds in part by way of adding commitments to S, the mechanism moves to step (3). But the dismissal may consist simply in removing commitments from S. Here lies a way to erotetic equilibrium: believing nothing, we entertain no questions we cannot answer and like the Pyrrhonist attain tranquility through nonacceptance. Insofar as erotetic completeness is a standard of completeness "internal to our cognitive horizons," it can be met by shrinking those horizons. But erotetic completeness so purchased "may well indicate poverty rather than wealth" (Rescher 1984, 34) of our conceptual apparatus, and it is no cause for self-congratulation.[3]*

3. *S'* furnishes presuppositions for questions hitherto unentertainable.

4. These new questions are *open* questions, which returns us to step (1).

Comment: *How do we know their answers are not already in S'? Suppose that the commitments added to S that function as presupposi-*

tions for the newly posable *questions do not also* answer *those questions.*[4] *Then in order for the answers to those questions to lie in* S′, *they must have been in* S *all along. Now suppose—a mighty assumption, to be sure—that* S *is closed under logical implication. Then whatever follows from a member of* S *is also a member of* S. *Ergo if answers to the newly posable questions were in* S, *their presuppositions were too (for a question's presuppositions are entailed by its answers). But then the questions were posable all along, and are not* newly *posable, as we have supposed. Modulo the presumption that the propositions rendering a question posable do not also answer it, newly posable questions are open questions.*

So running, Rescher's mechanism of erotetic propagation drives science without end.

The Tension

The mechanism of erotetic propagation posits that to answer a question genuinely open with respect to a set of scientific commitments we must add commitments to that set. But Rescher's emphasis on the plasticity of scientific explanation indicates another way to close an open question. Rather than adding to our set of commitments, we might instead add to our store of explanatory principles in such a way that some of the commitments *we already have* constitute an answer to the question. This means of answering a question does not alter our set of beliefs. It rather alters our notion of what it is to marshal some beliefs to explain others. (To say that even in this case our set of beliefs changes—it now includes the beliefs to the effect that *these* beliefs here support *those* beliefs there—is to flirt with enthymeme! Better to keep one set of books on our categorical commitments and another on the explanatory patterns we recognize.) We have already seen erotetic completeness (cheaply) achieved by the expedient of explanatory expansion: the "*n* explains *n* + 1" principle I derided earlier renders erotetically complete the state of science with respect to which it is articulated. The principle is shoddy, but the example suggests that novel explanatory principles can short-circuit Rescher's mechanism of erotetic propagation, by closing open questions without opening new ones. If so, what underpins Rescher's argument against insolubilia—a recognition of the plasticity of scientific explanation—undermines the perpetuity of his mechanism of erotetic propagation. A lacuna lurks in any argument appealing to that mechanism to establish that science, if ever incomplete, is forever incomplete.

Smolin's Cosmological Darwinism

Lee Smolin would pursue cosmology in that lacuna. Despite its dramatic empirical success, few physicists regard Einstein's general theory of relativity to be a *fundamental* theory of gravity because it is not a quantum theory, and the (perhaps bitter) lesson of twentieth-century physics seems to be that fundamental theories are quantum theories. Smolin pursues a quantum theory of gravity, and with it a "single, comprehensive picture of nature" (1997, 5). Drawing this picture he takes to be "the ultimate task of physics" (1997, 8). "Ultimate" here means "final": the picture drawn, physics would attain the completeness that Rescher's mechanism of erotetic propagation would postpone indefinitely.

A failed but noble candidate for the post of ultimate theory is the standard model of elementary particle physics. The standard model elegantly and efficiently unifies the electromagnetic, weak, and strong forces. But (as the editors of a recent collection of seminal particle physics papers warn) to dwell on its aesthetic virtues is to fail to "convey the standard model's real achievement, which is to encompass the enormous wealth of data accumulated over the last fifty years" (Cahn and Goldhaber 1989, ix) of scattering experiments. Its credentials notwithstanding, the standard model is not the ultimate theory Smolin seeks. For one thing, unifying three fundamental forces, the standard model leaves the fourth—gravity—out. And for another, it hosts around twenty free parameters—quantities (e.g., the mass of the electron) whose values the theory does not predict but instead leaves for physicists to determine on the basis of experiment. Smolin reckons that no *truly* fundamental theory would allow itself twenty parameters' worth of wiggle room. Part of what an ultimate theory *would* do, reckons Smolin, is tell us *why* these parameters take the values they do.

Rescher's mechanism of erotetic propagation is in motion. The standard model answers a barrage of questions about the nature and structure of light and matter. Its answers furnish presuppositions to questions hitherto unentertainable. One is Smolin's: only having embraced the standard model can he sensibly ask "why [must] nature . . . choose the masses and other properties of elementary particles to be as [they are in the standard model] and not otherwise[?]" (1997, 76). As Rescher's mechanism of erotetic propagation would lead us to expect, the standard model does not complete physics but rather makes available new ways to articulate its incompleteness.

To see how Smolin, having tripped the mechanism of erotetic propagation, hopes to escape it with an ultimate theory intact, we must examine the theory he proposes to answer the parameter attunement question. At the moment, this theory of cosmological Darwinism treats quantum gravity as a black box—Smolin can describe the results cosmological Darwinism requires of quantum gravity but cannot demonstrate that quantum gravity complies. Were the workings of the quantum gravity box revealed and compliant, and were the resulting theory verified, Smolin would count cosmological Darwinism as ultimate physics. If articulated cosmological Darwinism were both ultimate and correct, it would escape Rescher's mechanism of erotetic propagation.

Unaware of the mechanism, Smolin is nevertheless explicit about an escape strategy. His idea is to borrow a page from the biological playbook. Rather than explaining parameter attunement by deriving parameter values from some theory more fundamental (which theory sets the stage for more questions), Smolin invokes a statistical mechanism to explain why the free parameters have the values they do. He takes this explanation to be "historical" in character, like Darwinian explanations, and unlike what he terms the "principles"-based explanations endemic to physics. "This is how the values arose," the explanation runs, and *there ends the story.* So Smolin is gumming the mechanism of erotetic propagation with the plasticity of science. He aims to account once and for all for parameter attunement, not by taking on novel commitments (although we will see that he does do a bit of that) but by deploying novel (novel to cosmology, at least) explanatory principles. His strategy implies an answer to the question of how (if at all) the expansion of explanatory strategies can be undertaken responsibly. The implied answer is: explanatory expansion is responsible when the new conscripts are already wage-earning members of some legitimate scientific theory. *Scientific* explanations of the form Smolin offers for parameter attunement are commonplace. What is not is the use of such explanatory strategies in fundamental physics.

Now for Smolin's explanation: The parameter attunement Smolin would explain obtains every place we have looked. But we cannot look in every place. In particular, we cannot look inside black holes, remnants of collapsed stars so dense that not even light can escape their gravitational attraction. No signal of any sort can reach us from a black hole's interior. We might hope in general that our theories reveal what we cannot observe, and in particular that they reveal what transpires inside a black hole. Classical gravity dashes this hope, and dashes it dramatically. Classically, black holes contain *singularities,* regions where classical equations describing the coupling between space-time and matter-energy become nonsensical. What transpires inside black holes, classical gravity tells us, is classical gravity's own downfall. Here Smolin issues the first of many promissory notes funding cosmological Darwinism: he assumes that a properly *quantum* theory of gravity *will* smooth over the singularities classical gravity locates inside black holes. Quantum gravity having repaired these rents in the fabric of space-time, Smolin would put that fabric to extraordinary use. Inside a black hole is a region of ferocious density, a region that only a completed theory of quantum gravity can describe. We believe that a similar region lies some ten billion years in our past. It is the origin of our universe, and we call it the Big Bang. Emboldened by this similarity, Smolin hypothesizes that (quantum gravity will tell us that) what transpires inside a black hole is the birth of another universe, closely related to its progenitor in the sense that the free parameter values for the standard model characterizing the baby universe are close to those characterizing the parent universe.

Now if quantum gravity performs up to Smolin's specifications, and black holes are a sort of cosmic maternity ward, then we have the following picture: there is not one universe but a *population* of them, differing from one another only in the values of their free parameters. This population consists of *lineages* of universes descended from and closely related to one another. With this picture in place, all Smolin needs to explain the attunement of the free parameters in our universe is to *postulate* that *our universe is typical.* Let a *universe-type* be a set of universes whose free parameter values are pairwise similar. For Smolin, a universe is typical if it belongs to the same universe-type as most other universes in the population. Now imagine that the present (whatever that means in this setting) population of universes is many generations descended from some collection of seed universes, whose free parameter values are set arbitrarily. (Of course, at least one universe in this seed collection must admit a black hole, or else no real lineages result.) Each seed universe has as many baby universes as it has black holes, and the babies do the same. Eventually there is born a baby universe whose free parameters are extremely conducive to black hole formation.

Granting Smolin that "there are small ranges of parameters for which a universe will produce many more black holes than for other values" (1997, 95), we can take "conducive to black hole formation" to be a universe-type. Once a token of this universe-type appears on the scene, a population explosion occurs! For this universe gives birth to *many* others *much* like it, and they do the same. After a few generations, universes rich in black holes outnumber barren universes; they outnumber joyless universes bearing but few young; they outnumber every other universe-type. Therefore, a few generations after the black-hole-rich baby universe appears on the scene, *most* universes are of the black-hole-rich type. By hypothesis typical, our universe belongs to this type. But for our universe to belong to this type is just for our universe to have its parameters tuned to black hole formation, nuclear star formation, us formation. Cosmological Darwinism's explanation of parameter attunement is this: "the parameters . . . have the values we find them to because these make the production of black holes much more likely than most other choices" (Smolin 1997, 96).

The contention legitimating the explanatory expansion by which cosmological Darwinism would account for parameter attunement is that a "precise formal analogy" (Smolin 1997, 103) holds between cosmological Darwinism and the more mundane sort. This analogy opens by casting free parameters in the role of genes, passed from generation to generation, with only minor ("small and random") mod-ification. Just as Darwinian theory lay largely dormant until the re-discovery of Mendelian genetics provided it with a credible mechanism of inheritance, the heyday of cosmological Darwinism presumably awaits the articulation of a theory of quantum gravity that accounts not only for the birth of baby universes but also for their inheritance of free parameter values from their parents. The analogy continues: a set of free parameters is a genotype, inherited (perhaps with mutation) from a progenitor. Most free parameter sets are "uninteresting": they correspond to universes that fizzle swiftly out of existence or that sustain little order or structure. Just so, most genotypes code for un-viable organisms. In both the biological and the cosmic cases, fitness has to do with reproductive success: the better a type is at replicating itself successfully, the fitter it is; only a small range of possible types enjoy the success that is fitness. And "in both cases, what needs to be explained is why the parameters that are actually realized fall into these small sets" (Smolin 1997, 105); in both cases, the explanation involves a statistical mechanism rendering typical what is by some a priori measure improbable. Developing this analogy, Smolin develops a case for expanding the explanatory repertoire of fundamental physics to incorporate the explanations afforded by cosmological Darwinism. Changes in explanatory strategy can be undertaken responsibly, Smolin's account suggests, when they consist of borrowing established explanatory strategies from other sciences.

Smolin would urge that cosmological Darwinism also satisfies *Rescher's* criteria for responsible explanatory expansion, the pragmatic criteria of furthering science's aims of prediction and control. The explanation Smolin offers of parameter attunement contains the following prediction: in universes where the free parameters have values different from their actual values, far fewer black holes are produced than are actually produced. That he can make some form of this claim for eight of the free parameters Smolin takes to be the best evidence for his theory; evaluating the claim for the remaining parameters, he urges, could very well culminate in the theory's falsification.

Conclusion: Rescher's Responses?

I have little doubt that Rescher would deny that Smolin has completed physics, and much interest in the form Rescher's denial would take. Herewith, briefly and in what I take to be ascending order of interest, I consider four possible denials, and the light each would shed on Rescher's views. (In this last regard, the most interesting denial of all, of course, is Rescher's own, also in this volume.)

1. *Smolin has not completed physics because he has left quantum gravity a black box.* To open this box is to loose a Pandoran throng of questions[5]: How do black holes breed new universes? Are baby universes different in scale from their parents? If they are, need we revise our current physical theory to respect scale invariance? If they are not, where are they? By what principles are we to impose a probability measure on the universe set? How are we to understand cosmological evolutionary time? And so on. Not only are these questions manifestly open, they are also apparently the sort to resist clean closure. Hence cosmological Darwinism is neither physics complete nor physics completable.

All this is surely sensible, but it disarms the challenge Smolin's work poses to Rescher's without giving that challenge its due. For this denial consists of pointing to open questions Smolin hopes quantum gravity will close. If quantum gravity fails Smolin, we have a familiar example of science incomplete owing to recalcitrant open questions. But if quantum gravity comes through for Smolin, he claims to have a novel example of science completed by the expedient of diversifying explanatory strategies. (In a variant of the version of cosmological Darwinism presented here, it is the free parameters of the detailed theory of quantum gravity that Smolin would explain by a statistical mechanism. On this telling, closing the Pandoran throng of questions by adding a detailed theory of quantum gravity to our catalog of scientific commitments will not open new questions.) This strategy of explanatory diversification merits our attention. So, absent an argument that quantum gravity *cannot* come through for Smolin, let us allow that it might, and allowing that it might, ask whether cosmological Darwinism buttressed by a detailed and compliant quantum theory of gravity represents ultimate physics.

2. *Smolin has not completed physics because his novel (to physics) explanatory principles violate pragmatic criteria governing explanatory expansion.* Rescher has suggested that it is legitimate to take on novel explanatory principles when the adoption of those principles is conducive to science's aims of prediction and control. Although it is hard to see how the principles of cosmological Darwinism could facilitate control of anything, Smolin has suggested that they are predictive, and daringly so: should it transpire that universes with free parameter values different from ours have more black holes than does our universe, cosmological Darwinism fails. Still Rescher might deny that cosmological Darwinism's explanatory expansion thereby satisfies his pragmatic criteria. For one thing, it is not obvious that Rescher would join Smolin in reckoning cosmological Darwinism genuinely predictive. For the predictions it offers are not about which empirical commitments we will take on in virtue of experimental manipulation, but about which theoretical commitments we will take on in virtue of dogged ratiocination. For another thing, even if Rescher counts Smolin's prognostications as genuine predictions, he may not take the predictive capacities of cosmological Darwinism to compensate for its control inadequacies. Denying Smolin's success on these grounds, Rescher would shed light on his account of prediction, and on the force and nature of his pragmatic criteria for responsible explanatory expansion.

3. *Smolin has not completed physics because he has botched the strategy of legitimating his novel (to physics) explanatory principles by borrowing them from already legitimate science.* To sketch this denial —which relies on a controversial but not repudiated take on the explanatory umpf! of Darwinian theory—I will start with an untutored caricature of the theory in its mundane form. The caricature begins by distinguishing the roles of replication and interaction in the explanations the theory offers. The theory explains why certain types dominate or make up certain populations—why melanism is widespread among the moths of industrial England, for instance. The explanation invokes a statistical mechanism involving the replication of those types: dark moths have a history of replicative success superior to that of light moths, and so dark moths vastly outnumber light ones. What is characteristically *Darwinian* is not this bare mechanism, but its *explanation:* "If variations useful to any organic being do occur, assuredly individuals thus characterized will have the best chance of being preserved in the struggle for life and from the strong principle of inheritance they will tend to produce offspring similarly characterized. This principle of preservation, I have called, for the sake of brevity, Natural Selection" (Darwin 1964 [1859], 126–27). What explains the statistical mechanism of differential replication is the phenomenon of differential interactive success: dark moths enjoy replicative success superior to that of light moths *because* they are better suited to compete for scarce resources in their shared environment. (Compare this explanation of differential replicative success to one in terms of genetic drift to see the point of branding the former explanation distinctly Darwinian.)

This is only a caricature of mundane Darwinism, but it allows me to make the following point about Smolin's cosmological (pseudo?)-Darwinism: Smolin has cast universes in the role of replicators but has left the role of interactors empty. There is no shared environment for whose resources the universes in his population compete; and so no competition; and so no sense in which one universe-type is better suited to compete than any other universe-type; and so no explanation of differential replicative success by appeal to differential competitive success. In Darwin's theory, superior interactor adaptedness explains superior replicative success. But in Smolin's theory, the only "adaptation" is the tautologous one of fecundity. What replaces the explanatory work done in the Darwinian scheme by adaptation, competition, scarcity, and so on is Smolin's postulate of typicality. We understand our universe to be the type it is not because that type has prevailed in the struggle for existence, but because most universes are that type and we have assumed that our universe is typical. In Smolin's cosmological Darwinism, selection pressure has mutated into peer pressure. Borrowing some Darwinian resources, Smolin has neglected to borrow the *explanatory* ones. Cosmological Darwinism fails to complete physics because the "explanations" it offers of parameter attunement have been relieved of their explanatory umpf!

Although Smolin is aware that nothing in his theory answers to Darwinian notions of adaptation, competition, scarcity, and so on, he nevertheless regards the theory as genuinely explanatory.[6] Whether Rescher agrees, disagrees, or defers this question to the relevant community of inquirers will tell us something about the extent to which Rescher takes explanatory success to be situationally constituted.

4. *Smolin has not completed physics because the strategy of explanatory expansion (even when responsible) does not halt question propagation.* This denial comes in both a weak and a strong form. The weak form holds that in Smolin's case, the invocation of Darwinian explanatory patterns does not succeed in closing inquiry, because those patterns themselves contain elements requiring further investigation, perhaps even investigation of a sort that sets the mechanism of erotetic propagation back into perpetual motion. The strong form of the denial holds that no responsible strategy of explanatory expansion can defeat the mechanism of erotetic propagation, because what holds (in the weak version of the denial) of the Darwinian expansion holds of any responsible expansion: casting familiar commitments in a novel explanatory light requires further investigation of those commitments.

I take this denial in its strong form as the most provocative of the denials I have cataloged. Issuing and articulating it, Rescher would relieve us of detailed consideration of Smolin's theory and return us to his own subtle and principled consideration of the limits of science.

NOTES

1. These arguments are hardly the only ones Rescher marshals; limitations of space and time prevent me from considering others in detail. For some of these see (in addition to his 1984 work) Rescher (1978, viii–xi).

2. Hempel's analysis provides a counterpoint to Rescher's: declaring the question "a riddle which has been constructed in such a manner that makes an answer logically impossible" (1973, 200), he dismisses it as a legitimate scientific question.

3. In Chapter 9 of *The Limits of Science*, Rescher subjects putatively complete sciences (purchased cheaply or not) to standards of completeness decidedly *external* to our cognitive horizons. (One he considers is Hawking's: "a complete, consistent, and unified theory of physical interaction [would] describe all possible observations.") Rescher argues that we can never know that our science has met these standards. So at the end of the day, the mechanism of erotetic propagation he engineers is merely suggestive, for he backs down from the thesis that science can never be erotetically complete and from the thesis that science can never be (by some external standard) complete, to endorse the thesis that we can never determine that our science is (by some external standard) complete. If this thesis follows from the definition of "external standard," it is true but threatens to be trivial.

4. Rescher does acknowledge (1984, 138) that erotetic completeness is possible; the failure of this supposition is one way to attain it despite the mechanism of erotetic propagation.

5. I am grateful to Martin Carrier not only for emphasizing Smolin's commitment to substantive theory but also for pointing out some of the perplexing questions this commitment opens up.

6. It should be noted that I have taken Smolin to (at least terminological) task for leaving out of his cosmological Darwinism elements (scarcity, competition, adaptation) that some philosophers of biology argue are inessential to mundane evolution by natural selection. These philosophers may wish to turn the current point on its head: Smolin's theory achieves its explanatory success by deploying the folkloric Darwinian framework stripped of inessential elements; ergo the folkloric Darwinian framework stripped of inessential elements is explanatory!

REFERENCES

Belnap, N. D., and T. B. Steel. 1976. *The Logic of Questions and Answers*. New Haven, Conn.: Yale University Press.

Cahn, R. N., and G. Goldhaber, eds. 1989. *The Experimental Foundations of Particle Physics*. Cambridge: Cambridge University Press.

Darwin, C. 1964 [1859]. *On the Origin of Species*. Cambridge, Mass.: Harvard University Press.

Hempel, C. G. 1973. "Science Unlimited." *Annals of the Japan Association for Philosophy of Science* 14: 187–202.

Rescher, N. 1978. *Scientific Progress*. Pittsburgh: University of Pittsburgh Press.

———. 1984. *The Limits of Science*. Berkeley: University of California Press.

Sellars, W. 1962. "Philosophy and the Scientific Image of Man." In R. G. Colodny, ed., *Frontiers in Science and Philosophy*. New York: New York University Press, 35–78.

Smolin, L. 1997. *The Life of the Cosmos*. Oxford: Oxford University Press.

5

Nicholas Rescher on the Limits of Science

Jürgen Mittelstrass

Department of Philosophy, University of Konstanz

Examining the ideas of Nicholas Rescher is always intellectually profitable and enjoyable. No other thinker today has a broader philosophical horizon than Rescher, and virtually none can express his or her thoughts as precisely as Rescher does. We all—if we are lucky—know some things precisely and many things imprecisely, and thus in the case of philosophy have many opinions but few really tried and true insights. Nicholas Rescher apparently takes no satisfaction in mere opinions, which, as Kant (1942, 343) once put it, "pop up here and there" in philosophy, nor does he recognize any limits to knowledge. This attitude holds also for his thoughts about the limits of science. They are to be found preeminently in Rescher's classical works on these themes, *The Limits of Science* (1984) and *Scientific Progress* (1978), but also in *Cognitive Systematization* (1979) and *Empirical Inquiry* (1981).

Rescher marks out the philosophical framework for his analyses as follows: "On one side lies the exaggerated scientism of the 'science can do anything' persuasion that sees science in larger-than-life terms as an all-powerful be-all and end-all. On the other side lies the antiscientism or even irrationalism that sees science as a dangerous luxury, a costly diversion we would be well advised to abandon altogether." These are two extremes between which philosophical reflection not only scrambles back and forth but beats its head bloody. He who follows Rescher fares better, not only here but also in the dissection of the question of the limits of science into the three questions: "1. How far *might* science actually go: what are the *practical* limits on science? 2. How far *should* science go: what are the *prudential* and the *moral* limits on science? 3. How far *could* science go in principle: what are the *theoretical* limits on science?" (Rescher 1984, 1, 2).

In view of the clarity and cogency of his analyses, I should and will pass up this opportunity to offer inconsequential criticisms of Rescher. Instead, setting aside the second question, which opens up the problem of responsibility, I will discuss the first and the third questions under five key headings. For even the *theoretical* limits on which Rescher focuses his analyses are—and this is my guiding thesis—often *practically* grounded.

Scientific Progress

Wherever we examine the limits of science, we encounter questions about the *progress* of science as well. The principal question is whether, in view of the possible limits of knowledge, scientific progress still has a future. What I mean by this seemingly paradoxical formulation is this: Will scientific discoveries eventually reach a natural end at which there is nothing more to discover? Will technological developments one day reach a point at which there are no (new) purposes that such developments—scientific developments included—could pursue and at which there are no further possibilities of realization? Philosophy of science examines these questions, mainly in relation to the natural sciences, in the form of two theses, both of which are extensively presented in Rescher's *Scientific Progress* (1978, 6–15):

1. The thesis of the complete or asymptotically exhaustive survey of nature. According to this thesis, the future "history" of scientific discovery either is absolutely finite or at some point settles into an asymptotic approach to what can be known at all. Mere elaboration and further precision would then take the place of innovation. At some point science would no longer have a future: everything discoverable would have been discovered and everything in need of a scientific explanation would have been explained. Even the mopping-up operation—the calculation of additional decimal places and the classification of additional cases that add nothing essentially new—would gradually come to a close.

2. The thesis of the complete or asymptotic exhaustion of information capacities. According to this scenario, scientific information capa-

cities either are absolutely finite or at some point begin an asymptotic approach to absolute information limits. Here, too, elaboration and further precision would replace innovation. Science would have exhausted its own possibilities for research and articulation, and an "information barrier" would have arisen between science and nature irrespective of whether investigation of the latter had reached a point of exhaustion. The question "Does (scientific) progress still have a future?" thus expresses only an apparent paradox. But the question is unanswerable within the framework of the two cited theses—something Rescher shows by pointing to the unlimited character of scientific questions.

Questions and Purposes

Scientific progress is expressed not only by the multiplication of items of *knowledge* but also by the multiplication of the *questions* that undergird knowledge formation. In Rescher's words, "Cognitive progress is commonly thought of in terms of the discovery of new facts—new information about things. But the situation is in fact more complicated, because not only *knowledge* but also *questions* must come into consideration. Progress on the side of *questions* is also cognitive progress, correlative with—and every bit as important as—progress on the side of *information*. The questions opened up for our consideration are as definitive a facet of a 'state of knowledge' as are the theses that it endorses" (1984, 28). According to "Kant's principle of question propagation," which Rescher extracts from a passage in the *Prolegomena* (1911 [1783], 352), each new question in science implies new knowledge, especially when the question did not arise only from old knowledge.

However, it is not only questions that lead us beyond familiar limits. Just as the number of questions that science can pose is infinite—what sense does it make to assert that all questions have been answered?—so, too, is the number of *purposes* that scientific knowledge can pursue—if one should even want to speak of all possible scientific purposes. If research is characterized not only by means of the results that have been achieved within individual disciplines (say, with regard to answering scientific questions) but also by means of the (internal and external) purposes associated with it—and this is doubtless the case, as a comparison of Newtonian physics with Aristotelian physics, each of which pursued radically different purposes, makes clear—then the notion of an end to scientific progress would imply not only the assertion "we know everything (that we can know)" but also the assertion "we know all the purposes (that we could have)." The number of purposes, however, is unbounded, even if we take into account the limits to a scientific transformation of the world and of humankind. In other words, in order to answer the question "Does (scientific) progress still have a future?" we would already have to know what we do not yet know—what only progress or its failure to materialize could show us. In this sense, then, there are no limits to science.

Discoveries

If science—not least in regard to the questions and purposes that lie behind it—is a creative process—not something that, so to speak, runs on tracks—then its *unpredictability,* the circumstance that future knowledge cannot be foreseen, tells strongly against the assumption of any (theoretical) limits. Rescher also sees it this way: "We cannot tell in advance what the specific answers to our scientific questions will be. It would, after all, be quite unreasonable to expect detailed prognostications about the particular *content* of scientific discoveries. It may be possible in some cases to speculate *that* science will solve a certain problem, but *how* it will do so lies beyond the ken of those who antedate the discovery itself. If we could *predict* discoveries in detail in advance, then we could *make* them in advance" (1984, 97).

That science *makes* discoveries does not mean that it simply pursues what is to be discovered—something called *nature* in an older terminology—or that it considers its theoretical constructions themselves to be something natural that in turn determines science (in both cases predictability would be easy); rather it means that in scientific knowledge understanding and nature are constantly establishing new relationships and that the one cannot exist without the other—as Kant already knew. And this holds not only from the point of view of stability but also from that of instability. If nature has moods, the understanding loses its practical certainty; and if the understanding has moods, nature loses its theoretical definiteness.

In this context Rescher speaks of structural and substantial generality. We can certainly predict that future science "will be incomplete, that its agenda of availably open questions will be extensive, and so on.

But of course this sort of information tells us only about the *structure* of future science, and not about its *substance*. These structural generalities do not bear on the level of substantive detail: they relate to science as a productive *enterprise* (or 'industry') rather than as a substantive *discipline* (as the source of specific theories about the workings of nature)." Thus the unexpected itself remains a constitutive part of future science—for theoretical and practical reasons. For this reason, as Rescher appropriately concludes, present-day science cannot speak for future science. Or, formulated in the terminology of limits and stabilities, the very instability of science "is at once a *limitation* of science and a part of what frees it from having actual *limits*" (1984, 101, 102, 111).

The Mortality of Knowledge

In support of the now widely accepted view that scientific progress does not proceed cumulatively, Rescher cites Popper: "It is not the accumulation of observations which I have in mind when I speak of the growth of scientific knowledge but the repeated overthrow of scientific theories and their replacement by better or more satisfactory ones" (Popper 1972, 215; Rescher 1984, 68). Progress is, so to speak, displaced from things to theories, that is, to thought. And this shift is in its way boundless inasmuch as it does not progress empirically but *constructively*. In Rescher's words: "Scientific inquiry is a creative process of theoretical and conceptual innovation; it is not a matter of pinpointing the most attractive alternative within the presently specifiable range but of enhancing and enlarging the range of envisageable alternatives" (1984, 96). Or put in another way: Scientific thought is, so to speak, constantly reinventing itself, realizing itself in its constructions and destroying itself with its constructions. Fittingly, the phoenix is the symbol of science just as the owl is the symbol of philosophy. Science creates itself, just as philosophy contemplates itself and what it has seen. Science thrives on the mortality of knowledge, philosophy on the immortality (or, better, the limitlessness) of reflection, which therefore constantly encounters itself, while science forgets and discovers. Only the concept of construction holds the two, philosophy and science, together. For so long as it does not just reproduce itself hermeneutically, philosophical reflection, too, constructs and devises new worlds, but only to fill them again with its age-old experiences.

Perhaps these remarks themselves sound like philosophical reveries. I do not want to be caught dreaming, so let us return to the question of scientific understanding. Against the dissolution of this understanding in ideational discontinuities, which has been in vogue since Thomas Kuhn's (1970) classic work on the structure of scientific revolutions, Rescher recommends a sober realism—not a philosophical but a deeply pragmatic realism: Real progress

is indeed made, though this progress does not proceed along purely *theoretical* but along *practical* lines. Once one sees the validation of science as lying ultimately in the sphere of its applications, one also sees that the progress of science must be taken to rest on its pragmatic improvement: the increasing success of its applications in problem solving and control, in cognitive and physical mastery over nature. . . . Despite any *semantic* or *ideational incommensurability* between a scientific theory and its latter-day replacements, there remains the factor of the *pragmatic commensurability* that can (by and large) be formulated in suitable extrascientific language. (1984, 89–91)

This is the way out of the philosophical paradoxes; it is also the way that scientific progress chooses, even against many of its philosophical interpretations.

Perfected Science

If science knew everything that it could know, it would be perfect. As conditions for perfect science, Rescher (1984, 134–35) lists *erotetic completeness, predictive completeness, pragmatic completeness,* and *temporal finality.* That is to say, everything that is explicable with respect to its own questions must be explained; everything predictable with respect to its own cognitive base must be forecast; everything cognitively required with respect to its own cognitive intentions must be available as instruments; and nothing given with the relevant completeness may leave room for unexplained elements. These conditions, in turn, are at best dogmatically attainable or imaginable as regulative principles, as Rescher makes clear with an appeal to Kant's idea of a moral perfection: "The idea of 'perfected science' is the *focus imaginarius* whose pursuit canalizes and structures our inquiry. It represents the ultimate *telos* of inquiry, the idealized destination of a journey in which we are *still* and indeed are *ever* engaged, a grail of sorts that we can pursue but not possess" (1984, 152).

The reverse side of this Kantian idealism is Rescher's conception of *cognitive Copernicanism,* which maintains that no time is cognitively privileged in the sense of realized scientific perfection: even the "current state of 'knowledge' is simply one state among others, all of which stand on an imperfect footing." An equivalent formulation states that as a rule scientific knowledge must be taken to be imperfect, that is, incomplete and incorrect, not in the sense of a defect—such a notion would presuppose attainable perfection or completeness—but in the sense of an *in principle* openness of scientific knowledge (1984, 87, 86). Paradoxically, the boundlessness of science in the sense of an interminable progress of knowledge lies precisely in the limited character of knowledge, in its imperfection and corrigibility.

That science is interminable and so has no limits to its progress is not so much an epistemological thesis that appeals to the essence of knowledge as a thesis directed at practical unfeasibility, that is, the factual limitation of scientific progress effected by such external conditions as economical and technological conditions: "In a world of finite resources, this means that science must in the future progress ever more slowly—for strictly *practical* and ultimately *economic* reasons. Although natural science is theoretically limitless, its actual future development confronts obstacles and impediments of a strictly practical kind that spell its deceleration" (1984, 160). Scientific progress, in other words, is limited neither by an unattainable perfection of knowledge nor by absolute (theoretical) limits of knowledge. Epistemological jurisdiction in cases of the limits of science, therefore, lies not with metaphysics but with *"economic rationality."* Limits of science thus exist only in a practical sense, not in a theoretical one.

Conclusion

These observations on Nicholas Rescher's ideas about the limits of science have presented the fruits of his reflections without displaying their philosophical richness. Rescher is right when he says that "Science, like any other human enterprise, is inevitably limited, but there is no reason whatsoever to think that it must reach a dead end" (1984, 4). This insight can also be expressed in the following manner: the limits of science are either *error limits* (the scientific understanding gets stuck in its own inadequacy) or *economic limits* (scientific progress becomes unaffordable) or *moral limits* (scientific achievement is directed against humankind itself). In any case—and this was my point when I said at the beginning of this chapter that even theoretical

limits of science, if there be such, are in general practically grounded—
the standard or measure that sets limits to scientific progress is a
practical one, and precisely for this reason self-imposed.

REFERENCES

Kant, I. 1911 [1783]. *Prolegomena zu einer jeden künftigen Metaphysik die als Wissenschaft wird auftreten können* §57. In Königlich Preussische Akademie der Wissenschaften, ed., *Gesammelte Schriften* IV. Berlin: Georg Reimer, 253–383.

———. 1942. "Lose Blätter zu den Fortschritten der Metaphysik." In Königlich Preussische Akademie der Wissenschaften, ed., *Gesammelte Schriften* XX. Berlin: Georg Reimer, 333–51.

Kuhn, T. S. 1970. *The Structure of Scientific Revolutions,* 2nd ed. Chicago: University of Chicago Press.

Popper, K. R. 1972. "Truth, Rationality, and the Growth of Scientific Knowledge." In K. R. Popper, *Conjectures and Refutations: The Growth of Scientific Knowledge,* 4th ed. London: Routledge and Kegan Paul.

Rescher, N.1978. *Scientific Progress: A Philosophical Essay on the Economics of Research in Natural Science.* Oxford: Basil Blackwell.

———. 1979. *Cognitive Systematization.* Oxford: Basil Blackwell.

———. 1981. *Empirical Inquiry.* Totowa, N.J.: Rowman and Littlefield.

———. 1984. *The Limits of Science.* Berkeley: University of California Press.

6

Limits of Science: Replies and Comments

Nicholas Rescher

Department of Philosophy, University of Pittsburgh

These brief comments are responses—in alphabetic turn—to the critical commentary that Robert Almeder, Jürgen Mittelstrass, and Laura Ruetsche have offered regarding my published views about the limits of science. The brevity of these responses should not be taken to mean that I underestimate the importance of the issues being raised. However, it seems both necessary and sufficient for this occasion to indicate somewhat telegraphically the general direction—the overall tendency and general strategy—of my response.

Response to Almeder

Robert Almeder's discussion of my position is on the whole both informative and constructive. There are, however, a few issues about which we remain at odds.

Prominent among these is Almeder's critique of my contention that scientific progress can continue forever, so that natural science cannot be completed. The rationale that he attributes to this is the straightforwardly inductive view that since there have always been questions in science one should conclude that this will continue to be so. However, I do not think that this does justice to my position. For this does *not* really rest on straightforward induction from the history of science. Instead, it rests on the parametric space exploration model that I have sketched at great length in *Scientific Progress* (1978). Natural science is crucially dependent on data secured via technology, and the technological resources by which we explore nature's parametric space are ever improving but can never be perfected. But as we improve our technology of observation and experimentation we encounter the influence of resistance barriers. (We must propel particles ever closer to the speed of light, but we cannot reach it; we can push experimental temperatures ever closer to zero degrees Kelvin, but we cannot reach it; and so on.) And imperfection matters crucially here because the new data made available by technological improvements are bound to bring theory changes in their wake. Accordingly, our assurance that the fountain of scientific questioning will not dry up is neither naively inductive nor mysteriously intuitive—it rests, rather, on our understanding (our *best* understanding, I would say, were I not reluctant to act the judge in my own case) of how the process of empirical science in fact works. For the improvability of science is, in my view, rooted in the unprovability of technology, and there is (as I see it) no limit to this because we can always extend—in theory even if not in practice—the as-yet achieved levels of technical power and sophistication.

The crucial point is *not* that there are factual insolubilia—scientific questions that science cannot possibly answer. Rather it lies in the fact that science is a dynamical process—that as we answer our old questions we always change "the existing state of scientific knowledge" in such a way as to put new questions on the agenda—that in the case of scientific progress we open up new questions in the very cause of putting old ones to rest. And it is this dynamical, ever-changing aspect of the question agenda of natural science that assures that there is always a supply of as-yet-unresolved issues for the science of the day— irrespective of what the date on the calendar happens to be. Even though the question agenda of science may well always be finite, the fact remains that that question agenda itself is ever changing.

I will, however, concede to Almeder that the proliferation of new scientific questions might indeed come to a stop if one were to change the scoring system for new questions through a radical shift in the position of science on the generality/specificity curve. If science were to become content with vaguer, less definite contentions—so that what we would otherwise see as a new question now becomes seen as simply a matter of further detail regarding answers to old, preexisting questions—then scientific questions might indeed reach an end. But I see this as an implausible prospect, seeing that what is involved is a radical change not just in the processes of natural science but in its very nature.

A second important issue is raised by Almeder's query regarding the sort of "scientific realism" that I propose to adopt. An analogy may help to indicate how I view the relation of the theory claims of science to the actual descriptive nature of the world. Consider the attempt to represent a completely detailed, black-and-white, photographlike picture of a scene by devising another picture that blocks this picture into four rectangles as follows:

A given rectangle is to be rendered white, black, or gray depending on whether the original picture is predominantly light, dark, or mixed in that particular region. And let us carry this process further. Thus at step 2 we divide each of these four rectangles in the same manner and represent it according to the same rule. Then at step 3 we treat each rectangle in the step 2 picture in the same way. And so on. It is clear that at no stage whatsoever will we have a true, accurate representation of that original picture. None of those "theoretical claims" is ever true as it stands. But nevertheless we will gain an increasingly good grasp of the original scene—and eventually indeed one that is visually indistinguishable from it—given the inherent limitation of human vision.

Examples of this kind can serve to indicate by way of analogy that a thesis need not be true wholly *or even in part* to provide us with a workable way of coming to cognitive terms with reality. Success in dealing with the real world need not pivot on the truth of detail: our theory models can yield the truth at the level of aggregation (for example, as per "the book is on the table near the candlestick" in the case of the picture) without any of the individual *details* of the theory having to get matters exactly right—or anything near exactly right. Consider a barrel with ninety-nine McIntosh apples and one pippin. It would certainly not be a true characterization of the situation to say "There are one hundred McIntoshes in the barrel." But if this claim were "vagued up" to read "There are roughly one hundred apples in the barrel," then we would have a statement that is incontestably true. Our detailed theories in science do not actually need to be more or less correct descriptions of the external world to yield appropriate indications of the truth of things at the level of rough generalities. (That is why my *Scientific Realism* spoke of scientific realism's adequacy arising at the rough and approximative level of "schoolbook science.")

Response to Mittelstrass

I now turn to Jürgen Mittelstrass's very clear and cogent discussion of my position regarding the limits of science. It is difficult to know how to "respond" here because Mittelstrass is primarily concerned not with criticizing my views on the limits of science but with explaining them and placing them in the larger philosophical context of the theory of knowledge—with special reference to its relationship to the teachings of Kant. In reacting to this commentary, it must indeed be acknowledged that although the driving force of my views about science resides largely in an empirical, ex post facto concern for the actual practice of scientific inquiry, the position that results bears so many similarities to Kant's decidedly more aprioristic views that Mittelstrass is quite right to characterize the position as a "Kantian idealism."

However, an important issue in this connection pivots on the distinction between route and destination. The Kantian machinery of regulative principles and postulated ideals indeed plays a central role in my conception of the philosophy of science, as does the idea of limits and boundaries (*Grenzen* and *Schranken*) of knowledge. And so I too reach various Kantian destinations. But Kant arrives at them via the a priori route of transcendental argumentation. He is concerned for the theoretical "conditions under which alone" our human knowledge can achieve those features we are minded to claim for it. The idea of necessary conditions for the achievement of cognitive goals is pivotal for Kant. My own prime concern, by contrast, is with the ex post facto teachings of experience regarding the effective ways and means of successful inquiry. My position is thoroughly pragmatistic in nature, hinging upon the experientially learned conditions of effective cognitive praxis, and so my position has an empirical rather than a Kantianly necessitarian foothold.

As I see it, the goal of our philosophical theorizing is to achieve an explanatory closure that can account for how it is that a creature constituted as we are (and at this stage we require extensive empirical knowledge), operating in a world constituted as ours is (and here too there is a great need for empirical knowledge), can manage to conduct its inquiries in a way that can optimize its cognitive grasp on those context-mandated realities. Here we are, to be sure, dealing with a goal-driven inquiry into effective ways and means by which a successful praxis is possible, but the praxis at issue is one that is conducted by empirically characterized beings pursuing their cognitive objectives in empirically given circumstances. The line of reflection at issue is accordingly not one of Kantian aprioristic idealism but that of a remote cousin, which travels only with the lighter baggage of science-informed, ex post facto commitments.

I found particularly interesting and right-minded Mittelstrass's insightful stress on the *self-creativity of science,* the general idea that natural science ongoingly reconstitutes itself—reinvents itself as it goes along, so to speak. Present science differs from past science both in its questions and in its answers, and yet those new questions and answers emerge from the work that is done in the course of grappling with the old ones. This engenders an ongoing change not only in the findings of science—a change in *product,* so to speak—but also in the methods and procedures by which those findings are consolidated—that is, a change in *process.* In the course of scientific progress we continually learn not only more science but also more about how to do science. It is this dynamical aspect of science that, more than any other, prevents theorists from obtaining a secure cognitive grasp on what "science" is actually like. The target of their study is hard to hit because it is constantly on the move. Heraclitus might have been wrong about rivers, but he would have been right about science: no two generations deal with exactly the same sort of "science." And prediction is impracticable here, seeing that present science cannot speak reliably for future science. Mittelstrass, it seems to me, is entirely correct in stressing this crucial aspect of science as an ongoing process of self-reinvention.

Response to Ruetsche

My response to Laura Ruetsche will begin with the potential obstacle to question proliferation, which she formulates as follows: "Rather than adding to our set of [substantive] commitments, we might instead add to our store of explanatory principles in such a way that some of the commitments we already have constitute an answer to the question." Against this suggestion I would urge the process/product interaction inherent in the consideration that the explanatory principles of science are grounded in and emerge out of its substantive commitments, and indeed that just as there is no clear boundary between synthetic and analytic, so equally there is no clear boundary between substantive facts and explanatory principles in natural science, because those principles always hinge on substantive commitments. The pervasively dispositional nature of physical properties leaves no gap between its factual descriptions and its inferential if-thens. Moreover, in natural science there is an ongoing dialectic of feedback between process and product. Novelty on the one side must go hand in hand with innovation on the other. Accordingly there is, I would submit, no exploitable gap or lacuna between factual commitments and inference patterns into which Lee Smolin (or anyone else) can slip cosmological deliberations as question-proliferation stoppers.

And this brings us to Smolin's book, *The Life of the Cosmos* (1997). The sort of quasi-anthropic argument that Smolin favors pivots on the question "Why are the values of the free parameters in the cosmologically pivotal equations of astrophysics so fixed as to maximize the likelihood of black hole production by evolutionary processes?" This certainly does not seem to be a plausible candidate for a dead-end marker on the road of scientific cosmology.

On closer examination, the structural format of the question at issue appears as "Why are nature's fundamental descriptors (boundary value conditions, free parameters) so fixed that a certain type of result (stable subworlds, life-sustainable planets, organic beings, or whatever) will ensue under the aegis of nature's basic laws?" Two observations are in order here.

First and foremost is the consideration that to show *that* things could not have eventuated as they indeed have unless certain conditions were satisfied is very far from ending the explanatory project. What is residually left in mystery is just another version of the "Why?" question with which the deliberations began, with two big issues left unresolved: (1) the Leibniz question of why the laws are as they are, and (2) the explanatory bridging of the very wide gap between efficient and final causation. After all, the proliferation of black holes in the world is far from being its only characteristic feature. Why therefore should it be singled out as the ultimate pivot that provides the reason why everything else is as it is? A quasi-anthropic approach of the sort envisioned by Smolin seems to me not the end of question proliferation but rather the opening up of various new lines of questioning. After all, Smolin's "natural selection" is not driven, à la biology, by mere survival under conditions defined by an environment of external nature, seeing that at the level of the universe as a whole there is no identifiable restrictive externality that can provide the selective arena for an evolutionary process. (This in effect is the burden of Ruetsche's own discussion.)

There is accordingly no reason to think of Smolin's conjectures as bringing cosmology to an end. At best and most they would close off only one particular direction of questioning. After all, a Darwinian account takes the schematic form of explaining a presently realized uniformity through the selective elimination of alternatives that have proved unsuccessful in the competition for survival. But precisely because it is schematic, such an account has to be fleshed out in detail. It demands the telling of a detailed story about how it is only natural and to be expected that a struggle for existence occurs on the battlefield of certain particular conditions and is fought with certain particular weapons and strategies. A Darwinian account, in sum, does not bring question proliferation to an end—it simply puts a new range of questions on the agenda. (Here Ruetsche and I are substantially in agreement.)

A second main difficulty arises in relation to finalistic argumentation of the sort at issue in Smolin's argument. The argument envisions finality via a thesis of the form "This is how matters have to be if nature is to produce the existing end result it has in fact realized." But a critical premise has been suppressed here, namely, "This is how matters have to be if the existing end result is to be realized *by the operation of the fundamental laws of nature as we presently conceive of them.*" And this presupposes that our putative "laws of nature" are nature's law in actual fact—that is, that those public laws of nature are substantially final in the way of completeness and correctness. But in the present dialectal context this supposition becomes deeply problematic. One can scarcely invoke against a position that maintains the substantial incompleteness and imperfection of science, an argument that presupposes that science is substantially complete and correct.

I will, however, concede to Ruetsche that if we allow the prospect that future science might so gerrymander its question agenda that the register of questions acknowledged as legitimately possible comes to be identified with the inventory of questions that have in fact been answered, then science will indeed have attained a version of erotetic completeness. But that "future science" as thus conceived can hardly be acknowledged by us as constituting *science* as we understand it. And so my feeling about this concession is that it lies in the sphere of rather remote prospects. For it envisions an evolution of natural science into something very different from the project as we nowadays confront and understand it. True, I maintain that science is something plastic and changeable. But not *that* plastic and changeable.

Conclusion

In concluding I offer one further comment suggested by the general tenor of this volume. As I see it, the "limits of science" are not matters of cognitive boundaries set by insolubilia but rather result from the incompletability of natural science inherent in the circumstance that no matter how much we manage to do there always remains still more that could be done and can in principle be done by us. And such limits do not arise from defects of the human intellect but rather root in the fact that it cannot proceed in matters of scientific inquiry wholly on the basis of its own resources—that its exploration of nature's parameter space must inevitably proceed by means of instrumentation afforded by technology. These limits are accordingly bound to be both economic and physical—and are insuperable on both fronts.

To be sure, it is certainly possible that natural science might be completed in the sense that science-as-we-then-and-there-have-it manages to resolve (at some level of correctness and adequacy) all of the questions about the world that figure in our then-and-there question agenda, so that our question asking and science's question answering stand in cognitive equilibrium. But if this were to happen it could well reflect not so much the success of science as shortcomings on the side of inquiring intelligence. And even if this seemingly happy condition were to be realized, it could well betoken not so much the sufficiency of our answers through scientific power as an insufficiency of our questions through deficiencies on the part of inquiring intelligence.

REFERENCES

Rescher, N. 1978. *Scientific Progress*. Oxford: Blackwell.
Smolin, L. 1997. *The Life of the Cosmos*. New York: Oxford University Press.

7

How to Pile up Fundamental Truths Incessantly

On the Prospect of Reconciling Scientific Realism with Unending Progress

Martin Carrier
Department of Philosophy, University of Bielefeld

Progress is among the most cherished ideas in the communities of scientists and philosophers of science alike. It is widely believed that epistemic progress is unlimited and goes on forever. But there is another view, likewise dear to the hearts of most scientists and philosophers of science: the view that science manages increasingly to reveal the innermost secrets of nature. It approaches the truth about the universe. The problem is that these two beliefs do not harmonize easily. If the truth is discovered, the matter is settled; there seems to be no room for further progress. Conversely, if science moves on unabatedly, it seems to be in need of sustained improvement—which suggests that it has not gotten things right. This conceptual tension between realism and ongoing progress is the focus of my considerations. I begin by staking out the rival claims in, first, sketching scientific realism and the idea of a final theory, and, second, outlining the assumption of incessant progress in fundamental research. Subsequently, I examine options for resolving the conceptual contrast and for reconciling scientific realism with ongoing progress in basic science. My thesis is that the minimum demand for matching these views is the adoption of a mild form of emergentism.

Realism and the Final Theory

Scientific realism involves the commitment to the existence of entities that are not amenable to unaided observation but are assumed within successful scientific theories. Scientific realism is not supposed to apply to tables and chairs (whose existence is taken for granted anyway) but to objects such as photons or electrons. Such entities cannot be perceived directly. They are specified within a theoretical framework, and they are registered through the mediation of instruments whose operation relies on theories as well. The properties ascribed to them and the methods used for their identification thus crucially depend on scientific theories. Scientific realism claims that successful theories in the mature sciences are reliable enough to be trusted also in this ontological respect. Such theories are approximately true, and if a thus distinguished theory specifies a class of objects, these are likely to form part of the inventory of nature. To put it bluntly, if a true theory assumes certain entities, the latter exist in reality.

The overall commitment associated with scientific realism thus is that science approaches the independently existing reality. It manages to come closer to the truth about the nature of things. The theoretical terms of methodologically distinguished theories in the mature sciences do refer to real objects and processes. For instance, since the best available theories of the relevant phenomena essentially draw on photons or electrons, something of the kind actually exists. That is, such theories are supposed to be able to capture at least approximately the nature of the corresponding entities.[1]

Accordingly, scientific realism does not hold that there are real things "out there" whose constitution remains forever hidden—such as Kant's "things in themselves." Rather, the more ambitious contention is that science finally ventures into the realm of nature's innermost workings. Science eventually succeeds in unveiling the blueprint of the universe or reading the mind of God (as Stephen Hawking put it; Weinberg 1992, 242). Moreover, science is thought to be well under way on its approach to truth. That is, the presently accepted accounts in the mature sciences give a more or less accurate—albeit still incomplete—portrait of nature's collection of objects and processes.

Scientific realism suggests the vision of a final theory and the corresponding end of progress in fundamental research—although only a minority of realists actually face the idea. If science is supposed to approach successfully the truth about the universe, its task should be completed one day. It is hard to keep on deciphering the blueprint of the universe successfully. If you are successful, you are done sooner or later. It was not flattery but rather a simple consequence of realist beliefs when Lagrange called Newton the "luckiest of all mortals" since nobody could possibly ever match his achievement. After all, there is only one universe to be explained (Koyré 1965, 18).

To be sure, Lagrange was wrong. There was Einstein to come. And general relativity was joined by quantum theory and the standard model of elementary particle physics. But this is not the point. The issue rather is conceptual in nature: the view that science gets more and more things right entails—or appears to entail—the view that the endeavor of disclosing nature will be brought to completion at some time. Steven Weinberg is among those who recognize the conclusion. He observes a strikingly *convergent* pattern of explanation. Generalizations are explained by more comprehensive theories and again by deeper principles. And wherever you start asking "Why?" questions as to the cause of physical, chemical, or biological phenomena, the ultimate answer always draws on particle interactions and thus relies on the standard model of particle physics (Weinberg 1992, 19, 32). This convergence of "arrows of explanation" is paralleled by the structure of the history of science. It involves a succession of steadily more comprehensive and more unified theories, which most probably converge at the "final theory" (Weinberg 1992, 231–32, 235). The final theory "will bring to an end . . . the ancient search for those principles that cannot be explained in terms of deeper principles" (Weinberg 1992, 18). At this point of maximum depth the ultimate foundations of science are reached.

Nonrealism and Ongoing Revolutions

In striking contrast to this view of the history of science, Thomas Kuhn argued that scientific progress is *ontologically divergent*. The history of scientific thought is characterized by marked twists and turns that do not develop into a coherent picture (Kuhn 1962, 206–7). There is a continual displacement of disparate approaches; scientific revolutions keep on occurring. In the seventeenth century the conception of heat as internal vibration prevailed; it was succeeded by the view that heat is a weightless fluid, only to be taken up again in the nineteenth century by the kinetic theory of heat. Likewise, Newton assumed a particulate structure of light. This interpretation was first dropped in favor of the wave theory and then resumed (in a modified way) by the photon model. As Kuhn argued, the consideration of such cases suggests that progress cannot adequately be conceived as approaching reality. It rather consists of a constant improvement and refinement of the respective states of knowledge. A teleological understanding of progress is abandoned; science gets better without getting it right (Kuhn 1962, 170–73). This view combines nonrealism with the prospect of perpetual conceptual cataclysms. Science does not come nearer to reality, and its progress is unlimited for this reason.

Nicholas Rescher largely sides with Kuhn in this respect. He accepts the Kuhnian ideas of ongoing theoretical changes of undiminished import and of progressing without arriving. Science does not proceed by accumulating truths but by overthrowing formerly adopted principles and by replacing them with disparate ones. Theories are constantly rebuilt from the very foundations. As a result, scientific realism is not a viable option (Rescher 1984, 68–76, 153–56; 1987, 19, 23–25).[2] To be sure, the more developed a science is, the more difficult it is to accomplish further significant progress. Consequently, the pace of progress may slow down, but it is unlikely ever to come to a halt (Rescher 1978, 80–81, 122–23; 1984, 168–73).

Rescher admits that his picture of scientific progress is closely tied to the historiographic views of Kuhn and Paul Feyerabend (Rescher 1984, 66). However, these views have fallen into disfavor in many quarters lately; in the course of the past two decades a much more accumulative interpretation of the history of science has gained prominence. This shift of conventional wisdom does not imply, of course, that Rescher's historiographic view is out of focus. The point rather is that it would be unwise to connect the claim of unlimited progress too intimately with the theories of Kuhn and Feyerabend. Insisting on the fragile and transient nature of the fundamental principles would tie the unboundedness claim to the antirealist stance. Advocates of a final theory could simply join the realist mainstream, deny the premise of ongoing revolutions, and proceed undauntedly. Every viable defense of limitless progress should be compatible with scientific realism. Consequently, it would be worthwhile to have a more ecumenical argument in support of the unboundedness claim. In fact, Rescher gives at least two of them, one relying on "Kant's principle of question propagation," and another one drawing on the technology-dependence of observation and measurement.

First, underlying a question is a presupposition that is inherent in each of its explicit answers. Every question dealing with the melting point of lead assumes that there is such a stable temperature. As to scientific questions, it is the theoretical framework that furnishes the relevant commitments; it sets the corresponding "question agenda." The pivotal aspect of the principle of question propagation is that answering a question tends to change the relevant factual presuppositions and thus involves a modification of the question agenda. Answering a question is thus likely to allow for posing new questions—which in turn advances scientific progress (Rescher 1984, 18–22, 28–29).

Given the appropriateness of Rescher's mechanism of "erotetic propagation" (which is addressed more explicitly by Laura Ruetsche in this volume), the worry remains that it is insufficient for buttressing unending progress. The ability to pose new questions is certainly a necessary precondition for progress to occur; but it is not tantamount to progress. Scientific progress requires dependable answers as well. The abundance of questions is no safeguard against barren stagnation. A prominent example is the present state of particle physics. There are many pressing questions to be asked, for example, about the causes of symmetry breaking or the values of particle masses and coupling constants. The problem rather is that nothing even remotely resembling a reliable answer has been found for decades. There is a wide range of possible theoretical options but apparently no way of singling out one of them empirically. Theories in this field are instead judged by their mathematical elegance, which has led to the feeling that particle physics has stopped being an empirical science and is in danger of becoming part of aesthetics (Horgan 1996, 63, 70). It is by no means obvious that the contentious final state of fundamental science provides answers to all supposedly legitimate questions. It may rather look like the present impasse in particle physics, where scarcity lies with convincing answers, not with questions. The upshot is that erotetic propagation is good for science, to be sure, but not good enough for scientific progress.

Second, Rescher's argument from technology-dependence draws on the fact that scientific knowledge does not accrue from the unaided observation of phenomena, but rather from the use of sophisticated registering devices and experimentation techniques. Science is applied technology. Scientists interact with or sometimes even create the objects of their scrutiny. Consequently, new technological means represent new modes of interaction with nature, which generate new data

and eventually lead to new theories. Technological innovation underlies theoretical achievement, a proposition granted by all parties to the
discussion. So, if technological advancement engenders theoretical
progress by breaking new empirical ground, the latter will occur along
with the former (Rescher 1978, 45–46, 52, 134–36; 1984, 54–57).

This argument is all right as far as it goes, but it fails to underwrite
the option of significant theoretical progress. It does not follow from
the interactive and technology-dependent mode of data acquisition
that the application of novel instruments will unearth objects or processes of a kind *different* from those already known. Chemistry has for
a long time been in the business of synthesizing new molecules by
increasingly sophisticated technological means without any associated
theoretical breakthroughs. Decades of molecular design have brought
forth a multitude of designed molecules—but no genuinely new theoretical insights. Likewise, in spite of the application of ever more
advanced experimental equipment, electrons have tenaciously failed to
display any internal structure. The more advanced technology has
simply delivered more of the same. Consequently, Rescher's interactive
model of penetrating nature is not sufficient for supporting perpetual
fundamental progress; it is rather compatible with the prospect of a
final theory. The application of sophisticated technology only succeeds
in uncovering phenomena of a new kind if nature lends a helping hand.
New effects cannot be created at will by employing new instruments or
switching to another level of consideration. The prospect of limitless
progress in fundamental matters also hinges on preconditions on the
part of nature. I address this problem somewhat more systematically in
the following section.

Progress, the Exhaustion of Nature, and Emergent Laws

As explained previously, scientific realism at least prima facie suggests
the idea that progress in basic science sooner or later exhausts the
reservoir of laws of nature. If science approaches the truth about the
universe, it is to be expected that all fundamental truths will have been
revealed one day. Scientific realism thus suggests the "geography picture" of scientific progress: you chart the surface of the earth, and if
you are successful you inevitably reach a point at which no pristine
territories are left. As Richard Feynman put it, America could be

discovered only once (Feynman 1965, 172). In the end, all virgin islands are registered and integrated into the science empire. At this point fundamental research is completed. The claim is that realism and the assumption of unbounded progress do not go together.

A possible reply is that scientific realism does not entail the view that science approaches the truth, but only the more modest view that science gets an increasing number of things right. It accumulates solved problems and thus unceasingly improves its performance. And the idea of piling up truths in no way entails that there is a final goal or end-state in which all truths are in. The difficulty is, though, how to precisely conceive of this "nonteleological realism." One wonders where all the new fundamental issues might constantly arise, as they have to if science is supposed both to resolve them incessantly and to come across further ones continually. One might even suspect that nonteleological realism is an incoherent notion. But this would be rash. There are loopholes left for reconciling realism with unlimited basic progress.

The first option is the adoption of a "Russian-doll ontology." Blaise Pascal advocated the idea of an infinity of worlds within worlds. Each minute part of matter contains an entire universe with scaled-down versions of the earth, the sun, and the planets. And each material part of the latter system again comprehends an even smaller universe, and so on indefinitely. The world as a whole is constituted by an infinite sequence of nested universes (Pascal 1995 [1669], 34). However, this view in itself is not yet sufficient for grounding unending progress, for at each smaller scale the same structures are thought to be reproduced. If one level of the hierarchy is known, the others offer no surprise. So one has to adduce the idea that each layer of this infinite hierarchy is governed by a different set of laws. Consequently, upon penetrating ever deeper, new entities and properties turn up steadily. As ever smaller parts of matter are detected, different sets of laws are identified. Opening up a doll reveals a completely different doll. Variegated Russian-doll ontology entails, then, that there be a genuinely distinct set of entities and properties beyond every given horizon of theory.

This position indeed represents the sought-after coherent conjunction of realism and unlimited progress. It suggests that the presently assumed elementary structures, basically quarks and leptons, are composed of even more fundamental entities with new properties, which in turn consist of still more basic ones, and so on and so forth. Within this framework science could go on indefinitely in unveiling the true nature of things. So scientific realism and unlimited progress in fundamental research are compatible with one another. But adopting the variegated Russian-doll ontology involves quite a commitment.

One wonders whether the ontological burden can be lightened. At first sight, the "chess model" of progress, again suggested by Feynman, appears to present such an option. Its point is that even the discovery of all the fundamental laws of nature would leave ample room for further progress; we could still explore their consequences forever. After all, the rules of chess are well known in their entirety; nothing new will ever turn up at the fundamental level. And yet the number of new chess games that can be played is unlimited (Horgan 1996, 175–76).

The chess model is less restrictive than the geography picture in that it stresses that science is in no way bound to come to a halt after the discovery of the ultimate laws of nature. However, the chess model is clearly insufficient for buttressing the endurance of progress in fundamental matters. It serves to lighten up the somewhat gloomy geography picture of exhausted science by emphasizing that if science has touched its limits it has not yet reached its end. The chess model emphasizes that the completion of fundamental science would leave large chunks of pure science unaffected. Within its framework, one lasting challenge concerns the study of complex systems. The fundamental laws are to be employed for elucidating the intricate interactions among the constituents of tangled systems. Science would go on to devote itself to unraveling the knot of complexity.

So it seems that adopting a finite nature (i.e., rejecting the variegated Russian-doll ontology), along with scientific realism, implies the geography picture or the chess model of scientific progress. And the chess model is just as insufficient for grounding continual progress in fundamental matters as the geography picture. It entails that fundamental research, by increasingly revealing the secrets of nature, exhaust growing realms of nature and thereby undermine its own potential for keeping progress going. Fundamental progress, realistically conceived, is a self-destroying endeavor. It does not conform to the requirements of sustainable development; if it works, it will not last. On the other hand, constraints on fundamental science need not suffocate pure science as a whole. The completion of investigations into the ultimate structure of reality may well go along with perpetual progress in other disciplines.

However, there is another option for matching realism and unlimited progress besides adopting a Russian-doll ontology, namely the assumption of *emergent* laws. The concept of emergence is now used in a variety of ways. In its classical version, developed by C. D. Broad in 1925, the notion was supposed to refer to properties of a composite system that, owing to some in-principle limitation, cannot be derived from or explained by the properties of its constituents. To be sure, the properties of the composite system are determined by the features of the parts; there are no additional influences. Yet it is impossible to account for the former on the basis of the latter relying on overarching or comprehensive laws. One of Broad's examples is the chemical transformation of hydrogen and oxygen into water. The compound can be generated reliably from the elements, which shows that the properties of the latter are sufficient to bring forth the properties of the former. Nevertheless, there is—or in Broad's time there was—not a trace of a connecting mechanism that could account for the drastic alteration of properties. After all, it seems an amazing feat of nature that two gases produce a liquid. Emergent properties thus qualitatively differ from the properties of their respective bases but are still lawfully related to them. However, this specific lawful relation is inexplicable by comprehensive theoretical principles (Broad 1925, 58–66).

The critical feature of emergence is this explanatory gap. The occurrence of qualitatively different properties in complex systems only justifies an emergence claim, first, if these properties cannot be accounted for by relying on the properties of the parts along with their interactions, and, second, if this deficit arises from some in-principle limitation. A counterexample is the appearance of current oscillations in a properly equipped electric circuit. If either a capacitor or a coil is placed in a circuit, not a trace of oscillations is found. But if both are properly installed, oscillations in current intensity turn up. This is a qualitative difference by any measure. But it can be explained by drawing on the interactions between the two devices. The derivability of the novel type of behavior indicates that we are not dealing with a case of (classical) emergence.

The occurrence of emergence would invalidate the chess model of progress. Assume the existence of a final theory in the sense of a limited set of principles applying truthfully to the fundamental constituents of matter. These principles fail to extend to emergent properties, which are characterized by their nonderivability from the laws pertaining to the respective constituents. As a matter of definition, emergent phenomena are beyond the reach of a final theory in this sense. A law may be called *ultimate* if it cannot be explained by more comprehensive principles. The upshot of the emergentist scenario is that there are ultimate laws of composite systems. Consequently, these emergent laws are not obtained by applying the final theory to these systems and following its consequences. Basic research does not assume a chesslike structure since the final theory fails to give us all the rules of the game.

Emergentism entails a layered structure of reality with a particularly intricate relation among the layers. There may well be a finite set of fundamental laws of nature, to be sure, but the laws applying to the higher levels of the hierarchy of properties or processes cannot be derived from the fundamental ones. The principles governing the behavior of the basic constituents of matter fail to capture the behavior of complex objects—albeit the latter are "nothing but" aggregates of these constituents. The higher-level laws are in no way deducible from the fundamental ones; they are not derivative but ultimate themselves—in spite of applying to composite and structured aggregates. The distinctly holistic properties and laws of these aggregates also belong to the subject matter of *basic* science. They are ultimate in the same sense as the principles pertaining to the constituents. Complex structures constitute a field of basic research even after the most minute particles are detected and their interactions registered. Within the emergentist framework, and in contrast to the chess model, complexity vouches for the ongoing character of *basic* scientific progress after the discovery of the complete *fundamental* theory.

In fact, Rescher's argument from technology dependence, discussed previously, draws on emergence. The argument is that ongoing progress does not hinge on the "infinite complexity" of physical systems but merely on the features of the process of securing fuller information about them. Improvement of our conceptual devices and observational instruments suffices of itself to underwrite the prospect of a virtually endless series of new and significant discoveries (Rescher 1978, 40–42; 1984, 56–57). And the example is revealing. Assume, he argues, that we examine the behavior of a system over increasingly shorter periods of time. At each level of data acquisition the time average of a given quantity is measured. The repeated reduction of the time interval is likely to yield new insights into the system's mechanism, and the significance of these discoveries will probably remain the same. And here is the reason: "Given that the averages at each stage of the example are inferentially independent of one another, the discoveries one makes at each iteration are all major innovations" (Rescher 1978, 42). The argument is sound, but its premise involves an emergence claim. For how can it be that the averages at different time scales are inferentially disconnected? Barring the possibility that the measurements are simply unreliable, the answer is: emergence. Otherwise one could as well expect that the further decrease of the period of measurement might not reveal any changes. The value of the quantity, as determined during continually shorter time intervals, might simply remain constant. In this case, short-time averages would be inferentially dependent on long-time ones, and vice versa. In contradistinction, inferential decoupling, as assumed in the premise of Rescher's argument, is the constitutive element of emergence.

The upshot of this section is that the notion of nonteleological realism can be made coherent by assuming an intricate structure of reality. Both the adoption of a Russian-doll ontology and that of emergentism would be suitable for reconciling scientific realism with unending progress in fundamental research.[3] On the other hand, both views involve heavy ontological baggage, which some advocates of scientific realism may find troublesome to bear. This leads to the question of whether there are any indications in support of one of these views. I think there are. A mild version of emergentism is indeed suggested by present science.

The Emergence of Functional Properties

The critical issue is the status of explanations in the special sciences. Underlying this issue is a conceptual ambiguity as to which laws are to be considered basic. At bottom, there are two options. In the first understanding, basic laws apply to the fundamental constituents of all processes and objects. They are distinguished by their comprehensive character. A basic law in the second sense is one that cannot be traced back to deeper principles, that is, one that is ultimate. In his plea for a final theory, Weinberg endorses both conceptions (Weinberg 1992, 18,

27). However, if emergence occurs, the fundamental laws do not coincide with the ultimate laws (see the previous section). Scientific explanations resort to laws. Emergent properties are captured by ultimate laws of composite systems and thus transcend the scope of a completed fundamental theory. Such laws are not comprehensive but apply to a more restricted domain; they are specific and yet basic. This means that a final theory of the fundamental entities would not constitute the point of convergence of *all* scientific explanations. And this in turn implies that a final theory leaves room for ongoing *basic* research. In the case of emergence a completed fundamental theory would fail to bring to a close the ancient search for principles that cannot be explained by deeper principles, as discussed earlier.

I discuss the prospects of emergence by turning to biology and its relation to physical or chemical accounts. The general understanding of the difference between the two is that biological approaches rely on *functions* whereas physico-chemical explanations address the *mechanisms* for the realization of these functions. Classical genetics uses the concept of a gene for denoting the *capacity* of an entity to transfer organismic properties to the subsequent generation. By contrast, physics (or biochemistry for that matter) refers to DNA sequences and their processing. Functions express *that* a certain task is performed; mechanisms specify *how* it is achieved. The relation between biology and physics hinges on the relation between functions and their realizations. And this relation is truly intricate: the same function can be realized by several disparate, heterogeneous mechanisms, and, conversely, the same mechanisms can perform several disparate, heterogeneous functions. My claim is that this reciprocal multiplicity generates the explanatory gap that is the hallmark of emergence.[4]

Let me begin with a simple example from the world of human artifacts (where functions play an important role). The task of keeping a house warm can be performed using different mechanisms. Operating a heating device or insulating the walls and the roof are equally suitable for performing this function. Both are means to the same end; they are roughly equivalent functionally. But they are disparate physically. Regarding the mechanisms, there is no significant common ground among them. It is true, a roof and a gas heater are both material objects; but they share this feature with a multitude of other, functionally disparate objects. Conversely, one and the same object can perform different functions. A roof, for instance, in addition to its usefulness for thermal regulation, also provides shelter against the rain. These functions are disparate or even heterogeneous. They are disparate since keeping people warm is different, functionally speaking, from keeping people dry. And they are heterogeneous in that they miss a significant joint feature that could serve to demarcate these functions from others that are definitely not performed by a roof (like protecting people from hunger). There is a multiplicity of heterogeneous realizations of the same function; and there is a multiplicity of heterogeneous functions of the same object or process.

Biology furnishes a plethora of examples for these variegated connections between functions and mechanisms. The evolutionary path of the bird feather is instructive. According to one of the explanations presently on offer, the bird feather originated from peculiarities of the protein metabolism of reptiles. During the early stages of the evolution of birds the population of insects had increased drastically, and these insects provided a vast food resource for hungry reptiles. Insects contain a comparatively high amount of protein, which the metabolism of reptiles could neither use nor remove, at least initially. The evolutionary solution was excretion of the surplus protein through the skin. The excess protein was used to form elastic structures covering the skin, and this metabolic innovation marked the beginning of the bird feather (or so the account runs). The primary evolutionary cause for its development—or its initial function—was the removal of metabolic waste. But this is by no means the only function feathers perform. In addition, they serve the purpose of thermal insulation. And a third function is the enhancement of the capacity for flight.

This shows that one and the same object or mechanism may perform different functions. And it may do so under the same environmental conditions. That is, the difference between the functions is not due to the intrusion of additional influences. Feathers may fulfill all three functions at the same time in the same situation. Moreover, these functions are heterogeneous. There is no significant functional common ground among the regulation of protein metabolism, the preservation of body heat, and the ability to lift off the ground. Conversely, each of the functions could be realized differently. As polar bears and whales demonstrate, one needs no feathers to protect oneself against the cold, and as mosquitoes and butterflies reveal, flying is not dependent on feathers. These realizations are heterogeneous; they do not share a common physical feature that might serve to distinguish them from other physical properties.

On the whole, then, a multiplicity of heterogeneous functions may be performed by a given mechanism under given circumstances; and a multiplicity of heterogeneous realizations may be associated with a given function. There is a *reciprocal heterogeneous multiplicity* between functions and their realizations. The multiplicity of functions vitiates inferring the function from a given mechanism; after all, this mechanism may perform a number of distinct functions. And the multiplicity of realizations thwarts deriving the mechanism from a given function; after all, this function could be performed by many distinct mechanisms. There is an inferential gap between functions and mechanisms.

Let me make this gap more explicit. The first thing to be noted is that no explanatory hiatus appears in the individual cases. Nothing precludes each such case from being elucidated completely. No puzzle need be involved in the fact that a function is realized by a particular mechanism in a particular instance. The physiological process of protein excretion can certainly be clarified in all its minute detail. And the insulating effect of a layer of feathers poses no explanatory challenge whatsoever. Conversely, it is in no way enigmatic that a given mechanism can perform several functions at a time. There is no mystery involved in the fact that feathers protect against the cold and enhance the capacity for flight as well.

This shows that the obstacle to explanation has nothing to do with the individual instances. Rather, it concerns the relation of *being equal in kind*. First, multiple heterogeneous realization implies that differences in the mechanisms do not entail differences in their functions. Equivalently, sameness of function is compatible with differences in the mechanisms. Second, it follows from the existence of multiple heterogeneous functions of the same mechanism that differences in function do not imply differences in the associated mechanisms. Correspondingly, sameness of the mechanisms may go along with differences in function. The equivalence classes of functions and of their realizations are largely disjoint. This is where the explanatory hiatus turns up. Functional similarities cannot be explained on the basis of the physical similarities among their realizations—because the latter may be dissimilar after all. And physical similarities among mechanisms cannot be accounted for by the similarity of the functions they perform—because the mechanisms may perform distinct functions. The equivalence classes of functions and of their realizations are inferentially decoupled. Here is the place where emergence arises.[5]

It should be emphasized that the abyss between functions and their realizations really affects the explanatory power of theories on either level. Functions are explanatorily relevant since they generalize over physically disparate cases. Functional concepts group such physically distinct instances together and make overarching explanations possible. Biological accounts employ such functional concepts as "hereditary factor," "signal substance," or "pheromone." Such concepts express essential effects associated with these entities without paying attention to how these effects are brought about. Signal substances, by definition, are operative in the communication within or between cells. They are characterized by their function. Functional accounts address uniform effects of structurally distinct molecules and thus involve an explanatory unification. These functionally uniform kinds of objects or processes dissolve from the physical perspective and crumble into physically disparate pieces. From this perspective, one can no longer speak of biological signals in general but rather must address the disparate molecular details of each particular case. As a result, the unifying functional explanation is lost.

Conversely, the different functions of the same mechanism make it clear that recourse to the realizations allows for a particular type of unified explanation as well. Restriction to functions would fail to grasp that it is the same physical object or process that is responsible for a whole class of functionally disparate effects. One needs the physical level of consideration to understand that the function of protein excretion is realized by the same physical means as the function of thermal insulation. And only by drawing on the chemical nature of the substances involved does one realize that DNA, messenger RNA, and transfer RNA—in spite of the different roles they play in the process of heredity—have an important property in common: they are all nucleic acids. Abandoning the physical level of consideration likewise involves a loss of unification.

As a consequence, the level of functions and the level of mechanisms *each* miss important explanatory accomplishments of the other. Both are needed for a comprehensive account of the phenomena. The concept of a signal substance ties physically disparate processes together, and the concept of a nucleic acid connects functionally disparate cases with each other. Note the symmetry of the conceptual situation. There is an explanatory loss in either direction. But this implies that some explanations cannot be traced back to particle physics. *Pace* Weinberg (1992, 55), *not* all arrows of explanation converge in the vicinity of the standard model.[6]

Conclusion

The lesson that accrues from these considerations regarding the relation between scientific realism and the possible limitations of basic progress is as follows. In spite of initial appearances to the contrary, both views can be brought into harmony. But the minimum effort it takes to make them compatible is the assumption of emergent laws. Unlike a Russian-doll ontology, emergentism does not require an unending sequence of nested layers of reality. Within the emergentist framework, one may well assume a final level of entities and interactions. The point is that these entities and interactions, in contrast to the chess model, in principle fail to account for all the properties of every composite system they produce. Emergent laws are ultimate but not fundamental. They are ultimate in that they cannot be based on deeper principles, and they fail to be fundamental in that they do not apply to the final constituents of matter. By searching for emergent laws, basic scientific progress may survive the discovery of the fundamental laws.

Even if this amount of emergence is granted, it in no way *follows* that new problems pop up without limits and that progress goes on incessantly. The point is merely that this much of emergentism is the minimum ontological investment needed to make scientific realism and unending progress compatible with one another. For in this case it is *possible*, albeit by no means guaranteed, that new ultimate laws or properties arise continually from the study of complex aggregates. So there is at least a chance to advocate both realism and ongoing progress. The conjunction of the two claims becomes consistent; it is not conceptually ruled out that they both hold true. Philosophical considerations cannot vouch for the actual continual appearance of new issues. Philosophers cannot tell what the world is like. The argument merely concerns what the world needs to be like to allow for scientific realism and unlimited progress to obtain simultaneously. And the conclusion is, first, that possible worlds of this sort can be specified and that, second, judging from present science, the actual world *may* be one of the requisite kind.

The argument developed in the previous section suggests that the structure of special sciences that invoke functional properties in their explanations supports the emergentist position. The type of emergentism that need be adopted for underwriting ongoing progress in basic science is of a minimalist sort; it refers only to kinds of similarity, not to individual events (as Broad's classical emergentism has it). But this much of an inferential decoupling among different layers of properties is needed so as to allow for basic progress to continue indefinitely in the face of scientific realism and a finite set of fundamental constituents. Actually, something of the kind was stressed by Rescher. As he argues, it is *functional* complexity, not compositional complexity, that supports the prospect of unending scientific progress (Rescher 1978, 40, 42). The preceding considerations suggest that Rescher is quite right after all.

NOTES

1. For a sketch of this position see, for example, Boyd (1983, 45), Leplin (1984, 1–2), and Rescher (1987, 1–4). For a discussion of arguments adduced in its favor see Carrier (1991b, 1993).

2. Somewhat surprisingly, Rescher also assumes that some overarching or fundamental principles of science are exempt from revision and provide a basis for an attenuated version of scientific realism. "Schoolbook science" offers a home for a "science-indebted realism" (1987, 59–64).

3. Gerald Massey drew my attention to the fact that emergentism in no way entails literally "unending progress," that is, the emergence of an infinite number of "ultimate" laws or properties. In fact, the upper limit of the number of emergent properties is given, within this framework, by the number of combinations of the fundamental constituents of matter. Provided that this latter number is finite, the number of novel research areas is constrained as well. But it can be trusted to be large by any measure. Still, as a matter of principle, the series of new problems would be merely virtually endless in the emergentist picture, not infinite.

4. This issue was raised by Kincaid (1990, 576–83), who considered multiple realization and differences in the functions of the same entity or process as obstacles to a reduction of biology to chemistry. But in contrast to the present argument, Kincaid attributes functional differences solely to the intrusion of additional factors. Gasper (1992, 668) goes a step further by stressing that the same physical substrate can realize different higher-level properties. I conjoin these two approaches and link them to the notion of emergence.

5. An analogous disparity (along with the concomitant loss of explanatory power) is found among the kinds of states of physics and psychology (Carrier 1991a; 1998, §4).

6. In Weinberg's view, there is no emergence. The laws governing *all* complex systems that are not genuinely distinct arise from the interactions of the constituents (1992, 61–62). Weinberg admits that the introduction of a different conceptual perspective (he mentions thermodynamics) involves different generalizations. But he insists that each particular case can be accommodated by particle physics (1992, 39–41). This may be true, but it misses the point. The unification vanishes in turning to the disparate particulars of each case, and this constitutes an explanatory loss.

REFERENCES

Boyd, R. 1983. "On the Current Status of the Issue of Scientific Realism." *Erkenntnis* 19: 45–90.
Broad, C. D. 1925. *The Mind and Its Place in Nature.* Paterson, N.J.: Littlefield, Adams, 1960.
Carrier, M. 1991a. "On the Disunity of Science, or Why Psychology Is Not a Branch of Physics." In Akademie der Wissenschaften zu Berlin, ed., *Einheit der Wissenschaften.* Berlin: De Gruyter, 39–59.
———. 1991b. "What Is Wrong with the Miracle Argument?" *Studies in History and Philosophy of Science* 22: 23–36.
———. 1993. "What Is Right with the Miracle Argument: Establishing a Taxonomy of Natural Kinds." *Studies in History and Philosophy of Science* 24: 391–409.
———. 1998. "In Defense of Psychological Laws." *International Studies in the Philosophy of Science* 12: 217–32.
Feynman, R. P. 1965. *The Character of a Physical Law.* Cambridge, Mass.: MIT Press.
Gasper, P. 1992. "Reduction and Instrumentalism in Genetics," *Philosophy of Science* 59: 655–70.
Horgan, J. 1996. *The End of Science: Facing the Limits of Knowledge and the Twilight of the Scientific Age.* Reading, Mass.: Addison-Wesley.
Kincaid, H. 1990. "Molecular Biology and the Unity of Science." *Philosophy of Science* 57: 575–93.
Koyré, A. 1965. *Newtonian Studies.* Cambridge, Mass.: Harvard University Press.
Kuhn, T. S. 1962. *The Structure of Scientific Revolutions.* Chicago: University of Chicago Press.
Leplin, L. 1984. "Introduction." In L. Leplin ed., *Scientific Realism.* Berkeley: University of California Press, 1–7.
Pascal, B. 1995 [1669]. *Pensées.* Paris: Brooking International.
Rescher, N. 1978. *Scientific Progress: A Philosophical Essay on the Economics of Research in Natural Science:* Oxford: Basil Blackwell.
———. 1984. *The Limits of Science.* Berkeley: University of California Press.
———. 1987. *Scientific Realism: A Critical Reappraisal.* Dordrecht: Kluwer.
Weinberg, S. 1992. *Dreams of a Final Theory.* New York: Vintage.

8

Can Computers Overcome Our Limitations?

Nicholas Rescher
Department of Philosophy, University of Pittsburgh

Could Computers Overcome Our Limitations?

In view of the difficulties and limitations that beset our human efforts at answering our questions in a complex world, it is tempting to contemplate the possibility that computers might enable us to overcome our cognitive disabilities and epistemic frailties. We may wonder: Are computers cognitively omnipotent? If a problem is to qualify as soluble at all, will computers always be able to solve it for us?

Of course, computers cannot bear human offspring, enter into contractual agreements, or exhibit heroism. But such processes address *practical* problems relating to the management of the affairs of human life and so do not count in the present cognitive context. Then too we must put aside *evaluative* problems of normative bearing or of matters of human affectivity and sensibility: computers cannot offer us meaningful consolation or give advice to the lovelorn. The issue presently at hand regards the capacity of computers to resolve *cognitive* problems regarding matters of empirical or formal fact. Typically, the sort of problems that will concern us here are those that characterize the sciences, in particular problems relating to the description, explanation, and prediction of the things, events, and processes that make up the realm of physical reality. And to all visible appearances computers are ideal instruments for handling the matters of cognitive complexity that arise in such contexts. The question, then, is this: Is there anything in the domain of cognitive problem solving that computers cannot manage to do?

The history of computation in recent times is one of a confident march from triumph to triumph. Time and again, those who have affirmed the limitations of computers have been forced into ignominious retreat as increasingly powerful machines implementing increasingly ingenious programs have been able to achieve the supposedly unachievable. However, the question on the present agenda is not "Can computers *help* with problem-solving?"—an issue that demands a resounding affirmative and needs little further discussion. There is no doubt whatsoever that computers can do a lot here—and very possibly more than we ourselves can. But there is an awesomely wide gap between *a lot* and *everything*.

First some important preliminaries. To begin with, we must, in this present context, recognize that much more is at issue with a "computer" than a mere electronic calculating machine understood in terms of its operational hardware. For one thing, software also counts. And so does data acquisition. As we here construe computers, they are electronic information-managing devices equipped with data banks and augmented with sensors as autonomous data access. Such "computers" are able not only to *process* information but also to *obtain* it. Moreover, the computers at issue here are, we shall suppose, capable of discovering and learning, and thereby able significantly to extend and elaborate their own initially programmed modus operandi. Computers in this presently operative sense are not mere calculating machines, but general problem solvers along the lines of the fanciful contraptions envisioned by the aficionados of artificial intelligence. These enhanced computers are accordingly question-answering devices of a very ambitious order.

Given this expanded view of the matter, we must also correspondingly enlarge our vision both of what computers can do and of what can reasonably be asked of them. For it is the potential of computers as an instrumentality for universal problem solving (UPS) that concerns us here, not merely their more limited role in the calculations of algorithmic decision theory (ADT). The computers at issue will thus be prepared to deal with factually substantive as well as merely formal (logico-mathematical) issues. And this means that the questions we can ask are correspondingly diverse. For here, as elsewhere, added power brings added responsibility. The questions it is appropriate to ask thus can relate not just to matters of calculation but also to the things and processes of the world.

Moreover, some preliminary discussion of the nature of "problem solving" is required, because one has to be clear from the outset about what it is to *solve* a cognitive problem. Obviously enough, this is a matter of answering questions. Now "to answer" a question can be construed in three ways: to offer a *possible* answer, to offer a *correct* answer, and finally to offer a *credible* answer. It is the third of these senses that is of concern here. And with good reason. Consider a problem solver that proceeds in one of the following ways: it replies "yes" to every yes/no question; or it figures out the range of possible answers and then randomizes to select one; or it proceeds by pure guesswork. Even though these so-called "problem solvers" may give the correct response some or much of the time, they are systematically unable to resolve our questions in the presently operative credibility-oriented sense of the term. For the obviously sensible stance calls for holding that *a cognitive problem is resolved only when an appropriate answer is convincingly provided*—that is to say, when we have a solution that we can responsibly accept and acknowledge as such. Resolving a problem is not just a matter of having an answer, and not even of having an answer that happens to be correct. The actual resolution of a problem must be credible and convincing—with the answer provided in such a way that its cogency is recognizable. In general problem solving we want not just a dictum but an *answer*—a response equipped with a contextual rationale to establish its credibility in a way accessible to duly competent recipients. To be warranted in accepting a third-party answer we must ourselves have case-specific reasons to acknowledge it as correct. A response whose appropriateness as such cannot secure rational confidence is no answer at all.[1]

General-Principle Limits Are Not Meaningful Limitations

The question before us is the following: "Are there *any* significant cognitive problems that computers cannot solve?" Now it must be acknowledged from the outset that certain problems are inherently unsolvable in the logical nature of things. One cannot square the circle. One cannot comeasure the incommensurable. One cannot decide the demonstrably undecidable nor prove the demonstrably unprovable. Such tasks represent absolute limitations whose accomplishment is theoretically impossible—unachievable for reasons of general principle rooted in the nature of the realities at issue.[2] And it is clear that inherently unsolvable problems cannot be solved by computers either.[3]

Other sorts of problems will not be unsolvable as such but will, nevertheless, prove to be computationally intractable. For with respect to *purely theoretical* problems it is clear from Turingesque results in algorithmic decision theory that there will indeed be computer insolubilia—mathematical questions to which an algorithmic respondent will either give the wrong answer or be unable to give any answers at all, no matter how much time is allowed.[4] But this is a mathematical fact that obtains of necessity, so that this whole issue can also be set aside for present purposes. For in the present context of UPS, the necessitarian facts of Gödel-Church-Turing incompleteness become irrelevant. Here any search for *meaningful* problem-solving limitations will have to confine its attention to problems that are in principle solvable: *demonstrably* unsolvable problems are beside the point of present concern because an inability to do what is in principle impossible hardly qualifies as a limitation, seeing that it makes no sense to ask for the demonstrably impossible.

For present purposes, then, it is limits of *capability* not limits of *feasibility* that matter. In asking about the problem-solving limits of computers we are looking at problems that *computers* cannot resolve but that other problem solvers conceivably can. The limits that will concern us here are accordingly rooted neither in conceptual or logico-mathematical infeasibilities of general principle nor in absolute physical impossibilities, but rather in performatory limitations imposed specifically upon computers by the world's contingent modus operandi.

And in this formulation the adverb *specifically* does real work by way of ruling out certain computer limitations as irrelevant. Things standing as they do, some problems will simply be too large given the inevitable limitations on computers in terms of memory, size, processing time, and output capacity. Suppose for the moment that we inhabit a universe that, although indeed boundless, is nevertheless finite. Then no computer could possibly solve a problem whose output requires printing more letters or numbers than there are atoms in the universe. Such problems ask computers to achieve a task that is not "substantively meaningful" in the sense that no physical agent at all—computer, organism, or whatever—could possibly achieve it. The problems that concern us here are those that are not solution-precluding on the basis of inherent mathematical or physical impossibilities. To reemphasize: our concern is with the performative limitations of computers with regard to problems that are not inherently intractable in the logical or physical nature of things.

Practical Limits

Inadequate Information

Often the information needed for credible problem resolution is simply unavailable. Thus no problem solver can at this point in time provide credible answers to such questions as "What did Julius Caesar have for breakfast on that fatal Ides of March?" or "Who will be the first auto accident victim of the next millennium?" The information needed to answer such questions is just not available at this stage. In all problem-solving situations, the performance of computers is decisively limited by the quality of the information at their disposal. "Garbage in, garbage out," as the saying has it. But matters are in fact worse than this. Garbage can come out even where no garbage goes in.

One clear example of the practical limits of computer problem solving arises in the context of prediction. Consider the two prediction problems set out in figure 8.1. At first sight, there seems to be little difficulty in arriving at a prediction in these cases. But now suppose that we acquire some further data to enlarge our background information: pieces of information supplementary to—but in no way conflicting with or corrective of—the given premises:

Case 1   X *is extremely, indeed* inordinately, *fond of Trollope.*

Case 2   Z *also promised to repay his other neighbor the $7.00 he borrowed on the same occasion.*

Note that in each case our initial information is in no way abrogated but merely enlarged by the additions in question. But nevertheless in each case we are impelled, in the light of that supplementation, to *change* the response we were initially prepared and rationally well advised to make. Thus when I know nothing further of next year's Fourth of July parade in Centerville, USA, I shall predict that its music will be provided by a marching band; but if I am additionally informed that the Loyal Sons of Old Hibernia have been asked to provide the music, then bagpipes will now come to the fore.

Case 1

Data: *X* is confronted with the choice of reading a novel by Dickens or one by Trollope. And further: *X* is fond of Dickens.

Problem: To predict which novel *X* will read.

Case 2

Data: *Z* has exactly $10.00. And further: *Z* promised to repay his neighbor $7.00 today. Moreover, *Z* is a thoroughly honest individual.

Problem: To predict what *Z* will do with his money.

Figure 8.1

It must, accordingly, be recognized that the search for rationally appropriate answers to certain questions can be led astray not just by the *incorrectness* of information but by its *incompleteness* as well. The specific body of information that is actually at hand is not just important for problem resolution, it is *crucial*. And we can never be completely confident of problem resolutions based on incomplete information, seeing that further information can always come along to upset the apple cart. As available information expands, established problem resolutions can always become destabilized. One crucial practical limitation of computers in matters of problem solving is thus constituted by the inevitable incompleteness (to say nothing of the potential incorrectness) of the information at their disposal. And here the fact that computers can only ever ingest finite—and thus incomplete—bodies of information means that their problem-resolving performance is always at risk. (Moreover, this sort of risk exists quite apart from other sorts, such as the fact that computerized problem resolutions are always the product of many steps, each of which involves a nonzero probability of error.) If we are on a quest for certainty, computers will not help us get there.

Transcomputability and Real-Time Processing Difficulties

Apart from *uncomputable* (computationally unsolvable) problems, there is also the range of *transcomputable* problems: problems whose computational requirements exceed the physical bounds and limits that govern the concrete realization of theoretically designed al-gorithmic machines.[5] Because computers are physical devices, they are subject to the laws of physics and limited by the realities of the physical universe. In particular, since a computer can process no more than a fixed number of bits per second per gram, the potential complexity of algorithms means that there is only so much that a given computer can possibly manage to do.

Then there is the temporal aspect. To solve problems about the real world, a computer must of course be equipped with information about it. But securing and processing information is a time-consuming pro-cess, and the time at issue can never be reduced to an instantaneous zero. Time-constrained problems that are enormously complex—those whose solution calls for securing and processing a vast amount of data—can exceed the reach of any computer. At some point it always becomes impossible to squeeze the needed operations into the available time. There are only so many numbers that a computer can crunch in a given day. And so if the problem is a predictive one it could find itself in the awkward position that it should have started yesterday on a problem only presented to it today. Thus even under the (fact-contravening) supposition that the computer can answer *all* of our questions, it cannot, if we are impatient enough, produce those an-swers as promptly as we might require them. Even when given, an-swers may be given too late.

Limitations of Representation in Matters of Detail Management

This situation is emblematic of a larger issue. Any computer that we humans can possibly contrive here on earth is going to be finite: its sensors will be finite, its memory (however large) will be finite, and its processing time (however fast) will be finite.[6] Moreover, com-puters operate in a context of finite instructions and finite inputs. Any representational model that functions by means of computers is of finite complexity in this sense. It is always a finitely characterizable system: its descriptive constitution is characterized in finitely many information-specifying steps, and its operations are always ultimately presented by finitely many instructions. And this array of finitudes means that a computer's modeling of the real will never capture the inherent ramifications of the natural universe of which it itself is but a minute constituent. Artifice cannot replicate the complexity of the real; reality is richer in its descriptive constitution and more efficient in its transformatory processes than human artifice can ever manage to realize. For nature itself has a complexity that is effectively endless, so that no finite model that purports to represent nature can ever replicate the detail of reality's makeup in a fully comprehensive way, even as no architect's blueprint-plus-specifications can possibly specify *every* feature of the structure that is ultimately erected. In particular, the complications of a continuous universe cannot be captured completely via the resources of discretized computer languages. All endeavors to represent reality—computer models emphatically included—involve some element of oversimplification, and in general a great deal of it.

The fact of the matter is that reality is too complex for adequate cognitive manipulation. Cognitive friction always enters into matters of information management—our cognitive processing is never totally efficient; something is always lost in the process; cognitive entropy is always on the scene. But as far as knowledge is concerned, nature does nothing in vain and so encompasses no altogether irrelevant detail. Yet oversimplification always makes for losses, for deficiencies in cognition. For representational omissions are never totally irrelevant, so that no oversimplified descriptive model can get the full range of predictive and explanatory matters exactly right. Put figuratively, it could be said that the only "computer" that can keep pace with reality's twists and turns over time is the universe itself. It would be unreasonable to expect any computer model less complex than this totality itself to provide a fully adequate representation of it, in particular because that computer model must of course itself be incorporated *within* the universe.

Performative Limits of Prediction—Self-Insight Obstacles

Another important sort of practical limitation to computer problem solving arises not from the inherent intractability of questions but from their unsuitability for particular respondents. Specifically, one of the issues regarding which a computer can never function perfectly is its own predictive performance. One critical respect in which the self-insight of computers is limited arises in connection with what is known as the "halting problem" in algorithmic decision theory. Even if a problem is computer solvable—in the sense that a suitable computer will demonstrably be able to find a solution by keeping at it long enough—it will in general be impossible to foretell how long a process of calculation will actually be needed. There is not—and demonstrably cannot be—a *general* procedure for foretelling with respect to a particular computer and a particular problem "Here is how long it will take to find the solution—and if the problem is not solved within this time span then it is not solvable at all." No computer can provide general insight into how long it—or any other computer, for that matter—will take to solve problems. The question "How long is long enough?" demonstrably admits of no general solution here.

And computers are—of necessity!—bound to fail even in much simpler self-predictive matters. Thus consider confronting a predictor with the problem posed by the question

P_1 *When next you answer a question, will the answer be negative?*

This is a question that—for reasons of general principle—no predictor can ever answer satisfactorily.[7] Consider the available possibilities:

Answer given	Actually correct answer	Agreement?
Yes	No	No
No	Yes	No
Can't say	No	No

On this question, there just is no way in which a predictive computer's response could possibly agree with the actual fact of the matter. Even the seemingly plausible response "I can't say" automatically constitutes a self-falsifying answer, since in giving this answer the predictor would automatically make "No" into the response called for by the proprieties of the situation.

Here, then, we have a question that will inevitably confound any conscientious predictor and drive it into baffled perplexity. But of course the problem poses a perfectly meaningful question to which *another* predictor could give a putatively correct answer—namely, by saying "No: that predictor cannot answer this question at all; the question will condemn a predictor (predictor no. 1) to baffled silence." But of course the answer "I am responding with baffled silence" is one which that initial predictor cannot cogently offer. And as to that baffled silence itself, it is something that, as such, would clearly constitute a defeat for predictor no. 1. Still, that question which impelled predictor no. 1 into perplexity and unavoidable failure presents no problem of principle for predictor no. 2. And this clearly shows that there is nothing improper about that question as such. For while the question posed in P_1 is irresolvable by a *particular* computer, it nevertheless could—in theory—be answered by *other* computers, and so it is not irresolvable by computers in general.

However, there are other questions that indeed are computer insolubilia for computers at large. One of them is

P_2 *What is an example of a predictive question that no computer will ever state?*

In answering *this* question the computer would have to stake a claim of the form "*Q* is an example of a predictive question that no computer will ever state." And in the very making of this claim the computer would falsify it. It is thus automatically unable to effect a satisfactory resolution. However, the question is neither meaningless nor irresolvable. A *noncomputer* problem solver could in theory answer it correctly. Its presupposition, "There is a predictive question that no computer will ever consider," is beyond doubt true. What we thus have in P_2 is an example of an in-principle solvable—and thus "meaningful"—question that, as a matter of necessity in the logical scheme of things, no problem-solving computer can ever resolve satisfactorily. The long and short of it is that every predictor—computers included—is bound to manifest versatility incapacities with respect to its own predictive operations.[8]

However, from the angle of our present considerations, the shortcoming of problems P_1 and of P_2 is that they are computer irresolvable on the basis of theoretical general principles. And it is therefore not appropriate from the present perspective, as already explained, to count this sort of thing as a computer limitation. Are there any other, less problematic, examples?

Performative Limits: A Deeper Look

At this point we must contemplate some fundamental realities of the situation confronting our problem-solving resources. The first of these is that no computer can ever reliably determine that all its more powerful peers are unable to resolve a particular substantive problem (that is, one that is inherently tractable and not demonstrably unsolvable on logico-conceptual grounds). And this means that

T_1 *No computer can reliably determine that a given substantive problem is altogether computer irresolvable.*

This is to say that no computer can reliably determine that a particular substantive problem p is such that no computer can resolve it: $(\forall C) \sim C$ res p. We thus have

$$\sim(\exists C')(\exists P)C' \text{ det } (\forall C)\sim C \text{ res } P$$

or equivalently:

$$(\forall C')(\forall P)\sim C' \text{ det } (\forall C)\sim C \text{ res } P \text{ }^9$$

A brief explanation is needed regarding the use of *determine* that is operative here. In the present context this is a matter of so functioning as to be able to secure rational conviction for the claim at issue. As was emphasized previously, we want not just answers but *credible* answers.

Moreover, something that we are not prepared to accept from any computer is cognitive megalomania. No computer is, so we may safely suppose, ever able to achieve credibility in staking a claim to the effect that no substantive problem whatever is beyond the capacity (reach) of computers. And this leads to the thesis

T_2 *No computer can reliably determine that all substantive problems what-ever are computer resolvable.*

That is to say, no computer can convincingly establish that whenever a substantive problem p is at issue, then some computer can resolve it—in other words, that for any and every substantive problem p, $(\exists C)$ C res p. Thus

$$\sim(\exists C') \text{ } C' \text{ det } (\forall P)(\exists C) \text{ } C \text{ res } P$$

or equivalently:

$$(\forall C') \sim C' \text{ det } \sim(\exists P)(\forall C)\sim C \text{ res } P$$

A computer can neither reliably determine that an arbitrarily given substantive problem is computer irresolvable (T_1) nor reliably determine that no such problem is computer irresolvable (T_2).

We shall not now expatiate upon the rationale of these theses. Establishing their plausibility—which would at this point unduly interrupt the flow of the argument—is deferred to the appendix. All that matters at this juncture is that the principles in question merit acceptance—and do so not as a matter of abstractly logico-mathematical considerations, but owing to the world's practical realities.

The relationship between theses T_1 and T_2 comes to light more clearly when one considers their formal structure. The claims at issue are as follows:

T_1 *For all* C': *(∀P) ~C' det (∀C) ~C res* P

T_2 *For all* C': *~C' det (∀P) ~ (∀C) ~C res* P

Now let us also adopt the following two abbreviations:

- C-un p for: ~C det P ("C is unable to determine that p")
- X(p) for: (∀C)~C res P ("p is computer-unresolvable")

Then

$$T_1 = \text{For all } C\text{: } (\forall P)C\text{-un } X(P)$$

$$T_2 = \text{For all } C\text{: } C\text{-un } (\forall P)\text{~}X(P)$$

As this makes clear, both theses indicate a universal computer incapacity in relation to computer-unresolvability theses of the form $(\forall C)$~C res p. Thus T_1 and T_2 both reflect ways in which computers encounter difficulty in obtaining a credible grip on such universal incapacity. Fixing the bounds of computer solvability is beyond the capacity of any computer.

It should be noted that in his later writings Kurt Gödel himself took a line regarding mathematics analogous to that which the present discussion takes with respect to general problem solving. He maintained that no single particular axiomatic proof systematization will be able to achieve universality with respect to provability in general.[10] Even as Gödel saw algorithms as inherently incapable of doing full justice to mathematics, so the present argumentation has it that problem-solving computers cannot do full justice to science. Both theses implement the common idea that, notwithstanding the attractions and advantages of rigorous reasoning, the fact remains that in a complex world it is bound to transpire that truth is larger than rigor.

Contrast with Algorithmic Decision Theory

The unavailability of a universal problem solver in the setting of general problem solving has far-reaching theoretical implications. For it means that in UPS we face a situation regarding the capability of computers that is radically different from that of ADT.

In ADT we have Church's thesis: Wherever it is possible for computation to decide an issue, this resolution can be achieved by means of effective calculation. Thus computational resolvability/decidability (an informal conception!) can to all useful intents and purposes be equated with algorithmically effective computability (which is rigorously specifiable):

$$(C) \quad \text{sol } P \leftrightarrow (\exists C)(C \text{ res } P)$$

To this thesis one can adjoin Alan Turing's great insight that there can be a "universal computer" (a Turing machine)—a device T that can solve a computational problem if any calculating machine can:

$$(T) \quad (\exists C)(C \text{ res } P) \leftrightarrow T \text{ res } P$$

Combining these two theses, we arrive at the result that in the sphere of algorithmic computation solvability-at-large is tantamount to resolvability by a Turing machine:

$$(M) \quad \text{sol } P \leftrightarrow T \text{ res } P$$

Here one machine can speak for the rest: if a problem is resolvable at all by algorithmic calculations, then a Turing machine can resolve it. In ADT, there is thus an absolute, across-the-board conception of solvability.[11]

But when we turn our perspective to UPS, this monolithic situation is lost. Here the state of things is no longer Turingesque: there is not and cannot be a universal problem solver.[12] As we have seen, for any problem solver there will automatically be some correlatively unsolvable problems—problems that it cannot resolve but others can—along the lines of the aforementioned computer-embarrassing question P_1 ("When next you answer a question, will the answer be negative?"). Once we leave the calm waters of algorithmic computation and venture into the turbulent sea of problem-solving computation in general, it becomes impractical for any computer to survey all the possibilities. Here the overall range of computer-resolvable problems extends beyond the information horizon (the "range of vision," so to speak) of any given computer, so that no computer can make convincing claims about this range as a whole. In particular, these deliberations mean we would not—and should not—be prepared to take a computer's word for it if it stakes a claim of the form "Q is a (substantive) question that no computer whatsoever could possibly resolve."

A Computer Insolubilium

The time has come to turn from generalities to specifics. At this point we can confront a problem-solving computer (*any* such computer) with the challenging question

P_3 *What is an example of a (substantive) problem that no computer whatsoever can resolve?*

There are three possibilities here:

1. The computer offers an answer of the format "P is an example of a problem that no computer whatsoever can resolve." For reasons already canvassed we would not see this as an acceptable resolution, since by T_1 our respondent cannot achieve credibility here.

2. The computer responds "No can do. I am unable to resolve this problem: it lies outside my capability." We could—and would—accept this response and take our computer at its word. But the response of course represents no more than computer acquiescence in computer incapability.

3. The computer responds "I reject the question as being based on an inappropriate presupposition, namely that there indeed are problems that no computer whatsoever can resolve." We ourselves would have to reject this position as inappropriate in the face of T_2. The response at issue here is one that we would simply be unable to accept at face value from a computer.

It follows from such deliberations that P_3 is itself a problem that no computer can resolve satisfactorily.

At this point, then, we have realized the principal object of the discussion: we have been able to identify a meaningful concrete problem that is computer irresolvable for reasons that are embedded, via theses T_1 and T_2, in the world's empirical realities. For, to reemphasize, our present concern is with issues of general problem solving and not algorithmic decision theory.

The Human Element: Can People Solve Problems that Computers Cannot?

Our discussion has not, as yet, entered the doctrinal terrain of discussions along the lines of Hubert L. Dreyfus's *What Computers Still Can't Do* (1992).[13] For the project that is at issue there is to critique the prospects for "artificial intelligence" by identifying processes involving human intelligence and behavior that computers cannot manage satisfactorily. Dreyfus accordingly compares computer information processing with human performance in endeavoring to show that there are things that humans can do that computers cannot accomplish. However, the present discussion has up to this point considered solely problems that computers cannot manage to resolve. Whether *humans* can or cannot resolve them is an issue that has remained out of sight.

And so a big question remains as yet untouched: Is there any sector of this problem-solving domain where the human mind enjoys a competitive advantage over computers? Or does it transpire that wherever computers are limited, humans are always limited in similar ways?

In addressing this issue, let us be precise about the question that now faces us. It is

P_4 *Are there problems that computers cannot solve satisfactorily but people can?*

In fact what we would ideally like to have is not just an abstract answer to P_4, but a concrete answer to

P_5 *What is an example of a problem that computers cannot solve satisfactorily but people can?*

What we are now seeking is a computer-defeating question that has the three characteristics of (1) posing a meaningful problem, (2) being computer unsolvable, and (3) admitting of a viable resolution by intelligent noncomputers, specifically humans.[14]

This then is what we are looking for. And—lo and behold!—*we have already found it*. All we need do is to turn around and look back to P_3. After all, P_3 is, so it was argued, a problem that computers cannot resolve satisfactorily, and this consideration automatically provides us—people that we are—with the example that is being asked for. In presenting P_3 within its present context we have in fact resolved it. And moreover P_5 is itself also a problem of this same sort. It too is a computer-unresolvable question that people can manage to resolve.[15]

In the end, then, the ironic fact remains that the very question we are considering—regarding cognitive problems that computers cannot solve but people can—provides its own answer.[16] P_3 and P_5 appear to be eligible for membership in the category of "academic questions"— questions that are effectively self-resolving—a category that also includes such more prosaic members as "What is an example of a question formulated in English?" and "What is an example of a question that asks for an example of something?" The presently operative mode of computer unsolvability thus pivots on the factor of self-reference— as is the case with Gödelian incompleteness.

To be sure, their inability to answer the question "What is a question that no computer can possibly resolve?" is—viewed in a suitably formidable perspective—a token of the power of computers rather than of their limitation. After all, we see the person who maintains "I can't think of something I can't accomplish" not as unimaginative but as a megalomaniac—and one who uses *we* instead of *I* as only slightly less so. Nevertheless, in the present case this pretension to strength marks a point of weakness.

The key issue is whether computers might be defeated by questions that other problem solvers, such as humans, could overcome. The preceding deliberations indicate that there indeed are such limitations. For the ramifications of self-reference are such that one computer could satisfactorily answer certain questions regarding the limitation of the general capacity of computers to solve questions. But humans can in fact resolve such questions because, with them, no self-reference is involved.

Potential Difficulties

The time has now come to face up to some possible objections and difficulties.

An objection that deserves to be given short shrift runs as follows: "But the sort of computer insolubilium represented by self-reference issues like those of P_5 is really not the kind of thing I was expecting when contemplating the title of the chapter." But, alas, expectations do not really count for much in this sort of situation. After all, nobody faults Gödel for not having initially come up with the sorts of examples that people might have expected regarding the incompleteness of formalized arithmetic—some insoluble Diophantine problem in number theory.[17]

But of course other objections remain.

For example, do those instanced problems really lie outside the domain of trustworthy computer operation? Could computers not simply follow in the wake of our own reasoning here and instance P_3 and P_5 as self-resolving? Not really. For in view of the considerations adduced in relation to T_1-T_2 above, a computer cannot convincingly monitor the range of computer-tractable problems. So the responses of a computer in this sort of issue simply could not secure rational conviction.

But what if a computer were to establish its reliability regarding such supposedly "computer irresolvable" questions indirectly? What about a reliable black box? Could a computer not acquire reliability simply by compiling a good track record?

Well . . . yes and no. A black box can indeed establish credibility by compiling a good track record of correct responses. But it can do so only when this track record is issue-homogeneous with the matter at hand: when those correct responses relate to questions of the same type as the one that is at issue. Credibility is not transferable from botany to mathematics or from physics to theology. The only productively meaningful track record would have to be one complied *in a reference class of similar cases.* Now just how does type homogeneity function with respect to our problem? What is the "type of problem" that is at issue here? The answer is straightforward: it is *questions that for reasons of principle qualify as being computer intractable.* But how could a computer establish a good track record here? Only by systematically providing the responses we can reasonably deem to be correct on wholly independent grounds. The situation that arises here would thus be analogous to that of a black box that systematically forecasts tomorrow's headlines correctly. This sort of thing is indeed imaginable—it is a logically feasible possibility (and thereby, no doubt, an actuality in the realm of science fiction). But we would be so circumstanced as to deem that black box's performance miraculous. And we do not—cannot—accept this as a *practical* possibility for the real world. It is a fanciful hypothesis that we would reject out of hand until such time as actual circumstances confronted us with its realization—a prospect we would dismiss as utterly unrealistic. It represents a bridge that we would not even think about crossing until we actually got there—simply because we have a virtually ineradicable conviction that actually getting there is something that just will not happen.

"But surely the details of this discussion are not so complex that a computer capable of defeating grandmasters at chess could not handle them as well." There thus still remains a subtle and deep difficulty. One author formulated the problem as follows:

In a way, those who argue for the existence of tasks performable by people and not performable by computers are forced into a position of never-ending retreat. If they can specify just what their task involves, then they [must] admit the possibility of programming it on some machine. . . . [And] even if they construct proofs that a certain class of machines cannot perform certain tasks, they are vulnerable to the possibility of essentially new classes of machines being described or built. (Weinberg 1967, 173)

After all, when people can solve a certain problem successfully, then they can surely "teach" this solution to a computer by simply adding the solution to its information store. And thereafter the computer can also solve the problem by replicating the solution—or if need be by simply looking it up.

Well, so be it. It is certainly possible for a computer to maintain a solution registry. Every time some human solves a problem somewhere somehow, it is duly entered into this register. And then the computer can determine the person-solvability of a given problem by the simple device of "looking it up." But this tactic represents a hollow victory. First of all, it would give the computer access only to person-resolved problems and not to person-resolvable ones. But, more seriously yet, if a computer needs *this* sort of input for answering a question, then we could hardly characterize the problem at issue as computer solvable in any but a Pickwickian sense.

At this point the issue of the scoring system becomes crucial. We now confront what is perhaps the most delicate and critical part of the inquiry. For now we have to face and resolve the conceptual question of how the attribution of credit works in matters of problem solving.

Clearly if *all* of the inferential steps essential to establishing a problem solution as such were computer performed, and *all* of the essential data inputs were computer provided, then computers would have to be credited with that problem solution. But what of the mixed cases in which some essential contributions are made on both sides—some by computers and some by people? Here the answer is that *credit for mixed solutions lies automatically with people.* For if a computer "solves" the problem in a way that is overtly and essentially dependent on people-provided aid, then its putative "solution" can no longer count as authentically computer provided. A problem that is "not solvable by persons alone but yet solvable when persons are aided by computers" is still to be classed as person solvable, whereas a problem that is "not solvable by computers alone but yet solvable when computers are aided by persons" is not to be classed as computer solvable. (Or at any rate not until we reach the science-fiction level of self-produced, self-programmed, independently evolving computers that manage to reverse the master-servant relationship here.) For as matters stand, the scoring system used in these matters is simply not "fair." The seemingly table-turning question "Is there a problem that people cannot solve but computers can?" automatically requires a negative response once one comes to realize that *people can and do solve problems with computers.*[18] The conception of "computer-provided solutions" works in such a way that here computers must do not only the job but actually *the whole job.* And on this basis the difficulty posed by that subtle objection can be dismissed.

The crucial point is that although people use computers for problem solving, the converse simply does not hold: the prospect that computers solve problems by using people as investigative instruments is unrealistic—barring a technologico-cultural revolution that sees the emergence of functionally autonomous computers, operative on their own and able to press people into their service.

Does such a principle of credit allocation automatically render people superior to computers? Not necessarily. Quite possibly the things that computers cannot accomplish in the way of problem solving are things people cannot accomplish either—with or without computers. The salient point is surely that much of what we would ideally like to do, computers cannot do either. They can indeed diminish, but they cannot eliminate, our limitations in solving the cognitive problems that arise in dealing with a complex world—that is, in effect, a realm where *every* problem-solving resource faces some ultimately insuperable obstacles.

Appendix: On the Plausibility of T_1 and T_2

The task of this appendix is to set out the plausibility considerations that establish the case for accepting the pivotal theses T_1 and T_2.

A helpful starting point for these deliberations is provided by the recognition that the inherently progressive nature of pure and applied science ensures the prospect of continual improvements in the development of ever more capable instrumentalities for general problem solving. No matter how well we are able to do at any given state of the art in this domain, the prospect of further improvements always lies open. Further capabilities in information access, processing capacity, or both can always be added to any realizable computer, no matter how powerful it may be. And this suffices to substantiate the realization that there is no inherent limit here: for every particular problem solver that is actually realized there is (potentially) some other whose performative capability is greater.[19]

Now it lies in the nature of things that in cognitive matters, an agent possessed of a lesser range of capabilities will always underperform one possessed of a greater range. More powerful problem solvers can solve more problems. A chess player who can look four moves ahead will virtually always win out over one who can manage only three moves. A crossword-puzzle solver who can manage words of four syllables will surpass one who can manage only three. A mathematician who has mastered the calculus will outperform one whose competency is limited to arithmetic. This sort of thing holds for general problem solving as well.

One must also come to terms with the realization that no problem solver can ever reliably determine that all its more powerful peers are unable to resolve a particular substantive problem (that is, one that is inherently tractable and not demonstrably unsolvable on logico-conceptual grounds). The plausibility argument for this is straightforward and roots in the limited capacity of a feebler intelligence to gain adequate insight into the operation of a stronger. After all, one of the most fundamental facts of epistemology is that a lesser capacity can never manage to comprehend fully the operations of a greater. The untrained mind cannot grasp the ways of the expert. And try as one will, one can never adequately translate Shakespeare into Pidgin English. Similarly, no problem solver can determine the limits of what its more powerful peers can accomplish. None can reliably resolve questions of computer solvability in general: none can reliably survey the entire range. The enhanced performance of a more capable intellectual performer will always seem mysterious and almost magical to a less capable peer.

John von Neumann conjectured that computational complication is *reproductively* degenerate in the sense that computing machines can only *produce* others less complicated than themselves. The present thesis is that with regard to UPS, computational complication is *epistemically* degenerative in that computing machines can only reliably *comprehend* others less complicated than themselves (in the sense that one computer "comprehends" another when it is in a position to tell just what this other can and cannot do).

Considerations along these lines substantiate the proposition that no computer in the field of general problem solving can obtain a secure overview of the performance of its peers at large. And this means that

Thesis T$_1$ *No computer can reliably determine that a given substantive problem is altogether computer irresolvable.*

Furthermore, it is clear that the question "But even if computers had no limits, how could a computer possibly manage to determine that this is the case?" plants an ineradicable shadow of doubt in our mind. For even if it were the case that computers had no problem-solving limits, and even if a computer could (as is virtually inconceivable) manage to determine that this is so, the fact would nevertheless remain that the computer could not really manage to secure our conviction with respect to this claim. Trusting though we might be, we would not be—and could not reasonably be—*that* trusting. No computer could achieve credibility in staking so hyperbolic a claim. (We have the sort of situation that reminds one of the old Roman dictum "I would not believe it even were it told to me by Cato.")

The upshot is that no computer problem solver is in a position to settle the question of limits that affect its peers across the board. We thus have it that

Thesis T$_2$ *No computer can reliably determine that all substantive problems whatsoever are computer resolvable.*

But of course this thesis—like its predecessor—holds only in a domain that, like UPS, is totally open-ended. It does not hold for ADT, in which one single problem solver (the Turing machine) can speak for all the rest.

The following dialectical stratagem also deserves notice. Suppose someone were of a mind to contest the acceptance of theses T$_1$ and T$_2$ and proposed to reject one of them. This very stance would constrain that person to concede the other. For T$_2$ *follows from the denial of* T$_1$ (and correspondingly T$_1$ follows from the denial of T$_2$). In other words, there is no prospect of denying both these theses; at least one of them *must* be accepted.

The proof here runs as follows. Let $X(p)$ as usual represent $(\forall C)\sim C$ res P. Then we have

$$T_1: \text{For all } C: (\forall P)\sim C \text{ det } X(P)$$

$$T_2: \text{For all } C: \sim C \text{ det } (\forall P)\sim X(P)$$

We thus have

$$\sim T_1 = \text{For some } C: (\exists P) \ C \text{ det } X(P)$$

$$\sim T_2 = \text{For some } C: \ C \text{ det } (\forall P)\sim X(P) \text{ or equivalently } C \text{ det}\sim(\exists P)X(P)$$

Now the computers at issue are supposed to be truth-determinative, so that we stand committed to the idealization that C det P entails P. On this basis, $\sim T_1$ yields $(\exists P)X(P)$. And furthermore $\sim T_2$ yields $\sim(\exists P)X(P)$. Since these are logically incompatible, we have it that $\sim T_1$ and $\sim T_2$ are incompatible, so that $\sim T_1$ entails T_2 (and consequently $\sim T_2$ entails T_1).

It is a clearly useful part of the plausibility argumentation for T_1-T_2 to recognize that accepting at least one of them is inescapable.[20]

NOTES

I am grateful to Gerald Massey, Laura Ruetsche, and especially Alexander Pruss for constructive comments and useful suggestions on a draft of this chapter.

1. The salient point is that unless I can *tell* (i.e., be able to claim justifiedly myself) that you are justified in your claim, I have no adequate grounds to accept it: it has not been made credible to me, irrespective of how justified you may be in regard to it. To be sure, for your claim to be credible for me I need not know *what* your justification for it is, but I must be in a position to realize *that* you are justified. Your reasons may even be incomprehensible to me, but for credibility I require a rationally warranted assurance—perhaps only on the basis of general principles—that those reasons are both extant and cogent.

2. On unsolvable calculating problems, mathematical completeness, and computability see Davis (1958). See also Pour-El and Richards (1989) or, on a more popular level, Hofstadter (1979).

3. Some problems are not inherently unsolvable but cannot in principle be settled by computers. An example is "What is an example of a word that no computer will ever use?" Such problems are inherently computer-inappropriate, and for this reason a failure to handle them satisfactorily also cannot be seen as a meaningful limitation of computers.

4. On Gödel's theorem see Shanker (1988), a collection of essays that provide instructive, personal, philosophical, and mathematical perspectives on Gödel's work.

5. On this issue see Brennerman (1977).

6. For a comprehensive survey of the physical limitations of computers see Leiber (1997).

7. As stated this question involves a bit of anthropomorphism in its use of "you." But this is so only for reasons of stylistic vivacity. That "you" is, of course, only shorthand for "computer number such-and-such."

8. On the inherent limitation of predictions see my *Predicting the Future* (1997).

9. Note that T_1 is weaker than:

T_3: *No computer can reliably determine that there are substantive problems that are computer irresolvable:*

$$\sim(\exists C')\ C'\ det\ (\exists P)(\forall C)\ \sim C\ res\ P$$

This stronger thesis is surely false, but the truth of the weaker T_1 nevertheless remains intact.

The falsity of T_3 can be shown by establishing its denial:

$$(\exists C')\ C'\ det\ (\exists P)(\forall C)\ \sim C\ res\ P$$

First note that in a world of computers whose capabilities are finite there is for each computer C_i some problem P_i that C_i cannot resolve. The conjunctive compilation of all these problems, namely P_x, will thus lie beyond the capacity of all computers. So $(\forall C)\ \sim C\ res\ P_x$ and therefore $(\exists P)(\forall C)\ \sim C\ res\ P$. Now the reasoning by which we have just determined this can itself clearly be managed by a computer, and thus

$$(\exists C')\ C'\ det\ (\exists P)(\forall C)\ \sim C\ res\ P \qquad\qquad Q.E.D.$$

10. See Feferman et al. (1995), and especially the 1951 paper on "Some Basic Theorems on the Foundation of Mathematics and Their Indications."

11. On Church's thesis see Davis (1965) and Rogers (1967). On Turing machines see Hesken (1988).

12. But could one not simply connect computers up with one another so as to create one vast megacomputer that could thereby do anything that any computer can? Clearly this can (in theory) be done when the set of computers at issue includes "all currently existent computers." But of course we cannot throw future ones into the deal, let alone merely possible ones.

13. This book is an updated revision of his earlier *What Computers Can't Do* (Dreyfus 1972).

14. For some discussion of this issue from a very different approach see Penrose (1989).

15. We have just claimed P_5 as computer unresolvable. And this contention, of course, entails $(\exists P)(\forall C)\sim C\ res\ P$ or equivalently $\sim(\forall P)(\exists C)C\ res\ P$. Letting this thesis be T_3, we may recall that T_2 comes to $(\forall C)\sim C\ det\ \sim T_3$. If T_3 is indeed true, then this contention—that is, T_2—will of course immediately follow.

16. Someone might suggest: "But one can use the same line of thought to show that there are computer-solvable problems that people cannot possibly solve by

simply interchanging the references to 'computers' and 'people' throughout the preceding argumentation." But this will not do. For the fact that it is people that use computers means that one can credit people with computer-provided problem solutions via the idea that *people can solve problems with computers*. But the reverse cannot be claimed with any semblance of plausibility. The situation is not in fact symmetrical, and so the proposed interchange will not work. This issue will be elaborated in the next section.

17. Such examples were to come along only later, with the relating of Gödel's results to number-theoretic issues relating to the solution of Diophantine equations. For a good expository account see Davis and Hersh (1978). (I owe this reference to Kenneth Manders.)

18. To be sure, there still remains the question "Are there problems that people can solve *only* with the aid of computers?" But the emphatically affirmative answer that is virtually inevitable here involves no significant insult to human intelligence. After all, the same concession must be made with regard to reference aids of all sorts, the telephone directory being an example.

19. Certainly due care must be taken in construing the idea of greater performative capability. Thus let C be the computer in question and let s be a generally computer-undecidable statement. Then consider the question:

Q Is it the case that at least one of the following two conditions holds? (1) You (i.e., the computer at work on the question Q now being posed) are C. (2) s is true.

Since (1) obtains, C can answer this interrogation affirmatively. But since s is (by hypothesis) computer undecidable, no other computer whatever can resolve Q. It might thus appear that no computer whatever could have greater general capability than another. However, its essential use of "you" prevents Q from actually qualifying as "a question that C can answer but C' cannot." For in fact *different questions* are being posed—and different problems are thus at issue—when the interrogation Q is addressed to C and to C'. For the example to work its intended damage we would need to replace Q by one single fixed question that computer C would answer correctly with *Yes* and any other computer C' would answer correctly with *No*. And this is impossible.

20. Principles T_1 and T_2 also have close analogues in general epistemology. T_1's analogue is $(\forall x)(\forall t)\sim KxUt$ or equivalently $(\forall t)UUt$, where $Up = \sim(\exists x)Kxp$. And T_2's analogue is $U\sim(\exists t)Ut$, which follows at once from the assertability of $(\exists t)Ut$.

REFERENCES

Brennerman, H. J. 1977. "Complexity and Transcomputability." In R. Duncan and M. Weston-Smith, eds., *The Encyclopedia of Ignorance*. Oxford: Pergamon Press, 167–74.

Davis, M. 1958. *Computability and Unsolvability*. New York: McGraw-Hill. [Expanded reprint edition, New York: Dover, 1982.]

——, ed. 1965. *The Undecidable*. New York: Raven Press.

Davis, M., and R. Hersh. 1978. "Hilbert's Tenth Problem." In J. C. Abbot, ed., *The Chauvenet Papers*, vol. 2. Washington, D.C.: Mathematical Association of America, 554–71.

Dreyfus, H. L. 1972. *What Computers Can't Do*. New York: HarperCollins.
——. 1992. *What Computers Still Can't Do*. Cambridge, Mass.: MIT Press.
Feferman, S., and J. Dawson, eds. 1995. *Kurt Gödel, Collected Works,* vol. 3: *Unpublished Essays and Lectures*. Oxford: Oxford University Press.
Hesken, R., ed. 1988. *The Universal Turing Machine*. Oxford: Oxford University Press.
Hofstadter, D. 1979. *Gödel, Escher, Bach: An Eternal Golden Braid*. New York: Basic Books.
Leiber, T. 1997. "Chaos, Berechnungskomplexität und Physik: Neue Grenzen wissenschaftlicher Erkenntnis." *Philosophia Naturalis* 34: 23–54.
Penrose, R. 1989. *The Emperor's New Mind*. New York: Oxford University Press.
Pour-El, N. B., and J. I. Richards. 1989. *Computability in Analysis and Physics*. Berlin: Springer-Verlag.
Rescher, N. 1997. *Predicting the Future*. Albany: State University of New York Press.
Rogers, H. 1967. *Theory of Recursive Functions and Effective Computability*. New York: McGraw-Hill.
Shanker, S. G., ed. 1988. *Gödel's Theorems in Focus*. London: Croom Helm.
Weinberg, G. M. 1967. "Computing Machines." In P. Edwards, ed., *The Encyclopedia of Philosophy*, vol. 2. New York: Macmillan and Free Press, 168–73.

9

Limits to Self-Observation

Thomas Breuer

Department of Mathematics, Fachhochschule Vorarlberg

This contribution has two purposes. One is to show that for an internal observer it is impossible to measure exactly all past states of a system in which she is contained, if the time evolution of the observed system is Markov stochastic.

The other is to put this result into perspective. To this end I will introduce as an everyday example of self-observation the picture in the picture, then report earlier results that it is impossible to measure exactly from inside all present states, and to measure exactly from inside all past states if the time evolution is deterministic.

The Picture of the Picture in the Picture

Can a picture contain a full and precise picture of itself? Empirical evidence suggests that this is impossible: the resolution attained by the Old Masters, admirable as it is in some cases, does not suffice. For a picture to contain a full picture of itself one needs a picture of the picture in the picture, and a picture of the picture in the picture in the picture. On a finite canvas this is impossible (figure 9.1), except in the trivial case that the picture *is* the picture of itself. (Then the picture is also the picture of the picture of itself, and the picture of the picture of the picture of itself.) To exclude this trivial case I require that the picture contain the picture of itself and something else in addition. Then I say that the picture *strictly* contains a picture of itself. Even on a

Figure 9.1 On a finite canvas no picture O can strictly contain a perfect picture A of itself. The limit picture is in O.

spatially infinite canvas or on a canvas with infinite microstructure, it is impossible to have a perfect picture in the picture, as long as only finite resolution can be achieved. (By finite resolution I mean that only finitely many cells can be given different colors.)

How can we see that a picture with finite resolution cannot strictly contain a perfect picture of itself? Denote by O the canvas, by S_O the set of cells of the canvas that can carry color, by A the part of the canvas that is to be the picture of O, and by S_A the set of color cells of A. Since we required the picture strictly to contain a picture of itself, S_A is a strict subset of S_O. Having a finite canvas or finite resolution on an infinite canvas amounts to S_O being finite.

By its coloring each cell of the canvas represents something. Define a map I describing this representation: for a color cell $x \in S_x$, $I(x)$ is the set of color cells of O represented by x. I obviously varies with the painting we consider. Since A is to be a picture of O, I is a map from S_A to the power set of S_O. Now we want A to carry a *perfect* picture of O. So for each color cell $y \in S_O$ there should be a color cell $x \in S_A$ representing it, that is, for which $I(x) = \{y\}$. This implies that to I there corresponds a surjective map from S_A onto the set S_O of color cells of O. But this is impossible since S_A is a strict subset of S_O and S_A is finite.

Figure 9.2 A picture O can strictly contain a perfect *partial* picture A of itself. The limit picture can be outside O.

The result breaks down if we have a canvas with an infinite number of color cells. Such a canvas may carry a perfect picture of itself. What we can have with finite resolution is a picture containing an *approximate* picture of itself. We can also have a picture strictly containing a perfect picture of a part of itself (figure 9.2).

Exact Immediate Self-Measurements Are Impossible

All this is very similar to the situation we have when a system is measured by some subsystem. One can show along the lines of the picture in the picture argument that not all states of a system with a finite state space can be measured exactly by a proper subsystem.

The argument goes like this. A measurement performed by an apparatus A on some observed system O establishes certain correlations between the states of A and the states of O. If A could measure every state of O exactly, these correlations would give rise to a surjective map from the states of O to the states of A. But since A is a proper subsystem of O, its state space is strictly smaller than the state space of O, and thus for a finite state space of O such a map cannot exist.

This result is not very useful since most interesting systems have an infinite state space. (Even the state space of a single particle moving in Euclidean space has infinitely many points.) In this chapter I derive similar results for systems with infinite state space. This requires some slightly more sophisticated formalism.

To describe the inference that is made after the measurement from information about A to information about O, let us use a map I from the power set $P(S_A)$ of the set S_A of apparatus states into the power set $P(S_O)$ of the set S_O of system states.

I *assigns to every set* X *of apparatus states (except the empty set) the set* I(X) *of object states compatible with the information that the apparatus after the experiment is in one of the states in* X.

This defines the inference map I, which depends on the kind of measurement we are performing. The main reason for defining the inference map as a map of sets of states rather than as a map of states is that after the measurement the experimenter usually does not know the exact state of the apparatus but only the pointer reading. Furthermore, experiments in general do not determine exactly one state of the observed system but rather the (perhaps approximate) value of a physical quantity. Usually measurement results only specify a set of states, not a state.

I is different in different measurement situations. But when the observer chooses the experimental setup, she also chooses a map I describing how she is going to interpret the pointer reading after the experiment. This map is fixed throughout the measurement.

To simplify notation I will from now on write $I(s)$ instead of $I(\{s\})$ when I is applied to a set consisting of only one element. (Still, I is taken to be a map of sets of states, not a map of states.) I will say that an experiment with inference map I can *measure a state* $s \in S_O$ *exactly* if there is a set X_s of apparatus states referring uniquely to s, $I(X_s) = \{s\}$.

All this can be done irrespective of whether the apparatus A is part of O, or outside O, or partially part of O. Now consider the special case in which A is fully contained in O. So the observed system O is composed of the apparatus A and of a residue R. We assume that the observed system has strictly more degrees of freedom than the apparatus and contains it. This can be formulated in an *assumption of the strictly internal observer*:

$$(\exists s, s' \in S_O) : R(s) = R(s'), s \neq s'.$$

Here $R(s)$ denotes the state of A that is determined by restricting the state s of O to the subsystem A. So R is a map from S_O into S_A. (Later on I will, by a slight abuse of notation, also denote by R the map from $P(S_O)$ into $P(S_A)$ defined by $R(X) := \{R(s) : s \in X\}$ for $X \subset S_O$.) In classical mechanics, for example, the map R is defined by discarding the coordinates that refer to degrees of freedom of O that are not in A. In quantum mechanics, one can take R to be the partial trace over R. For our purposes it is enough to take an arbitrary but fixed map.

Note that the self-measurements described by Albert (1987) are not measurements from inside in the sense discussed here, because they use an apparatus only partially contained in the observed system. In Albert's self-measurements, the assumption of the strictly internal observer is not satisfied.

The states of the apparatus after the measurement are self-referential: they are states in their own right, but they also refer to states of the observed system that in turn determine the state of the apparatus. This leads to a *meshing property* that is satisfied by all consistent inference maps of internal measurements.

Lemma 1   *For every apparatus state* s $\in$ S$_A$, *the restriction of the system states* I(s) *to which* s *refers is again the same apparatus state* s. *So every consistent inference map* I *satisfies*

$$R(I(s)) = \{s\} \qquad (1)$$

for all s $\in$ S$_A$.

Theorem 1   *Not all present states of a system can be measured exactly by a strictly internal observer.*

For a proof of Lemma 1 and Theorem 1 see Breuer (1995). Note that the lemma and the theorem hold only because both s_A and $R(I(s_A))$ refer to the state of A at the *same* time.

The basic idea of this first result is the following: If an internal observer could measure all states exactly, then for every state s of O there would be a state s' of A with $I(s') = \{s\}$. Therefore exact measurability of all states implies the existence of surjective map $s' \to s$ from S_A to S_O. If there is no such map A cannot measure exactly all states of O. The assumption of the strictly internal observer and Lemma 1 prevent the existence of such a surjective map. This is easy to see. It follows from the fact that the following three requirements are incompatible: (1) there is a surjective map I from S_A to S_O, (2) the assumption of the strictly internal observer, and (3) the meshing property of I (Lemma 1). By Lemma 1 $R \circ I$ is the identity map on S_A, and therefore R is the inverse map of I. Now R is surjective, but by the assumption of the strictly internal observer it cannot be injective. Obviously there cannot be a surjective map I such that I^{-1} is simultaneously a surjective map but not injective.

Note that, unlike the result about the picture in the picture, Theorem 1 also holds for systems with infinite state space.

Exact Delayed Self-Measurements Are Impossible for Deterministic Time Evolutions

Let us turn to the case in which an observer wants to measure the *past* state of a system in which she is contained. From information about the state of A at a certain time t_1 after the measurement, she infers information about the state of O at an earlier time t_0. Denote by I_A^1 the set of states of A at t_1 and by S_O^0 the set of states of O at t_0. (So the upper index of S_O^0 or S_A^1 refers to the time, the lower to the system. By introducing the upper index I do not want to imply that there are many interesting cases where $S_O^0 \neq S_O^1$ or $S_A^1 \neq S_A^0$. It simply makes it easier to indicate to which times states and sets of states refer.) I now is defined in the following way:

I associates with every set $X \subset S_A^1$ *of apparatus states at time* t_1 *the set* I(X) *of those object states at time* t_0 *that are compatible with the information that at* t_1 *the apparatus is in one of the states in* X.

I have to assume that the time evolution of O from t_0 to t_1 is *deterministic*:

The state of O *at* t_0 *determines uniquely the state at* t_1.

This implies that there is a map $T : S_O^0 \rightarrow S_O^1$ associating with every state $s \in S_O^0$ the state $T(s) \in S_O'$ into which s evolves. T is surjective because every possible state of O at t_1 must have evolved from some state at t_0. T can be extended to the power set $P(S_O^0)$ by putting $T(X) := \cup_{s \in X} T(s)$. For the extended map I will use the same letter T.

Now assume that the observer A is strictly contained in the observed system O. In fact I require that

There are states s, s' *of* O *at* t_1 *whose restriction to* A *coincides.*

This *assumption of the strictly internal observer* implies that there are states $s \neq s' \in S_O^1$ for which $R(s) = R(s')$.

The meshing property (1) mentioned previously holds only for the measurement of present states, but there is a different meshing property for consistent inference maps in measurements of past states from inside.

Lemma 2   *Every consistent inference map* I *describing a measurement of past states from inside fulfills*

$$R(T(I(s_A))) = \{s_A\} \tag{2}$$

for all apparatus states $s_A \in S_A^1$ *with* $I(s_A) \neq \varnothing$.

A map $P(S_A^1) \rightarrow P(S_O^0)$ not fulfilling (2) cannot be used as an inference map because it would describe the inference of paradoxical conclusions from the pointer reading. In the special case that A wants to measure the *present* state of O, we have $t_0 = t_1$ and T is the identity map. Then the meshing property of Lemma 2 reduces to the one of Lemma 1.

Lemma 3   *The assumptions of the strictly internal observer and of determinism imply that there are states* s_1, s_2 *of* O *at* t_0, $s_1 \neq s_2$, *for which*

$$R(T(s_1)) = R(T(s_2)). \tag{3}$$

Theorem 2   *If a system* O *evolves deterministically from* t_0 *to* t_1, *then an apparatus that is strictly contained in* O *cannot measure exactly at time* t_1 *all states of* O *at* t_1.

For a proof of the lemmata and Theorem 2 see Breuer (1998). Why did I have to make the assumption of determinism? It would have been possible to define T not as a map $S_O^0 \rightarrow S_O^1$ but more generally as a two-place relation on $S_O^0 \times S_O^1$. This would also be appropriate for stochastic time evolutions.

But for stochastic time evolutions Lemma 2 does not hold; (2) can be violated without a contradiction arising. For example, if there is an $s \in I(s_A)$ with $R(T(s)) \ni s_A' \neq s_A$, then (2) is violated but $s \in I(s_A)$ still does not contradict $R(T(s)) \ni s_A' \neq s_A$. Assume A at t_1 is in s_A; then O at t_0 is in one of the states in $I(s_A)$, possibly in s. Admittedly, s could have evolved into a state whose restriction to the apparatus is not s_A, but this does not contradict $s \in I(s_A)$ because it could have happened—and actually it did—that s evolved into a state whose restriction to the apparatus in fact is s_A.

Neither does Lemma 3 hold for stochastic time evolutions. The proof of Lemma 3 breaks down because $T(T^{-1}(s))$ just contains s but is not equal to $\{s\}$ for stochastic time evolutions.

Exact Delayed Self-Measurements Are Impossible for Stochastic Time Evolutions

For the results of the previous section it is essential to assume that O evolves deterministically. Now we will see which kind of restrictions hold for the measurement of past states from inside a stochastically evolving system. I proceed in three steps: First, stochastic Markovian time evolutions are related to maps of probability distributions over pure states. Second, I describe the inference we make in a statistical experiment from the distribution of the pointer readings after the measurement to the distribution of the observed quantity before the measurement. Then I prove two lemmata and the main result.

First I explain how stochastic Markovian time evolutions lead to a map not of pure states but of probability distributions over pure states. To describe stochastic time evolutions it is essential to keep apart two state concepts: the pure states and the statistical states. The pure states are the states in which *one* system can be at a time. Denote by S_O the set of pure states of O. For a quantum system S_O consists of the rays of the Hilbert space; for a classical system it is the phase space.

A statistical state of O is a probability distribution on S_O. It may describe an ensemble of identical noninteracting systems, or the pure states of one system subject to repeated measurements, or our subjective knowledge about some true pure state. In classical mechanics, the statistical states are the probability distributions on phase space. In quantum mechanics, the statistical states are the probability distributions on the rays of the Hilbert space. Associated with each of these is a density matrix. But this relation is not one-to-one: many probability distributions lead to the same density matrix. For the purpose of prediction, all probability distributions with the same density matrix are equivalent. But still, different probability distributions describe different ensembles, even if they have the same density matrix. For the purpose of this chapter it is convenient to take as statistical states in quantum mechanics the probability distributions on the pure states, not the density matrixes.

Even if O does not evolve deterministically, its time evolution can be described by a stochastic process (see, e.g., Arnold 1974). A general stochastic process is characterized by infinitely many probabilities. Denote by $s_1, s_2, \ldots \in S_O$ pure states of O. Let $\mu_t(s)$ be the probability that the state of O at time t is s. Let $\rho_2(s_2, t_2 | s_1, t_1)$ be the conditional probability that the state of O at time t_2 is s_2 given that the state at time t_1 is s_1. And for arbitrary $n \geq 3$ let $\rho_n(s_n, t_n | s_{n-1}, t_{n-1}, s_{n-2}, t_{n-2}, \ldots, s_1,$ $t_1)$ be the conditional probability that the state of O at time t_n is s_n given that the states at times $t_1, \ldots, t_{n-1}$ are $s_1, \ldots, s_{n-1}$. The specification of all these probabilities defines a stochastic process. A stochastic process is called a Markov process if the conditional probabilities satisfy

$$\rho_n(s_n, t_n | s_{n-1}, t_{n-1}, s_{n-2}, t_{n-2}, \ldots, s_1, t_1) = \rho_2(s_n, t_n | s_{n-1}, t_{n-1})$$

for all $n \geq 3$. For a Markov process the probability that at t_n the system is in s_n depends only on the state at t_{n-1} but not on its history previous to t_{n-1}. Therefore these processes are memory-free. Such a process is fully specified by the probabilities μ, ρ_2 alone.

Now the decisive point is that for a Markov process a probability distribution μ_{t_0} over S_O at some time t_0 uniquely determines a probability distribution μ_{t_1} for each later time t_1 by

$$\mu_{t_1}(s_1) = \int_{S_O} ds_0 \rho_2(s_1, t_1 | s_0, t_0) \mu_{t_0}(s_0)$$

The differential equation governing the evolution of μ is the master equation. For some arbitrary but fixed time $t_1 \geq t_0$ denote the map $\mu_{t_0} \rightarrow \mu_{t_1}$ by T.

All this can be done for quantum and for classical systems. The only point to keep in mind is that in quantum mechanics a density matrix can result from many different probability distributions on the pure states. Therefore the stochastic time evolution of a quantum system cannot be represented as a map of density matrixes but only as a map of probability distributions on the pure states (Breuer and Petruccione 1995).

The second step is to describe the inference we make in a statistical experiment from the distribution of the pointer readings at some time t_1 after the measurement to the distribution of the observed quantity at some earlier time t_0. Making a statistical experiment presupposes either that the measurement is performed on many independent copies of the system or that it is often repeated. In the second case the times at the beginning of each single run are identified, and so are the times at the end of each run. So t_0 and t_1 now refer to certain stages in each run and not to an absolute time in the laboratory.

Now consider an experiment in which we want to measure the statistical state of some system O at an earlier time t_0. How do we make the inference to the distribution of O? Identifying stochastic models from observation data is a common technique in statistical physics (Honerkamp 1998, 357). Most systems are described by stochastic models that are not observed directly. For each type of model one has to formulate a so-called observation equation, which relates the observable pointer quantity p to the system quantities used in the model. The observation equation also encodes the setup of the experiment. In our case the system quantity is the state s of the observed system O. Assuming that the time dependence of s is described by a Markov process and thus follows a master equation, the observation equation can be written as

$$p(t_1) = f(s_{t_0}) + \sigma(t)\eta(t)$$

where $\eta(t)$ is a white noise term and $\sigma(t)$ some real number. The second term on the right-hand side represents a stochastic error of measurement, which usually is assumed to be normally distributed. The value of the pointer quantity at some time t depends on the state of the apparatus by some function $P: p(t) = P(s_t^A)$. Using P and the observation equation, the usual tools of statistical estimation theory give us a probability distribution for the state of O at t_0 as a function of the state of A at t_1: $\mu_{t_0}(s|s_{t_1}^A)$. This a posteriori distribution gives an estimator for the state of O from observation data about A, under the condition that the function P and the observation equation are given. Let us denote by J this function associating to a distribution over S_A at t_1 a distribution over S_O at t_0.

If $J(\mu^a) = \mu$ I will say that the apparatus distribution μ^a *refers* to the system distribution μ. We can say that the distribution μ over S_O at t_0 can be *measured* in a statistical experiment with inference map J if there is a distribution μ_A on S_A at t_1 with $J(\mu_A) = \mu$. This terminology is sensible. If for some μ there is no s_A referring to μ, there is no series of observation data that we would interpret as indicating that the state of O is distributed according to μ. A different experiment would be described by a different observation equation and a different estimator. This would lead to a different inference map J, for which there perhaps is a μ'_A with $J(\mu'_A) = \mu$. In this other experiment the statistical state μ would be measurable.

Now finally we can turn to prove the main result. If A is part of the observed system O every distribution μ over S_O determines a distribution $R(\mu)$ over S_A. In classical as well as in quantum mechanics, $R(\mu)$ is the marginal of μ over the degrees of freedom of O that do not belong to A. If A is part of the observed system O and O has some degrees of freedom outside A, there are different distributions $\mu \neq \mu'$ on S_O at t_1 whose restrictions $R(\mu)$, $R(\mu')$ to A coincide. This is the *assumption of the strictly internal observer*. It will be indispensable in the following argument.

Lemma 4   *Every distribution μ_A over S_A at t_1 can only refer to distributions μ over S_O at t_0 which evolve into distributions whose restriction to A is again μ_A. Thus every consistent inference map J used in a statistical measurement of past states from inside satisfies*

$$R(T(J(\mu_A))) = \mu_A \tag{4}$$

for all distributions μ_A over S_A at t_1 with $J(\mu_A) \neq \varnothing$.

To prove this, assume (4) is not satisfied for some μ_A with $J(\mu_A) \neq \varnothing$. So we have $R(T(J(\mu_A))) \neq \mu_A$. This will lead to a contradiction. If after the measurement the states of A are distributed according to μ_A, one infers that before the measurement the states of O were distributed according to $J(\mu_A)$. Then the states of O after the measurement are distributed according to $T(J(\mu_A))$, and the states of A after the measurement according to $R(T(J(\mu_A)))$. Because of $R(T(J(\mu_A))) \neq \mu_A$, this contradicts the assumption that at t_1 the states of A are distributed according to μ_A. This ends the proof of the lemma.

Would it really be a contradiction if one ascribed to A after the measurement two different distributions μ_A, $R(T(J\mu_A))$? After all, no finite series of observations can justify with certainty the conclusion that the states are distributed according to one distribution rather than the other. Still, I think if a contradiction can arise in a statistical description at all, the violation of (4) is one. Finitely many single measurements suffice for determining the distribution with arbitrarily high probability (instead of absolute certainty). However high you set your confidence threshold $1 - \varepsilon$, as long as $\varepsilon > 0$, the statistical methods of test of a hypothesis specify how many single experiments need to be made in order to exclude with probability $1 - \varepsilon$ a certain distribution. Contradictions in statistical descriptions will not get clearer than this.

Denote by M_O the set of probability distributions over S_O and by M_A the set of distributions over S_A.

Lemma 5   *If at t_1 the assumption of the strictly internal observer is satisfied, there are distributions $\mu_1 \neq \mu_2$ over S_O at t_0 with $R(T(\mu_1)) = R(T(\mu_2))$.*

To prove this define the inverse T^{-1} as the map from M_O into the power set of $M_A : \mu \to \{\mu_1 \in S_O : T(\mu_1) = \mu\}$. T^{-1} associates to each distribution μ at t_1 the set of distributions at t_0 that evolve into μ. By the assumption of the strictly internal observer there are distributions μ, μ' of O at t_1, $\mu \neq \mu'$, for which $R(\mu) = R(\mu')$. Since every possible distribution at t_1 has evolved from some distribution at t_0, neither $T^{-1}(\mu)$ nor $T^{-1}(\mu)$ are empty. Furthermore $T^{-1}(\mu) \not\subset T^{-1}(\mu')$; if this were not the case we would have $\{\mu\} = T(T^{-1}(\mu)) \subset T(T^{-1}(\mu')) = \{\mu'\}$, which contradicts $\mu \neq \mu'$. For the same reason we have $T^{-1}(\mu') \not\subset T^{-1}(\mu)$. The two together imply that there are states $\mu_1 \in T^{-1}(\mu)$, $\mu_2 \in T^{-1}(\mu')$ with $\mu_1 \neq \mu_2$, $\mu_1 \notin T^{-1}(\mu')$ and $\mu_2 \notin T^{-1}(\mu)$. Because of $\mu_1 \in T^{-1}(\mu)$ and $\mu_2 \in T^{-1}(\mu')$ we have $R(T(\mu_1)) \in R(T(T^{-1}(\mu)))$ and $R(T(\mu_2)) \in R(T(T^{-1}(\mu')))$. Now since $R(T(T^{-1}(\mu))) = \{R(\mu)\}$, we have $R(T(\mu_1)) = R(\mu)$. For the same reason we have $R(T(\mu_2)) = R(\mu')$. Finally $R(\mu) = R(\mu')$ implies $R(T(\mu_1)) = R(T(\mu_2))$. This ends the proof of the lemma.

Theorem 3   *No strictly internal observer can measure from inside all distributions of a system whose time evolution is described by a Markov stochastic process.*

To prove this assume A could measure both of two distributions μ_1, μ_2 over S_O at t_0 satisfying $\mu_1 \neq \mu_2$ and $R(T(\mu_1)) = R(T(\mu_2))$. (That there are such distributions follows from Lemma 5.) This will lead to a contradiction. It implies that there are two distributions μ, μ' of A at t_1 with $J(\mu) = \mu_1$, $J(\mu') = \mu_2$. By repeated application of (4) we arrive at

$$\mu_1 = J(\mu) = J(R(T(J(\mu)))) = J(R(T(\mu_1))) =$$
$$J(R(T(\mu_2))) = J(R(T(J(\mu')))) = J(\mu') = \mu_2$$

which contradicts $\mu_1 \neq \mu_2$. Thus for a strictly internal observer there is no consistent inference map J that can measure exactly both of two arbitrary distributions μ_1, μ_2 over S_O at t_0 satisfying $R(T(\mu_1)) = R(T(\mu_2))$. This ends the proof of the theorem.

Some final words about the relevance of Theorem 3. Are statistical measurements from inside possible at all? Admittedly, it will usually be impossible for an experimenter to bring a system in which she is contained into some fixed initial state. But for the theorem to apply it is not always necessary to bring the observed system into the same initial state. Usually the initial state is characterized by only some few parameters; all other quantities may have arbitrary values. This can be taken into account by a proper choice of initial distribution. Therefore it is essential in statistical measurements from inside to fix only a probability distribution over the states, not the precise initial state.

Do restrictions on measurability apply only to internal observers? No, even experiments on an external system in general cannot measure all distributions because background knowledge about the measurement setup excludes some distributions as possible initial distributions of O and others as possible final distributions of A. This is described by an inference map J whose image does not contain every distribution over S_O. So in general not even a measurement from outside can measure all distributions.

What are the specific restrictions on measurements from inside? Although measurements from outside in general will not be able to measure all distributions over S_O, for two arbitrary distributions one can find an inference map J that can measure both. (It is a different issue whether experimental ingenuity and resources are sufficient to perform in the laboratory an experiment with such an inference map.) In contradistinction to this, for measurements from inside there can be no inference map that can measure both of two distributions μ_1, μ_2 with $R(T(\mu_1)) = R(T(\mu_2))$.

Some Limits of Science

A theory of everything (TOE) seems to be the ultimate goal of scientific inquiry. If we had a TOE, would we be able to explain everything in physics? If we would not, we touch the limits of science: the limits of a possible TOE are limits of science.

There are traditional arguments that even with a TOE we cannot explain everything: 't Hooft (1994, 30) thinks that "complexity may very quickly reach the very limits even of the most powerful computers, and so, even if we have the full equations, there would be uncertainty when we try to apply them." Barrow (1994) argues that due to symmetry breaking a TOE need not account for everything: it may well be that the laws possess a symmetry that the solutions do not possess in Stöltzner and Thirring (1994). So there seems to be agreement that one must not take literally the label "theory of everything." A TOE does not explain everything. However, this does not necessarily preclude a TOE from being true of everything.

148   Thomas Breuer

Theorems 1, 2, and 3 have implications for the experimental relevance of a TOE. The biggest system a TOE supposedly can describe is a system without an external observer, namely the universe. Applying the theorems we conclude that (1) no observer can distinguish all states of the universe and (2) not even all observers together can achieve this.

Conclusion (1) is almost obvious. Admittedly, for any two states there is some part of the universe where they differ; an observer located there is not precluded from distinguishing the two states. But it is reasonable to assume there is some part of the universe that is not part of any observer. States that differ on this part but coincide on all observers will be indistinguishable for every observer.

Conclusion (2) is not quite so obvious. Perhaps two observers could cooperate to measure the state of the universe. Each might be able to measure the state of his outside world exactly, and the union of the two outside worlds is the whole universe. But this does not work either. The knowledge of the two observers has to be brought together somewhere. The system where the knowledge is brought together is again a part of the universe. So the theorems apply: this system again cannot distinguish all states of the universe. Here is the physical reason why this attempt fails: bringing together the knowledge in one system changes the state of the system and thus the state of the universe. So the information provided by the two observers will always be outdated.

I therefore conclude that even if we had a TOE, problems of self-reference prohibit any exact measurement of the state of the universe. A TOE is thus experimentally not fully accessible. However, this does not imply that a TOE does not exist or that we cannot find it by chance. But even if we had it, we would not be able to say for sure that we had found the One True Story of the world.

REFERENCES

Albert, D. Z. 1987. "A Quantum Mechanical Automaton." *Philosophy of Science* 54: 577–85.
Arnold, L. 1974. *Stochastic Differential Equations.* New York: Wiley.
Barrow, J. 1994. "Theories of Everything." In Hilgevoord 1994, 38–60.
Breuer, H.-P., and F. Petruccione. 1995. "Stochastic Dynamics of Quantum Jumps." *Physical Review E* 52: 428–41.
Breuer, T. 1995. "The Impossibility of Accurate State Self-Measurements." *Philosophy of Science* 62: 197–214.
———. 1998. "Ignorance of the Own Past." *Erkenntnis* 48: 39–46.

Hilgevoord, J., ed. 1994. *Physics and Our View of the World*. Cambridge: Cambridge University Press.

Honerkamp J. 1998. *Statistical Physics*. Heidelberg: Springer-Verlag.

Stöltzner, M., and W. Thirring. 1994. "Entstehen neuer Gesetze in der Evolution der Welt." *Naturwissenschaften* 81: 243–49.

't Hooft, G. 1994. "Questioning the Answers or Stumbling upon Good and Bad Theories of Everything." In Hilgevoord 1994, 16–37.

10

Blinded to History?

Science and the Constancy of Nature

Alfred Nordmann
Department of Philosophy, University of South Carolina

> *Tout change, tout passe,*
> *il n'y a que le tout qui reste.*
>
> —Diderot, Reve d'Alembert

When Ptolemeian astronomers studied the skies, they did so as spectators who, from their fixed vantage point, watched celestial motions. Kant and Lichtenberg taught us that the Copernican Revolution began by questioning this relationship between inquirers and their domain of inquiry. When Copernicus, Galileo, or Kant watched the skies, they saw only how things *appeared* to them as earthbound observers whose vantage point is implicated and participates in the motions of the heavenly bodies. The epistemological lesson of the Copernican Revolution is therefore that knowledge becomes more objective when the implicit conditions under which observations are made and knowledge is obtained are explicitly acknowledged as constraints on inquiry: objectivity increases when subjectivity is taken into account and controlled for, when *Erkenntnisbedingungen* become *Erkenntnisvoraussetzungen.*

This chapter reflects on a precondition of all scientific knowledge that sets a limit to science: science is based on the constitutive counterfactual that nature and its representations have no history. The explicit recognition of that limit advanced the production of objective knowledge and allowed for a discovery of time and the ahistorical treatment of vast evolutionary processes. However, the founding myth of an eternally fixed nature that alone can underwrite eternally true scientific theories also creates an unresolvable, productive tension. The paradigmatic conflict between Priestley and Lavoisier shows that a gain in objectivity may involve a loss of truth. In its present reincarnation, the conflict between Bruno Latour and a "Lavoisian" philosophy of science shows that the very conditions for successful scientific inquiry prevent science from knowing itself.

The Denial of History and the Possibility of Scientific Knowledge

The scientific philosopher Charles Sanders Peirce became painfully aware that science does not and cannot know itself, and moreover that it cannot do so because it must deny the history of reality and thereby its own historicity. According to "The Fixation of Belief," scientific inquiry proceeds from the hypothesis of reality (Peirce 1992, 120–21); however, the very content of that hypothesis systematically blinds science to the epistemological and ontological process of postulating, articulating, and establishing the truth of that hypothesis.

The content of the scientific hypothesis of reality is familiar enough: reality is independent of what any one person thinks of it; instead, it is thought to be the cause of our sensations or perceptions and thus precedes our knowledge of it. In Peirce's terminology, this conception of reality is nominalistic (Peirce 1992, 87–88), and it has been articulated by philosophers as diverse as Plato, Hume, Meyerson, or Wittgenstein. If truth is to be eternal, Plato might argue, there must exist a realm of eternal and unchanging forms, a true reality behind the appearances of change, becoming, and decay. In order to draw any conclusions from experience at all, Hume points out, we must rely or (adds Reichenbach) wager on the unprovable assumption that nature does not change and that the future will be like the past (Hume 1955, 40–53; Reichenbach 1961, 348–63). Meyerson shows that all claims concerning causality reduce to claims of identity and thus that science works out in which respects nature stays always the same (Meyerson 1930). And Wittgenstein elaborates that all attempts to represent nature in a descriptive language presuppose a static world that is an aggregate of given states of affairs: science cannot give meaning to statements like "everything is in flux," and this is one of the reasons why Wittgenstein's later philosophy questions so deeply the linguistic paradigm provided by the descriptive language of science (Wittgenstein 1981; 1993, 188–92; see also Nordmann 1999).

If these philosophers articulate the ontological commitment implicit in the scientific method of fixing belief, the philosophy of science has developed its epistemological correlate, namely the distinction between contexts of discovery and justification, between the space of facts and the space of reasons. In order to determine truth or falsity, science assumes that its theories and hypotheses can be evaluated separately from their parochial origin in the minds of particular people at particular times. Scientists and philosophers therefore share the common concern for sustaining, defending, and explicating this "separability assumption" (Feyerabend 1989, 394).

Science treats nature as if it has no history, as if it has always been there already, standing ready to be discovered and understood. This *objective* or ontological aspect of science's conception of history is complemented by its *subjective* or epistemological correlate: science treats hypotheses and theories as if their content can be dissociated from the history of their production, as if the historical agency of particular scientists will not be reflected and preserved in certified textbook knowledge (see Engelhardt 1979, 10, 223). Taken together, these objective and subjective preconceptions yield the promise of empirical knowledge that holds at all times and in all places, such as true knowledge of eternal laws of an immutable nature.[1]

To the extent that the assumption of these various *as ifs* is necessary for the pursuit of this promise, they are constitutive of science. To the extent, however, that their truth is in question, they may also be counterfactual. Indeed, from Kant to Popper, from Poincaré to Kuhn, philosophers and scientists have understood such presuppositions as constitutive counterfactuals, regulative ideas, useful fictions, founding myths, or self-fulfilling prophecies that shape the conduct of scientists and prefigure the objects and results of inquiry. Science anticipates nature as an object of theoretical representation: if one wants to take a picture of the world, one must hold it still and hold still oneself.[2]

On the subjective side, the truth of the separability assumption is most likely to be contested by historians, sociologists, and philosophers of science. Initially, the division between contexts of discovery and justification was called into question by debates about internal versus external determinants of scientific development. A second line of questioning explores the tension between experimental and mathematical traditions within physical science. It shows how the experimental undercurrent of scientific inquiry subverts the fiction of a syntactically closed theoretical science that moves from one universal generalization to another via a process of testing and revision. The experimental tradition asks whether the development of scientific knowledge is a chapter in the history of nature, whether natural and human history intersect in the performance of experiments, whether the progress of science effects change in the natural and human order, whether novel experimental facts are constructed rather than merely discovered or revealed. Once the epistemological division between contexts of justification and discovery is rejected, science appears as a historically creative endeavor, producing novel facts, novel actors, and a new world.

On the objective side, many scientific controversies revolve around the ontological assumption of an essentially unchanging nature. The scandal and ambition of alchemy consisted of its attempt to effect transubstantiation, that is, to change a thing's nature by turning base earth into pure gold. Catastrophist geologists denied as a matter of revealed and observed truth the idea of a conservative and unchanging nature. Instead, some viewed nature as an agent of divine will and aligned the history of volcanic activity in Italy with apocalyptic texts and the naturally ordained destruction of Roman Catholicism. The historicity of nature is also at issue just as soon as one wonders about an origin of species, challenging the definition of species as immutable substances with essential natures. Contemporary ecologists recommend the notion of nature as a living organism, author of and actor in a social history of nature (see, e.g., Böhme and Schäfer 1983). In this conception, nature might be endowed with intrinsic value, purpose, or will; it can be violated and, for example, get angry. Physicians and medical scientists question their own success at dehistoricizing disease, which began with *The Birth of the Clinic* (Foucault 1975): recognizing that so-called medical histories are mere records of successive states of health and disease, they call for true patient histories in order to better identify, specific to the patient, the relevant causes and most promising treatments of disease. Finally, futurologists model nature on the assumption that present trends continue, with the hope and expectation that they will not.

Especially in times of crisis, the sciences themselves are therefore asking to what extent the immutable laws of nature define a narrowly fixed arena in which events unfold, or whether this arena can or should be widened to admit genuinely historical processes.

Taken together, these epistemological and ontological challenges to the constitutively counterfactual denial of history undermine the promise that science might gain true knowledge of eternal laws of an immutable nature.

In the current "science wars," most philosophers argue that it must be possible to maintain the separability assumption, and many sociologists of scientific knowledge underscore its counterfactuality. The philosophy of Peirce holds special promise in that it motivates and justifies science's denial of history, at the same time historicizing and contextualizing that denial within a realist, rather than a nominalist, ontology and epistemology. According to Peirce, reality is the "normal product of mental action"; it does not precede inquiry but is gradually articulated or fixed as inquiry moves toward a final opinion. Reality is that which corresponds to true belief (Peirce 1992, 91), and the nominalist hypothesis of reality becomes realized only in the pursuit of a realist epistemology of scientific world-making (Peirce refers to this as "objective idealism"; see Peirce 1992, 293). Committed from necessity to the nominalist hypothesis of a fixed, ahistorical reality, which is the ultimate cause or warrant of true belief, science systematically blinds itself to its constructive procedures, to the historical process by which ideas are generated and their corresponding reality shaped, concretized, articulated.[3]

Copernican Revolutions in the Discovery of Time

In the spirit, though not the letter, of Peirce's proposal, a somewhat schematic account of scientific development can now be offered: the dynamic interplay between ahistorical and historicized perspectives on science and nature explains a particular kind of scientific progress, one that required *The End of Natural History* (Lepenies 1976) for a *Discovery of Time* (Toulmin and Goodfield 1965), but which decidedly did not involve the anticipated move *From Natural History to the History of Nature* (Lyon and Sloan 1981).

Much science seeks to understand change. Theories of motion deal with change of location over time; alchemy and chemistry explore changes in the composition of matter; biology, geology, and astronomy consider change on a large temporal scale with their theories concerning the origin and development of species, of the earth, and of the universe. All talk of change, however, presupposes a fixed frame of reference that preserves the identity of the something that undergoes change and persists through it. What persists through change is how things are, that is, the nature of things, that is, nature; and this conception of an unchanging nature sets the previously discussed epistemological condition or limit of science. As in the case of the Copernican Revolution, however, new and more objective knowledge becomes possible once a mere precondition of knowledge is problematized and taken as an explicit point of departure. This revolution begins with the question concerning the history of nature, that is, with the negotiation of science's fixed frame of reference: can one conceptualize that which persists through change as itself susceptible to change? If not, how can one at least allow for those things to change that were hitherto thought to persist?

The unchanging and immutable nature of things is most readily attributed to the things themselves, to their essential or substantial being. Therefore, perhaps, biology began with a substantial conception of species: an organism is what it is and does what it does by virtue of being an organism of such nature or kind. Just as earth cannot change into gold, a lizard cannot by its nature change into a bird.

Buffon may have been the first to question seriously the fixity and suggest the historicity of species (Lyon and Sloan 1981, 23, 170). When he defines a species as a "chain of successive existences of individuals," he undermines not only the substantial conception of species but all hope of gaining true knowledge of the world. While the "first causes of things will remain ever hidden from us," the most productive question concerning secondary causes "gives liberty of enquiry and admits the employment of imagination." In "a world of infinite combinations, some harmonious and some opposed," he will choose those hypotheses that establish "the greatest analogy with other phenomena of nature" (Lyon and Sloan 1981, 101, 176). However, such hypotheses are not underwritten by a conception of a stable and unchanging nature, but hold only by the pleasure of God. For example, in the transmission of internal molds consists "the essence and the continuity of the species . . . and [it] will so continue while the great Creator permits their existence" (Lyon and Sloan 1981, 177, 202). Buffon's

General and Particular History of Nature thus fails to yield a "physical system for the understanding" (Kant 1977, A7) and is "more of a romance of nature than a true history of nature."[4]

Similar complaints have been brought by his contemporaries against Cuvier's catastrophism. As Martin Rudwick points out, Cuvier himself runs up against the contradiction implicit in his desire on the one hand to be "a new species of antiquarian," on the other hand to call for a Newton who will "burst the limits of time" (Rudwick 1997, 183, 185; 1998). Cuvier's catastrophism admits radical discontinuities, resists universal laws of nature, and promotes his orientation toward archaeology: he studies geological strata as if they were foreign cultures, each with its own flora and fauna. In the absence of a theory that governs the temporal succession of these strata, they might as well be contemporary rather than stacked in a chronological order. This antiquarian exploration of strange worlds is therefore confusedly at odds with the question "So why should not natural history also have its Newton one day?" (Rudwick 1997, 185).

If a "Newton of natural history" were possible, that title would be claimed by Darwin, who succeeded where his predecessors had failed precisely because he avoids Buffon's imaginative eclecticism as well as Cuvier's catastrophism. Darwin specifies exactly how an unchanging nature can persist through evolutionary change. Relying on the assumptions of actualism or uniformitarianism (there are no causes but those that are currently in operation) and gradualism (nature makes no leaps), Darwinian biology proves to be anything but a historical or historicized science. Although it treats of the origin and transmutation of species, it succeeded as a science because it posits the uniform operation at all time of an evolutionary mechanism.

Darwin did not suggest that substantial species with essential natures can change, but that the term *species* does not designate substantial natures at all, that it is ontologically continuous with varieties and rather more pragmatically picks out groups of organisms that have become sufficiently separate from other such groups. He thereby effected the retreat of a substantial or essentialist conception of an unchanging nature in favor of a principled or nomological conception. Darwin's actualism and gradualism explicitly specify nature as the wide arena in which evolutionary processes unfold over vast stretches of time—as the conceptually fixed space that frames the succession of time-indexed contiguous states, that is, as governed by eternal physical laws that describe the mechanisms of geological, biological, and perhaps social change within the general model of mechanics with its laws of motion that describe changes of place.

Darwin's "Copernican Revolution" opened a new branch of objective knowledge by transforming an epistemological constraint on science into a productive and explicit presupposition. Is the constancy of nature simply a given, Darwin asked, or is science to determine whether it obtains at the level of species or at the level of general mechanisms? It has not escaped scientific notice, however, that Darwin's discovery of time did not thereby overcome that epistemological limit of science and that, like its predecessors, Darwinian biology continues to be a thoroughly ahistorical science.[5]

Criticizing the (neo-)Darwinian adaptationist program, Stephen Jay Gould and Richard Lewontin historicize evolutionary thought by highlighting the role of contingency. To the extent that they succeed, however, they also question the objectivity and even the possibility of evolutionary biology as a (nomological) science. Aside from Gould's qualified return to a "catastrophist" conception of punctuated equilibria, they jointly oppose the mechanicism inherent in the strictly adaptationist program, which reifies the opposition between organism and environment. According to the adaptationist view, the organism is merely responsive to its environment and submits random variations for consideration by natural selection. Gould and Lewontin propose that organisms are more creative and innovative than that, producing apparently nonadaptive spandrels that can promote contingent evolutionary developments down the road (Gould and Lewontin 1979). Lewontin argues explicitly that his dialectical conception of the organism as object and subject of evolution questions the scientific standing of biology: "Darwin's alienation of the environment from the organism was a necessary step in the mechanization of biology. . . . But what is a necessary step in the construction of knowledge at one moment becomes an impediment at another. . . . 'Adaptation' is the wrong metaphor and needs to be replaced by a more appropriate metaphor like 'construction'" (1995, 131).

In order to speak, dialectically, of the interdependent construction or coevolution of organism and environment, Lewontin denies that organisms are universally governed by external forces or laws of nature, even gravitation (1995, 136). Indeed, the founding myth of science, that there is an unchanging nature that persists through change, is rejected as bourgeois and alienated thought:

Even in evolutionary theory, the quintessential study of change, we saw the deep denial of change. Evolution was merely the recombination of unchangeable units of idioplasm; species endlessly played musical niches; the seemingly sweeping changes through geological time were only prolongations of the microevolution observed in the laboratory; and all of it was merely a sequence of manifestations of the selfish gene in different contexts of selfishness. . . . The dialectical view insists that persistence and equilibrium are not the natural state of things but require explanation, which must be sought in the actions of opposing forces. The conditions under which the opposing forces balance and the system as a whole is in stable equilibrium are quite special. (Levins and Lewontin 1985, 275, 280)

To be sure, in order to provide local explanations of the unique conditions under which organisms and their environments change or persist together, the "dialectical biologist" will draw on relevant bits and pieces of science (physics, genetics, ecology), and those sciences still presuppose that nature remains constant in some relevant respects. But the demand for local rather than global explanations negates a general scientific theory of the origin, divergence, and development of species as envisioned by Darwin. If we wish to take evolutionary *history* seriously, Lewontin argues, we must incorporate the model of social or political history, refraining from the temptation to construe a dialectical science of the laws of history that govern all actions of opposing forces (Levins and Lewontin 1985, 267–68).[6]

    Another response to the limitation of evolutionary theory redefines the constancy of nature once again, retreating from Darwin's principled or nomological conception of an unchanging nature even further to a structural or parametric conception. If there are ahistorical laws of nature and an evolutionary mechanism, one might ask, have not these laws and mechanisms themselves evolved? Charles Sanders Peirce viewed the history of the universe as an "unfolding argument" and sought the law of mind that describes and predicts the evolution of natural law (see, e.g., Peirce 1997, 201). He discovered a principle of self-organization that has been interpreted as a bold anticipation of proposals by physicists John Barrow, Paul Davies, and John Wheeler (Brent 1993, 174–76) and by chemist Ilya Prigogine (Prigogine and

Stengers 1984, 302–3; Pape 1991).[7] However, where Lewontin and Gould invoke the recognition of "contingency" as their criterion for a historical treatment of nature, Prigogine and Stengers propose the far weaker criterion of "irreversibility": under far-from-equilibrium conditions, no laws of nature lead from initial states to final states and back, since the relevant information content of the initial conditions depends on the dynamics of the system. When science treats of such conditions, it "moves away from the repetitive and the universal to the specific and the unique" and thus it might appear that the division of science and the humanities, of an unchanging nature and history, finally breaks down. However, Laplace's demon is defeated only insofar as science discovers "a new kind of order" (dissipative structure) and a "new unity" that arises from "irreversibility [as] the mechanism that brings order out of chaos." This *mechanism* can explain, for example, why "a universe far from equilibrium is needed for the macroscopic world to be . . . a living world" (Prigogine and Stengers 1984, 280, 13, 292, 300–301). However, to recognize that the irreversible evolution of informational order implies randomness is a very general and ahistorical fact, law, or definition of nature. Its discovery, at an even greater remove than Darwin's from the essentialist conception of an unchanging nature, required yet another "Copernican Revolution": "Demonstrations of impossibility, whether in relativity, quantum mechanics, or thermodynamics, have shown us that nature cannot be described 'from the outside,' as if by a spectator. Description is dialogue, communication, and this communication is subject to constraints that demonstrate that we are macroscopic beings embedded in a physical world" (Prigogine and Stengers 1984, 299–300).[8]

Like Buffon and Cuvier, Gould and Lewontin ask for a conceptually impossible historicized science. They are running up against the limits of science and find them unacceptably limiting, impeding the construction of knowledge. Like Darwin, in contrast, Prigogine and Stengers attempt to push back the limits of science, recovering greater segments of time within an ahistorical framework of theory. Thus—whether we remain evolutionary thinkers in the Darwinian mold, whether we embrace Gould's and Lewontin's dialectical emphasis on evolutionary contingencies, or whether we follow Prigogine and Stengers into more abstracted theories that admit hitherto unacknowledged temporal processes and dimensions—we are staying within the constitutive interplay of epistemological constraint and cognitive opportunity, within ahistorical conceptions of science and nature that yield powerful accounts of the succession of natural states.

Contested Trade-Offs: Objectivity and Truth

Implicitly and explicitly, science invokes a great variety of principles by which to postulate the constancy of nature. They range from actualism and gradualism ("there are no causes but those that are currently in operation," "nature makes no leaps") to predictive posits ("if present trends continue") to *ceteris paribus* clauses ("everything else being equal") to inductive and statistical principles ("the future is conformable to the past," "in the long run, relative frequency converges on objective probability") to principles of conservation (of, e.g., weight and energy). For each of these it may be possible to write a history of Copernican Revolutions, showing that the development of science saw an increasingly explicit, perhaps increasingly generalized, reliance on these principles. Such histories of apparent progress might obscure, however, the fact that the advance toward greater objectivity may involve a costly sacrifice: these accounts of progress tend to render us oblivious to the real possibility that nature has a history, that inquiry consists of a productive collaboration between human and nonhuman agents in the laboratory, that organism and environment, and scientist and nature, are involved in local, contingent interactions between opposing forces, simultaneously subject and object of a social history of nature.

Joseph Priestley and Antoine Lavoisier disagreed not only about chemical facts but on the very conceptions of science, knowledge, and nature. According to Priestley, truth is eternal, but scientific truths are established in the course of a simultaneously historical and natural process. Nature works through the mind toward the formation of a true system of belief; the mind works through nature toward the exhibition of novel facts and the true system of nature. In contrast, Lavoisier recommends an advancement of objective knowledge by making explicit the relevant features of a fixed and constant nature: formal constraints on reasoning and practice would allow for his experiments to advance a theoretical agenda (see Holmes 1989). Since Lavoisier's limitations on chemistry created the conditions for its success, the term "Lavoisian science" was introduced to characterize in general the ahistorical orientation of modern science (Meyerson 1991).[9] However, Priestley or his latter-day successors, such as Bruno Latour, resist Lavoisier's "Copernican Revolution" on a variety of levels, defending their profoundly historicized conceptions of nature and science in the name of truth.

Any review of Priestley's and Lavoisier's attitudes on the relation of science and history might begin with a passage from the 1791 edition of the *Experiments and Observations on Different Kinds of Air,* in which Priestley contrasts himself with Lavoisier, reflecting on the ironic circumstance that Lavoisier's famous new system of chemistry was erected upon Priestley's discovery of oxygen:

> When, for the sake of a little more reputation, men can keep brooding over a new fact, in the discovery of which they might, possibly, have very little real merit, till they think that they can astonish the world with a system as *complete* as it is *new,* and give mankind a high idea of their judgment and penetration; they are justly punished for their ingratitude to the fountain of all knowledge, and for their want of a genuine love of science and mankind, in finding their boasted discoveries anticipated, and the field of honest fame preoccupied, by men, who, from a natural ardour of mind engage in philosophical pursuits, and with an ingenuous simplicity immediately communicate to others whatever occurs to them in their inquiries. (1970 [1790], 1:xvii–xviii)

Priestley's polemic can be illuminated in three steps, each engaging a particular aspect of Lavoisier's denial of history. A first step involves the *history of science* and how it shapes the practices and self-understanding of both scientists.

When Priestley praises himself for his natural ardor of mind and for the ingenuous simplicity with which he communicates whatever occurs to him in his inquiries, he does not distinguish between science and its history, and he recommends himself as a historian of inquiry in contrast to Lavoisier, the ingenious theoretician. Priestley is, after all, the author of numerous histories: aside from a two-volume *History of the Corruption of Christianity* and four volumes on the *History of Early Opinions Concerning Jesus Christ,* there are, more significantly, two volumes on the *History and Present State of Electricity* and a *History of Discoveries Relating to Vision, Light and Colours.* Priestley's emphasis on histories is not a variant of an otsogian and ultimately self-promoting modesty (Merton 1965), nor does it express an antiquarian compulsion to detail the prehistory of his researches if only to highlight the novelty and originality of his own contributions.

Science, according to Priestley, is the cumulative history of revealed facts, of the world becoming filled at the fountain of all knowledge, which is fed by the inexhaustible plenitude of nature.

Lavoisier's chemistry, in contrast, was consciously based on the idea of revolution, on historical discontinuity, on devising a system of chemistry that gets the relationship between the known facts right. In his published writings Lavoisier does not dwell on the history of chemical researches: his introduction to the *Traité Élémentaire de Chimie* explicitly warns that a historical presentation would lead one to lose "sight of the true object I had in view" (Lavoisier 1952 [1789], 6; see also Engelhardt 1979, 98). Indeed, in an important sense, chemistry begins with him. One of his posthumously published manuscripts lends credence to Priestley's portrayal of Lavoisier as a boastful scientist when it asserts that "the theory is not, as I have heard it said, the theory of the French chemists, it is *mine* [elle est *la mienne*], and it is a possession that I claim before my contemporaries and before posterity. Others no doubt have added new degrees of perfection" (quoted by Siegfried 1988, 39). All previous discoveries are simply present in that they constitute the stock of known facts; their order of discovery has been superseded by their new systematic ordering: they are meaningful and considered, or meaningless and ignored. Future discoveries add degrees of perfection, confirm and articulate Lavoisier's system of chemistry.

Priestley's history of chemistry is no history of ideas, which (like Fourcroy's) "points out the errors of our predecessors, and indicates the path that leads to success" (Siegfried 1988, 43). Instead, Priestley's history is a record of researches, including his own (Priestley 1970 [1790], 1:xvi). This record repeatedly exemplifies the experience of astonished surprise, of experimental facts overwhelming the rudiments of prejudice that encumber even the most ingenuous of experimenters. By readying himself to be surprised, Priestley deliberately inserts himself into a natural process. His extensive account of the discovery of oxygen or dephlogisticated air "encourages philosophical investigations." It shows that "more is owing to what we call *chance*, that is, philosophically speaking, to the observation of *events arising from unknown causes*, than to any proper *design*, or preconceived *theory* in this business" (Priestley 1970 [1790], 2:102–3; see also 113–14). Accordingly, Priestley refers to "surprise" as an epistemological category that relates to the history of science ("unexpected in light of background knowledge") and as an ontological category that relates to the history of nature ("a chance affectation of matter and that property of matter called mind").

In contrast, Lavoisier encourages philosophical investigations by "adopting a new arrangement of chemistry . . . consonant to the order of nature." He can do so on the assumption that the order of nature is simultaneously the "order as shall render it most easy for beginners in the study of chemistry thoroughly to understand" the facts and the conclusions drawn from them (Lavoisier 1952 [1789], 2).

Here emerges a second dimension in the relation between science and history: how do Priestley and Lavoisier conceive of *process*, be it a chemical process (e.g., calcination or combustion) or the epistemological process leading to the formation of knowledge or belief?

According to Priestley, discovery is effected not by design but by the cumulative workings of "slight and evanescent" motives. In tune with his millenarian convictions, Priestley viewed all historical processes as the gradual revelation of a truth that displaces all human error and prejudice, asserting the system of nature as "superior to any *political* system upon earth" (Priestley 1970 [1790], 1:xxvi). His discoveries of numerous kinds of air, of photosynthesis, and of a method for making soda water proceeded from the observation of processes like fermentation, calcination, and putrification; he did not privilege in his researches the investigation of sudden or even violent reactions (see Gibbs 1967, 57–60). Similarly, the history of knowledge is conceived on the model of growth and maturation. Priestley's various tracts on education testify to this, as does his active endorsement of Hartley's associationist theory of the human mind. Indeed, Priestley recommends Hartley's system precisely because it departs from the crude mechanicism of earlier theories and agrees with "the more subtle and important laws of matter exhibited in chymical operations" (Priestley 1973 [1775], xviii; see also xxxii). Chemical operations teach us "that the laws and affections of matter are infinitely more complex than we had imagined" (xviii). Since the affectations of matter are infinitely complex, they alone suffice for an account of history and epistemology, a separate entity like "spirit" or "mind" need not be invoked, and thus the complexity of matter leads us to admire the "simplicity of nature, and the wisdom of the great author of it, in this provision for the *growth of all our passions,* and propensities, just as they are wanted" (xxxii).

Growth from affectations of matter issues in genuine novelty; something new is created in processes of fermentation, maturation, and growth; along with heat and light, the passage of time itself is productive in these processes and brings it about that in the end there is more than there was before. This applies to the growth of true ideas as well as to the "gradual formation of the ideas of moral right and wrong" (xliii); it also applies to the productions of nature (Priestley 1970 [1790], 1:xviii, xli). Whereas the true system of nature is historically final (xxxvi–xxxvii), the history of science is not only an epistemological history of coming to know that system, but also the ontological history by which the system of nature establishes itself ("discovered and directed by human art"): "the idea of continual rise and improvement is conspicuous in the whole study [of experimental philosophy], whether we be attentive to the part that nature, or that which men are acting in the great scene" (Priestley 1775, 1:iv). "New worlds may open to our view" (xv) in the pursuit of experimental philosophy and "by these sciences also it is, that the views of the human mind itself are enlarged, and our common nature improved and enobled" (xxi). And while the artifice of the experimental scientist provokes nature to reveal novel facts from its infinite storehouse of facts, Priestley describes this sometimes as "discovery" and sometimes as a qualitative change of nature itself: "[W]hen electricity began to show itself in a greater variety of appearances, and *to make itself sensible* to smell, the sight, the touch, and the hearing . . . , electricians were obliged to make their systems more complex, in proportion as the facts were so" (Priestley 1775, 2:18, emphasis added).

If Priestley considers "process" a transmutation of our common nature, Lavoisier's "Copernican Revolution" conceives chemical and epistemological processes as transitions between states within an ahistorical framework of a conservative and unchanging nature. "We may lay it down as an incontestable axiom," Lavoisier writes, "that, in all the operations of art and nature, nothing is created; an equal quantity of matter exists before and after the experiment; the quality and quantity of the elements remain precisely the same and nothing takes place beyond changes and modifications in the combination of these elements. Upon this principle the whole art of performing chemical experiments depends" (Lavoisier 1952 [1789], 41; see also Siegfried 1989). In tune with the principle of conservation of weight, Lavoisier's chemistry celebrates the symmetry of composition and decomposition, of analytic and synthetic procedures; it is a chemistry of reversible processes. Chemical and epistemological change consists of the transitions between contiguous chemical or cognitive states: since they add nothing new to the world, we observe nothing but the reconfigurations of known quantities, redistributions of matter, rearrangements of the facts. Although the quantity of matter needs to be tracked with utmost diligence, time does not need to be accounted for: even though he knew that reactions can be substantially different at high and low heat (1952 [1789], 66), Lavoisier treats equally a sudden and violent reaction (as in his gun barrel experiment) and a protracted chemical process. In both cases, he balances the quantity of matter before and after the experiment. Like other eighteenth-century sciences, Lavoisier's new chemistry asserts "that there is no process of evolution in which duration introduces new events of itself" (Foucault 1975, 12).[10]

This difference between Priestley and Lavoisier plays into their larger conceptions of history. Priestley's science is devoted to the production of novel facts. Together with novel social and political facts (especially the French Revolution) they provide evidences for his radical millenarianism (see Garrett 1973; Fruchtman 1981; Fitzpatrick 1987). "We are, as it were," Priestley proclaims, "laying gunpowder, grain by grain, under the old building of error and superstition, which a single spark may hereafter inflame, so as to produce an instantaneous explosion" (quoted by Gibbs 1967, 173). Later on, Priestley extends this metaphor into the sphere of science when he suggests that "the English hierarchy (if there be anything unsound in its constitution) has . . . reason to tremble even at an air pump, or an electrical machine" (1970 [1790], 1:xxiii).

According to Priestley, chemical and epistemological processes aim at a new, radically altered world. Indeed, enlightenment is such a process: darkness and ignorance are violently displaced as the bright (and blinding?) light of revelation streams in.

Lavoisier's conception of process, including the process of enlightenment, is more in keeping with his role as a liberal administrator in (pre-)revolutionary France (see Donovan 1993; Poirier 1996). Even art cannot create anything; however, evenly bathed in the light of reason, social and natural relations can be clarified, ordered, arranged.

These profound differences influence, third and finally, Lavoisier's and Priestley's *experimental and communicative practice.*

Priestley presents himself as one of those scientists "who, from a natural ardour of mind engage in philosophical pursuits, and with an ingenuous simplicity immediately communicate to others whatever occurs to them in their inquiries." He aims to be a transparent medium of nature (see Nordmann 1994), always worried that the force of prejudice "biasses not only our *judgments,* properly so called, but even the perceptions of our senses" (1970 [1790], 2:103).

When he experiments, he favors instruments that compare to real history rather than works of fiction: "Works of fiction resemble those machines which we contrive to illustrate the principles of philosophy, such as globes or orreries, the use of which extend no further than the views of human ingenuity; whereas real history resembles the experiments by the air pump, condensing engine and electrical machine, which exhibit the operations of nature and the nature of God himself" (Priestley 1999 [1788], 27–28; see also Priestley 1775, 1:xii–xiv.). As a modest witness to the operations of nature, Priestley's job is simply to wait and see, to produce striking phenomena, to chronicle and communicate the course of events. Indeed, "speedy publication" (Priestley 1970 [1790], 1:xvii) is one of the hallmarks of the ingenuous scientist.

Antoine Lavoisier, in contrast, does not share the ideal of speedy communication: "Among the facts that chemists discover are many that have little significance until they are connected together and assembled into a system. Chemists may withhold facts of this sort and ponder in the quiet of their studies how to connect them to other facts, announcing them to the public only when they have been pulled together as far as possible" (quoted by Perrin 1986, 665). Facts become meaningful only once they are fitted into his system of chemistry; before then their communication would be meaningless: "every branch of physical science must consist of three things: the series of facts which are the objects of the science, the ideas which represent these facts, and the words by which these ideas are expressed. Like three impressions of the same seal, the word ought to produce the idea, and the idea to be a picture of the fact" (1952 [1789], 1).

Whereas Lavoisier's chemical universe is controlled by an aesthetic and philosophical ideal of systematic closure, Priestley's chemical universe involves him in an open-ended course of experimentation: as a solitary investigator he typically moved through a series of states until his time had run out or until there was nothing left to experiment on in his receivers and jars. He encountered only a few elements and principles, each a fundamental constituent of matter, each creative and powerful. Phlogiston, for example, was conceived in analogy to "gravity," "electricity," "magnetism," or, indeed, the human soul. Chemical change involves the passage of qualities from one body to another.

In contrast, Lavoisier recognizes many elements, and in chemical change a material body is changing place (Meyerson 1991, 62–63, 206–7).[11] The task of chemical experimenters is therefore twofold. First, they must produce meaningful facts that can speak unambiguously to theoretical questions concerning the redistribution of the given quantity of matter.[12] Second, they must read nature's answers to their questions, and since their questions or predictions are framed in terms of quantities of matter, the balance becomes their paradigmatic instrument, with other instruments (like the calorimeter) modeled on the balance:

In the act of weighing Lavoisier sought to create an experimental space that was entirely under the experimenter's control. Once balanced with weights on Lavoisier's scale, substances were transformed from objects of nature to objects of science. The balance divested substances of their natural history. Their geographical and geological origins, their circumstances of production made little difference. They were transformed into samples of matter made commensurable by a system of standardized weights. (Bensaude-Vincent 1992, 222–23; see also Roberts 1991; Wise 1993)

Lavoisier's questions constitute objects of science and for the determination of answers forged a culture of experimentation that is maintained by a community of reasoners, *such as* the Academy of Science. This is how the same Lavoisier who earlier claimed the new chemistry for himself can make the following allowance: "[I]f at any time I have adopted, without acknowledgment, the experiments or opinions of . . . any of those whose principles are the same as my own, it is owing to this circumstance that frequent intercourse, and the habit of communicating our ideas, our observations, and our way of thinking to each other, has established between us a sort of community of opinions in which it is often difficult for every one to know his own" (Lavoisier 1952 [1789], 6). The sharing of principles (like the principle of conservation of weight) constitutes a community of opinion and of experimental practice; it creates a condition under which consensus becomes possible (see Nordmann 1999). The solution to a chemical problem can be recognized by any member of the community: it requires that the quantities of matter before and after the experiment be entirely accounted for and found to be equal. Lavoisier's ahistorical conception of science and nature thus enables the envisioned scientific community to *decide* the truth or falsity of theoretical propositions. It is a characteristic of Kuhnian "normal" science that such decision procedures are available, and Lavoisier's "Copernican Revolution" therefore created the opportunity for the normal production of objective chemical knowledge.

Excluded from Lavoisier's normal science are those who believe, as a matter of fact, that something can be created in an experiment or that the duration of the reaction represents an input of time that also must be matched by some output.[13] For the sake of a "system as complete as it is new" Lavoisier's community banished from chemistry the powerful and enigmatic principle "phlogiston." Priestley had to reject this restrictive move. His epistemology and his historicized conception of chemistry demanded that human ingenuity not censor or delimit the unfolding powers of nature. His task was to exhibit more and more facts from the infinite storehouse of nature, and only "when all the facts belonging to any branch of science are collected [will] the system . . . form itself" (Priestley 1775, 2:172–73; see also 1:xvii–xviii).

This, then, is the trade-off between Lavoisier's and Priestley's conceptions of science and nature: The founding myth of an ahistorical science and nature creates the hermetic space in which universal claims about objects of science become decidable, that is, the space of reasons or the context of justification. Although Priestley's more historically oriented science can provide paradigms or exemplars in its own right, it falls short of providing such a formal space of reasons.[14] Lavoisier's science thus emerges as superior to or more successful than Priestley's precisely in that its "Copernican Revolution" better fixed the relevant ways in which nature is thought to be conservative and unchanging. Although he acknowledges Lavoisier's successful revolution of chemistry (Priestley 1929 [1796], 19), Priestley never considered nature to be conservative and unchanging. In turn, Lavoisier knows that light and heat contribute to chemical reactions, that light is absorbed by matter, that reactions can be substantially different at high and low heat (1952 [1789], 10, 57, 66): he does not and cannot consider as literally and empirically true his incontestable axiom of conservation of weight.[15] Lavoisier's success therefore involved a costly sacrifice in that it systematically blinded chemistry to the very real possibility that novelty can be created. Priestley was only one of the first to warn against this.

The Metaphysical Burden of Lavoisian Science

Even after Lavoisier's undisputed triumph, Priestley's epistemology is echoed in expressions of ambivalence regarding the question of the historicity of nature and science. We are reminded of Priestley whenever there is emphasis on the creative dimensions of experimental science, on the notion that, along with the growth of knowledge, science effects a growth of concepts and habits and the growth or articulation of a corresponding reality.[16]

Bruno Latour is perhaps the most eminent among Priestley's successors. In his paper on "The Force and the Reason of Experiment" he questions the view that experimentation is a form of discovery: "If an experiment be a zero-sum game, if every output must be matched by an input, then nothing escapes from a laboratory that has not been put into it. Whatever the philosopher's list of the inputs in a setting, it always features the *same* elements before and after: the same [experimenter], the same [chemical elements], the same colleagues, or the same theory."[17] "Unfortunately," he continues, "there is always more in the experiment than was put in" (Latour 1990, 65). Indeed, standard philosophy and historiography of science testify to the prima facie inadequacy of the zero-sum view of experiment. The hallmark of a rigorous test or a well-designed experiment is supposed to be, after all, that its immediate outcome and long- term repercussions could run afoul of theoretical expectations. Something is ascertained, established, found out in an experiment.

According to Latour, such acknowledgments do not ordinarily lead philosophers, historians, or sociologists of science to abandon their view of experimentation as a zero-sum game. While agreeing that "*explaining* the outcome of the experiment by using a list of stable factors and actors will always show a *deficit*," philosophers and sociologists have devised strategies to make up for this deficit. Philosophers are likely to appeal to "nature 'out there'" and argue that it is a hidden stable factor, another input that was merely revealed by the output. What the experiment found, they insist, must have been there already (since experimentation, to them, is a mode of finding or discovery). Some sociologists do not refer to nature as their favorite stock to make up the deficit. Instead, they suggest that social structures are revealed in the experiment (since experimentation, to them, is a mode of affirming social identity, of projecting social values onto nature). According to both, philosopher and sociologist, "[n]othing new has happened. Nature (or society, or theory, or x) has simply been *revealed* at the occasion of an experiment" (Latour 1990, 65).

Latour acknowledges that these are successful explanatory strategies, indeed that, of the two, the philosopher's may well be an inference to the better explanation. And yet he recommends an entirely different perspective on experimentation: *"an experiment is an event."* Novelty arises, experiences are made, new qualities are acquired in historical events. In the course of experimentation Pasteur becomes a successful scientist whose theoretical predictions are borne out by rigorous tests; yeast proves that it distinguishes left and right tartrate acid; the Academy gains new techniques, new questions, new textbook knowledge. Latour's various human and nonhuman actors *"gain* their definition through the very trials of the experiment": "It is always admitted that science grows through experiment: the point is that Pasteur is also modified and grows through the experiment, as does the Academy, and, yes, as does the yeast. They *all* leave their meeting in a different state than which they went into. In other words, there is a history of science, not only of scientists and there is a history of things, not only of science" (Latour 1990, 66).

So far, Latour has not dismissed by argument the philosophical appeal to nature "out there" that allows us to speak of experimentation as a mere discovery or "the actualization of a potentiality." He simply points out that "the list of inputs does not *have to be* completed by drawing on any stock resource" like nature "out there" (1990, 65–66, emphasis added) and then offers his alternative account as at least equally plausible.

Latour's largely implicit argument consists of the recognition that the standard view of experimentation as mere discovery in a zero-sum game is less parsimonious and involves greater metaphysical commitments than his historicized alternative.

In 1775 Immanuel Kant calls for a natural history that is true to its name and gives a scientific account of physical developments or evolutionary processes, one that "transforms the currently so multifarious scholastic system for a description of nature into a physical system for the understanding" (1977, A7). In particular, he wants to understand the evolution of the various races, noting the admirable *prudence* [*Fürsorge*] of nature, which "equips its creatures for their preservation with hidden inner provisions for all kinds of future situations [Vorkehrungen auf allerlei künftige Umstände]":

During the migration and resettlement of animals and plants [this prudence] brings forth what appear to be [dem Scheine nach] new species but which are nothing but variations and races of the same species whose germs and natural dispositions have only occasionally developed in different ways over long periods of time. Chance or general mechanical laws cannot bring about such adaptations. We must therefore consider such occasional unfoldings [Auswickelungen] as *preformed*. (1977, A6–A7)

In the absence of mechanical law, Kant suggests, we can still account for the apparent development of novelty within a physical system for the understanding: there is no real novelty at all and nature does not change; apparent change consists of the unfolding of the preformed or the actualization of natural disposition, a mere transformation (from latent to manifest) of a physical state.

The philosopher's ahistorical account of experimentation ("nothing new has happened; nature has simply been revealed") thus mirrors the scientist's ahistorical explanation of irreversible (nonmechanical) physical processes, such as the dissolution of sugar or the unfolding of preformed adaptations ("nothing new has happened; a disposition has merely been realized"). All these accounts multiply properties well beyond necessity, attributing to nature dispositions to behave under any or all circumstances, some yet to come, many never to be realized.

Familiar critiques of dispositional explanation suggest that one must populate the world with hypothetical properties in order to regard physical changes as manifestations of latent states (see Tuomela 1978). Like Peirce before him, Latour shows that one need not incur this metaphysical burden.[18] Moreover, in light of our productive and creative exertions in the laboratory, it seems counterintuitive if not counterfactual to regard experimentation as anything but a constructive engagement with nature.

Given that one need not draw on the dispositional stock resource of an unchangeable nature "out there" and might choose otherwise, it becomes a characteristic of Lavoisian modern science (see page 161 and note 9 above) that it *does* choose to view experimentation as a zero-sum game, that it *does* explain all physical change ahistorically in terms of (irreversible) transitions between contiguous states, that it *does* assume a considerable metaphysical burden (the hypothesis of reality) in order to maintain its ahistorical orientation. Latour's argument accentuates this characteristic; he is right to ask why science would shoulder such a burden, and thereby sets an agenda for a philosophical theory of science. The metaphysical burden of Lavoisian science is justified by its ability to transform parochial objects of nature into exemplary objects of science, to render theoretical propositions decidable within an ahistorical space of reasons.

If experiments are not zero-sum games, they contravene Lavoisier's axiom of the conservation of mass. Lavoisier himself does not consider as literally and empirically true the incontestable axiom of conservation of weight. And yet "[u]pon this principle the whole art of performing chemical experiments depends." It constitutes Lavoisier's community of opinion; it provides a criterion and procedure for judging chemical explanations adequate and thus goes a long way toward making scientific consensus and objective knowledge possible.[19]

Nothing is created in nature or in the course of a chemical process, and nothing is created in art or in the design and performance of an experiment. Lavoisier's principle thus encompasses the objective and subjective aspects of science's conception of history. The assumption that nature is unchanging issues in objectifying procedures for capturing, weighing, and defining objects of science. And the assumption that the experiment is not a creative event but uncovers the facts separates the context of justification from the context of discovery: the adequacy of a scientific claim can be judged objectively and without reference to the scientists who perform the experiment. Lavoisier's principle successfully separates history and science. Any attempt to historicize Lavoisian science effectively undoes it—which is precisely why modern science justifiably resists constructivist deconstructions of its assumptions.

Latour notes that "[i]t is always admitted that science grows through experiment." He insinuates that one should willingly move on and concede that the scientists, the Academy, and nature itself also grow through experiments. After all, they all undergo experimental trials together, and they all take on new qualities.

It is a characteristic of Lavoisian science that it resists this move. To understand this characteristic is to understand why science chooses to limit itself explicitly against the history of nature and its own history, why it must refuse to know itself.

NOTES

1. As its epistemological correlate, the separability assumption safeguards and maintains the ontological commitment to an ahistorical nature. It does not presuppose this commitment, however. One may well imagine a non-nomological, historical science (e.g., History) that posits for its own reasons the separability between contexts of discovery and justification.

2. Giora Hon's contribution to this volume (Chapter 17) explores just this tension: experimenters encounter the cell as a living organism, but they must fix and observe it in particular states. Science cannot know the cell as a historical being or living organism.

3. Peirce is frustrated that his scientific epistemology goes unrecognized by science itself, that is, that science tends to underwrite the nominalist tendencies of what Peirce considers bad philosophy (see Peirce 1992, 104).

4. Marginal note by Kant to his *Lectures on Physical Geography*, quoted by Lyon and Sloan (1981, 27).

5. The history of evolutionary theory does not agree therefore with Toulmin and Goodfield's all too simple and confident declaration that "the picture of the natural world we all take for granted today has one remarkable feature . . . : it is a *historical* picture" (1965, 17). In contrast, it agrees with Lepenies's somewhat cryptic but far more fruitful suggestion that the end of natural history inaugurated an interplay of temporalization (*Verzeitlichung*) and dehistoricization (*Enthistorisierung*) (Lepenies 1976).

6. Note, however, that Levins and Lewontin dedicate their book to "Frederick Engels, who got it wrong a lot of the time but who got it right where it counted."

7. Peirce's "law of mind" characterizes the work of the irreversible principle of habit-taking, which, once evolved through random sporting, cannot but re-enforce itself, leading to the consolidation of habit and law, gradually crystallizing into a completely articulated reality that consists only of rigid relations of matter, that is, of effete mind. Limits of time and space prevent me from discussing John Barrow's intriguing inversion of the argument presented here. While nature and its constants may have a history, Barrow claims, the particularities of our universe make science possible and lock it into an ahistorical conception of the world: "Science exists only because there are limits to what Nature permits. The laws of Nature and the unchanging 'constants' of Nature define the borders that distinguish our Universe from a host of other conceivable worlds where all things are possible" (1998, 248, see 188).

8. Prigogine and Stengers's remarks on cause and effect, organism and environment, science and society resonate strikingly with those of Levins and Lewontin. Indeed, one might imagine Prigogine and Stengers attempting to close the theoretical gap identified by Levins and Lewontin. The latter would reject that attempt, however, precisely because it fails in its ambition to historicize nature: just as time cannot be equated with history, randomness should not be identified as contingency. (Incidentally, according to Prigogine and Stengers, macroscopic organisms are implicated in far-from-equilibrium systems only to the extent that they can scientifically observe them—which excludes most of organic nature from irreversible processes.)

9. The term "Lavoisian science" appears in the context of Gaston Bachelard's criticism of Meyerson. Bachelard credits Georges Matisse with the introduction of the term (Bachelard 1968, 54).

10. See also Bachelard: "With regard to substances in particular, conditions of stability having been given value, it was believed that structural conditions decided everything, the idea being, no doubt, that one masters time when one is well

organized in space, with the result that all the temporal aspects of chemical phenomena came to be neglected. There was no appreciation of the fact that time was
itself structured; no pains were taken to study rates, unfoldings, operations,
transformations—along these lines, therefore, there is new knowledge to be
gained" (1968, 72).

11. Although Lavoisier defines the term *element* operationally as that which
affords no further decomposition, his system of nomenclature turns each element
into a substance *sui generis,* a substance with an indestructible and immutable
nature (see Jeremy Bentham quoted by Beretta 1993, 326). The interactions of
these substances obey the incontestable axiom of conservation of weight. For
Bachelard (1968), "Lavoisian" science is defined by the intimate connection between its "substantialism" and its reliance on principles of conservation.

12. Accordingly, Frederick Holmes (1989) points out that Lavoisier changed
chemistry as an investigative enterprise by developing chemical apparatus for
increasingly specific theoretical purposes. Philip Kitcher (1993, 272–90) provides
a particularly compelling reconstruction of the steps leading to the gun barrel
experiment.

13. Cf. Bachelard (1968, 57) on absorption of light in regard to Lavoisian
chemistry. For his discussion of temporal aspects, see pages 58–59 and 71, and
also note 10 above.

14. The exemplary series of experiments on metallification or calcination testifies to the workings of phlogiston and indicates that other chemical problems can
be solved in an analogous manner, consistent with various assumptions about
phlogistic processes and principles. However, this implicit injunction to analogy
and consistency pales against Lavoisier's exemplary analysis and synthesis of water, which enjoins chemists to attend to composition and decomposition, to the
conservation of weight, allowing not only for the construction of possible solutions to chemical problems, but also for the immediate and univocal recognition of
correct solutions.

15. Bensaude-Vincent (1992) also points out "that Lavoisier was not attempting to demonstrate the principle of conservation" and that its "heuristic virtues
. . . should not blind us to its limiting characteristics": "In translating the conservation of elements' qualities into quantitative and ponderal terms, the principle
dodges the question of how—in what form—the elements are conserved. How do
they subsist in compounds and how do they move from one compound to another
during a reaction? Leaving this an occult question was the price Lavoisier paid for
being able to redefine elements as non-decomposable substances" (224–25). See
Lavoisier (1952 [1789], 10) on caloric and pages 57 and 62 for his reflections on
the problems posed by light and caloric, including the defining question by which
he sets his chemistry apart from phlogistic chemistry: "Are the heat and light
which are disengaged during the different species of combustion furnished by the
burning body or by oxygen which combines in all these operations?" Finally, the
"incontestable axiom" does not appear anywhere near the top of his systematic
and hierarchical exposition of the new chemistry. Instead, it is introduced in a
chapter on the decomposition of vegetable oxides, that is, when it guides and
justifies a particular course of experimentation.

16. See Schelling, Hegel, Peirce or Bachelard, Latour, Pickering.

17. Since Latour's writing tends to be somewhat cryptic, the following is an extended paraphrase, considering its implications only as far as they concern the present argument.

18. Latour (1996, 1999) revisits the claims of his 1990 paper and the dispositional account of discovery (which is "carried along by the 'propensity of things'"). Following Deleuze (and Whitehead), he distinguishes between the realization of a disposition and the actualization of a potentiality: experimental interventions serve as initiating events in the history of a thing and that of the experimentalist. However, Peirce's objective idealism renders unnecessary the precarious Deleuzian distinction and all reference to antecedent states, dispositions, or potentialities.

19. The notion of an unchangeable nature, in which all possible and actual behaviors and events—past, present, and future—already lie ready at hand, offers the tremendous opportunity to conceive all theory as simple description, which pictures or corresponds to a fixed reality; this idea of correspondence yields transparent criteria of objective knowledge or truth (Nordmann 1999). Heidegger recognized this is in his "Question Concerning Technology": "Modern science's way of thinking summons nature as a calculable coherence of forces. Modern physics is experimental physics not because it applies apparatus to the questioning of nature. The reverse holds true. Because physics—even taken as pure theory—summons nature to present itself as a pre-calculable coherence of forces, it therefore summons the experiment to question whether and how nature has presented itself upon the summons" (my translation; see Heidegger 1977, 21).

REFERENCES

Bachelard, G. 1968. *The Philosophy of No: A Philosophy of the New Scientific Mind.* New York: Orion.

Barrow, J. D. 1998. *Impossibility: The Limits of Science and the Science of Limits.* Oxford: Oxford University Press.

Bensaude-Vincent, B. 1992. "The Balance: Between Chemistry and Politics." *The Eighteenth Century* 33(2): 217–37.

Beretta, M. 1993. *The Enlightenment of Matter: The Definition of Chemistry from Agricola to Lavoisier.* Canton, Mass.: Science History Publications.

Böhme, H., and W. Schäfer. 1983. "Towards a Social Science of Nature." In W. Schäfer, ed., *Finalization in Science: The Social Orientation of Scientific Progress.* Dordrecht: Reidel, 251–69.

Brent, J. 1993. *Charles Sanders Peirce: A Life.* Bloomington: Indiana University Press.

Donovan, A. 1993. *Antoine Lavoisier: Science, Administration, and Revolution.* Oxford: Blackwell.

Engelhardt, D. von. 1979. *Historisches Bewußtsein in der Naturwissenschaft: Von der Aufklärung bis zum Positivismus.* Freiburg: Alber.

Feyerabend, P. 1989. "Realism and the Historicity of Knowledge." *Journal of Philosophy* 86: 393–406.

Fitzpatrick, M. 1987. "Joseph Priestley and the Millenium." In R.G.W. Anderson and C. Lawrence, eds., *Science, Medicine, and Dissent: Joseph Priestley.* London: Wellcome Trust/Science Museum, 29–37.

Foucault, M. 1975. *The Birth of the Clinic.* New York: Vintage.

Fruchtman, J. 1981. "Politics and the Apocalypse: The Republic and the Millennium in Late-Eighteenth-Century Culture." *Studies in Eighteenth-Century Culture* 10: 153–64.

Garrett, C. 1973. "Joseph Priestley, the Millennium and the French Revolution." *Journal of the History of Ideas* 34: 51–66.

Gibbs, F. W. 1967. *Joseph Priestley: Revolutions of the Eighteenth Century.* Garden City, N.Y.: Doubleday.

Gould, S. J., and R. Lewontin. 1979. "The Spandrels of San Marco and the Panglossian Paradigm: A Critique of the Adaptationist Programme." *Proceedings of the Royal Society of London, Series B: Biological Sciences* 205: 581–98.

Heidegger, M. 1977. *The Question Concerning Technology and Other Essays.* New York: Garland.

Holmes, F. L. 1989. *Eighteenth Century Chemistry as an Investigative Enterprise.* Berkeley: Office of History of Science and Technology, University of California.

Hume, D. 1955. *An Inquiry Concerning Human Understanding.* New York: Bobbs-Merrill.

Kant, I. 1977. "Von den verschiedenen Rassen der Menschen." In W. Weischedel, ed., *Werkausgabe,* vol. 11. Frankfurt: Suhrkamp, 11–30.

Kitcher, P. 1993. *The Advancement of Science.* New York: Oxford University Press.

Latour, B. 1990. "The Force and the Reason of Experiment." In H. Le Grand, ed., *Experimental Inquiries.* Dordrecht: Kluwer, 49–80.

———. 1996. "Do Scientific Objects Have a History?" *Common Knowledge* 5(1): 76–91.

———. 1999. *Pandora's Hope: Essays on the Reality of Science Studies.* Cambridge, Mass.: Harvard University Press.

Lavoisier, A. 1952 [1789]. *Elements of Chemistry.* In *Great Books of the Western World,* vol. 45. Chicago: Encyclopaedia Britannica, 1–159.

Lepenies, W. 1976. *Das Ende der Naturgeschichte: Wandel kultureller Selbstverständlichkeiten in den Wissenschaften des 18. und. 19. Jahrhunderts.* Munich: Hanser.

Levins, R., and R. Lewontin. 1985. *The Dialectical Biologist.* Cambridge, Mass.: Harvard University Press.

Lewontin, R. 1995. "Genes, Environment, and Organisms." In R. B. Silvers, ed., *Hidden Histories of Science.* New York: New York Review of Books, 115–39.

Lyon, J., and P. Sloan. 1981. *From Natural History to the History of Nature: Readings from Buffon and His Critics.* Notre Dame, Ind.: University of Notre Dame Press.

Merton, R. K. 1965. *On the Shoulders of Giants: A Shandean Postscript.* New York: Harcourt, Brace, Jovanovich.

Meyerson, E. 1930. *Identity and Reality.* London: Allen & Unwin.

———. 1991. *Explanation in the Sciences.* Dordrecht: Kluwer.

Nordmann, A. 1994. "Der Wissenschaftler als Medium der Natur." In J. Zimmermann and K. Orchard, eds., *Die Erfindung der Natur: Max Ernst, Paul Klee, Wols und das surreale Universum*. Freiburg: Rombach, 60–66.

———. 1999. "The Limited World of Science: A *Tractarian* Account of Objective Knowledge." Unpublished manuscript.

Pape, H. 1991. "Einleitung." In Charles Peirce, *Naturordnung und Zeichenprozeß*. Frankfurt: Suhrkamp, 11–109.

Peirce, C. S. 1992. *The Essential Peirce*, vol. 1: *1867–1893*. Nathan Houser and Christian Kloesel, eds. Bloomington: Indiana University Press.

———. 1997. *Pragmatism as a Principle and Method of Right Thinking: The 1903 Harvard Lectures on Pragmatism*. Patricia Ann Turrisi, ed. Albany: State University of New York Press.

Perrin, C. E. 1986. "Lavoisier's Thoughts on Calcination and Combustion." *Isis* 77: 647–66.

Poirier, J.-P. 1996. *Lavoisier: Chemist, Biologist, Economist*. Philadelphia: University of Pennsylvania Press.

Priestley, J. 1775. *The History and Present State of Electricity*. 2 vols. London: C. Bathurst et al.

———. 1929 [1796]. *Considerations of the Doctrine of Phlogiston, and the Decomposition of Water*, W. Foster ed. Princeton, N.J.: Princeton University Press.

———. 1970 [1790]. *Experiments and Observations on Different Kinds of Air, and Other Branches of Natural Philosophy Connected with the Subject*. 3 vols. New York: Kraus Reprints.

———. 1973 [1775]. "Introductory Essays." In his *Hartley's Theory of the Human Mind, on the Principle of the Association of Ideas*. New York: Arno Press.

———. 1999 [1788]. *Lectures on History and General Policy*. The Theological and Miscellaneous Works of Joseph Priestley, vol. 24, J. T. Rutt, ed. Bristol, U.K.: Thoemmes.

Prigogine, I., and I. Stengers. 1984. *Order Out of Chaos*. Toronto: Bantam.

Reichenbach, H. 1961. *Experience and Prediction*. Chicago: University of Chicago Press.

Roberts, L. 1991. "A Word and the World: The Significance of Naming the Calorimeter," *Isis* 82: 199–222.

Rudwick, M. 1997. *Georges Cuvier, Fossil Bones, and Geological Catastrophes: New Translations and Interpretations of the Primary Texts*. Chicago: University of Chicago Press.

———. 1998. "The First Historical Science of Nature." History of Science Society Distinguished Lecture, October 24, Kansas City, Missouri.

Siegfried, R. 1988. "The Chemical Revolution in the History of Chemistry." *Osiris* (second series) 4: 34–50.

———. 1989. "Lavoisier and the Conservation of Weight Principle." *Bulletin for the History of Chemistry* 5: 18–24.

Toulmin, S., and J. Goodfield. 1965. *The Discovery of Time*. New York: Harper.

Tuomela, R., ed. 1978. *Dispositions*. Dordrecht: Reidel.

Wise, M. N. 1993. "Mediations: Enlightenment Balancing Acts, or the Technologies of Rationalism." In P. Horwich, ed. *World Changes: Thomas Kuhn and the Nature of Science*. Cambridge: MIT Press, 207–56.

Wittgenstein, Ludwig. 1981. *Tractatus Logico-Philosphicus*. London: Routledge & Kegan Paul.

——. 1993. *Philosophical Occasions, 1912–1951*. J. C. Klagge and A. Nordmann, eds. Indianapolis: Hackett.

11

On Kinds of Timelessness

Comment on Nordmann

Richard Raatzsch
Department of Philosophy, University of Leipzig

In his chapter Alfred Nordmann considers a great variety of problems. I will not comment on all of them; instead I restrict myself to what I take to be one main line of his argumentation, which can be summarized in three propositions:

1. Lavoisier's chemistry is—contrary to that of Priestley—ahistorical.

2. Lavoisier's exclusion of time from chemistry marks an important step in the direction of objectivity in science.

3. In Priestley's as well as in Lavoisier's case, views on the nature of chemical processes go hand in hand with certain metaphysical, ethical, and political views.

In my comment on Nordmann's stimulating chapter I rely on insights from Goethe. Because prejudices about the fruitfulness of his views for the philosophy of science still linger, I am grateful for the opportunity to offer some suggestions that might help correct these.

Lavoisier's Ahistorical Chemistry—Contrary to that of Priestley

Suppose Miller's old school class is having its fiftieth reunion. Everyone tells a bit about him- or herself. The man next to Miller introduces himself with, "Hi, I'm Smith." Miller interrupts him with the words, "This cannot be true; you have totally changed!" To this, Smith retorts: "What *I* say is true, but what *you* say cannot be true, since, if I have totally changed, I am no longer the one I used to be. But if I am not the one I used to be, then I cannot be the one of whom you say 'he has totally changed.' So what you say is wrong. But if I were the one of whom you say 'he has totally changed,' then I would be the same as I used to be. But if I am the same as I used to be, then I cannot have totally changed. So what you say is wrong again. Since there are only these two possibilities, what you say *cannot* be true. Therefore, the contrary must be true—namely, that I have not changed at all—and this entitles me to say, 'Hi, I'm Smith.'"

How can Miller respond here? I propose that he says, "Miller, you have become a philosopher!" If Miller were to agree to this, then it would be easier for him to recognize in what sense there might be a problem in Nordmann's approach, according to which science "treats nature as if it has no history, as if it has always been there already, standing ready to be discovered and understood," so that nature remains that which persists through change rather than being subject to change itself.

I try to follow Smith when commenting on this in the following way: "Let us take a closer look at the process you are calling 'change of nature itself.' Let us call nature at the beginning of this process 'first nature' and nature at the end of it 'second nature.' Now there are two possibilities. Either first and second nature have something in common or they do not. If they have something in common, then this may serve as a justification for our calling both beginning and end by similar names, just as we are allowed to call both the first and the last man in the row 'man in the row' because of their having something in common, namely standing in the very same row. Now we can also answer the question '*What* is that thing at the beginning and at the end of this process called "change of nature itself"?' Well, it is that which we find both at the beginning and at the end of this process, something that persists through change. But if first and second nature have nothing in common, how can we justify our calling both by similar names? Here we could also give them totally different names, say 'Iggydiggy' and 'Abadaba.' But what are we supposed to say when confronted with the question: '*What* is that thing at the beginning and at the end of this process called "change of nature itself"?' Are we supposed to say 'Iggydiggy' or rather 'Abadaba'? However we decide, the problem reappears, now starting with, say, 'Iggydiggy,' having a break in 'first Iggydiggy,' and ending in—well, choose yourself a new word."

It seems as if we are only allowed to say that if nature changes itself, then it stays the same instead of being itself subject to change. In short: "Whenever something changes itself, then it remains as it is." But if something stays the same, then it does not change. On first hearing, that statement *sounds* at least as paradoxical as the following proposition from Nordmann: "Darwinian biology proves to be anything but a historical or historicized science. Although it treats of the origin and transmutation of species, it succeeded as a science because it posits the uniform operation at all time of an evolutionary mechanism."

So is science paradoxical? I think it *sometimes* is—but only when and as long as it involves itself with philosophy! When does this normally happen? When (a certain discipline of) science is either in the state of coming into existence or, as many philosophers of science since Thomas S. Kuhn have termed it, in a "revolutionary situation," when it undergoes "revolutionary changes." And these last until the new "paradigm" is found, that is, as long as the procedures of searching for one last. The moment the (new) way of looking at things is found, or the decision to accept one known way as the *standard* way (or *norm*) is made, the paradox is gone—since it is no longer a question of how things are, but a frame for all possible questions about how things are, and here the frame itself is no longer in question. It is exactly this which happens at the point quoted by Nordmann from Lavoisier's "Elements of Chemistry": "every branch of physical science must consist of three things: the series of facts which are the objects of the science, the ideas which represent these facts, and the words by which these ideas are expressed. Like three impressions of the same seal, the word ought to produce the idea, and the idea to be a picture of the fact" (Lavoisier 1964, xiv).

This idea is of fundamental importance for Lavoisier. He refers to it when pointing out and justifying that which is *new* about his chemistry. It sounds a bit like an excuse when he writes that at the beginning his only objective was to write on the "necessity to reforming and completing the Nomenclature of Chemistry":

[But while] engaged in this employment, I perceived, better than I had ever done before, the justice of the following maxims of the Abbé de Condillac, in his System of Logic, and some other of his works. "We think only through the medium of words.—Languages are true analytical methods. . . . —The art of reasoning is nothing more than a language well arranged." . . . Thus, while I thought myself employed only in forming a Nomenclature, . . . my work transformed itself by degrees, without my being able to prevent it, into a treatise upon the Elements of Chemistry. (Lavoisier 1964, xiii–xv)

When reading this, one gets the impression that Newton is only half right in his famous warning that physics should stay away from metaphysics. He is right insofar as the physicist who engages with metaphysics is destined to produce paradoxical statements, at least to some extent. But if the scientist is not willing to engage with metaphysics he will hardly get a theory. The extent to which this insight is itself a *philosophical* insight is exactly the extent to which Newton's warning itself (ironically) turns out to be of a *metaphysical, or philosophical,* nature—as well as Lavoisier's (nonironical) quoting of Condillac's remarks on the relation between facts, ideas, and nomenclature.

Giving Lavoisier's remarks a name can be only a first step, although not an unimportant one. Giving a name to something is a part or a result of classifying it, and different classifications go hand in hand with different questions and answers. So let us have a closer (and philosophical) look at Lavoisier's remarks. Again: ideas, facts, and words *ought* to be like three impressions of the same seal. When can we say that they *really* are? Suppose two people are trying to describe a game of chess using only symbolic notation. But one of them does not know the sign for the bishop—let us say he has forgotten what it looks like. Here we can say that the language of one of the two observers fits the world better than the language of the other. This is a case in which we can justify a proposal for a new or extended language, or a reminder respectively, with reference to the facts.

But the difference between Priestley and Lavoisier can hardly be of this kind. It is not that Lavoisier sees something that Priestley also sees, without having a word for it. If we want to talk about "seeing" in this case at all, then we can perhaps say that Lavoisier sees the same things as Priestley, but Lavoisier sees them differently, in a new way. This new way of looking goes hand in hand with the introduction of a new language. But *what* is seen here, now in one way and then in another? Here temptation says: What it really is that we both see, we will perhaps never know; but we can come closer and closer to this knowledge. That is, our way(s) of looking at things is (are) going to be more and more *objective.* This leads us to Nordmann's second main clause.

Lavoisier's Exclusion of Time from Chemistry: An Important Step toward Objectivity in Science

Let us first consider time. "Suppose," Miller says to Smith, "I am pouring 2 liters of this fluid into this measuring jug and then 2 liters of that fluid, so that there will be 4 liters in the jug, because 2 and 2 = 4." No sooner said than done. Yet the measuring jug does not show 4 liters, only 3. What now? Well, Smith says that Miller made a prediction that turned out to be wrong. But suppose Miller answers, "Are you going to debate that 2 + 2 = 4?" Of course, Smith retorts, "Your proposition '2 + 2 = 4' is not wrong, but your proposition '2 liters of this substance and 2 liters of that one result in a 4-liter mixture' is. You are confusing a *mathematical* with an *empirical* proposition. The truth of the mathematical proposition does not depend on how things are or are not, only that of the empirical one does. The mathematical proposition is *timeless;* the empirical one is *temporal.* When someone says, 'If one pours 2 liters of this fluid and 2 of that fluid together, one will get a 4-liter mixture, *and* I want to call '4 liters' whatever volume the measuring jug reveals,' he is not making a prediction. You did not make a *mathematical* discovery—that 2 + 2 is sometimes 3—but an *empirical* one—that some fluids result in less volume when mixed than when their individual volumes are added. And this is something you can only discover because what '4 liters' means and the result of 2 + 2 are fixed. The timeless proposition makes the temporal possible."

But let us suppose that there are *always* 3 liters in the measuring jug, no matter which fluids we use. Here the philosopher will perhaps say that this indicates a characteristic feature of all fluids compared with solid bodies. But what are we supposed to say if it turns out that *whenever* we put 2 solid bodies in an empty room and then 2 further ones, there are only 3 when we count them again? Shall we say that sometimes solid bodies disappear or amalgamate, although we cannot find any cause for this?

In a real case we would probably say that we must have miscounted and count again, several times. At some point, we would file this kind of case under "wonders" and let it rest. We would get on with things again—if there were still something to get on with *in the usual way.* For if that which is now the rule became the exception and that which is now the exception became the rule, we would not know what to do. In this sense, then, can we indeed say also that the timeless propositions of mathematics depend on what is going on in time? But until this happens, $2 + 2 = 4$ is fixed for us. That is the ground on which we move (see Raatzsch 1995). And this too sheds some light on the principle of conservation of mass. We say, "If you measured a weight after the reaction (say a burning) different from what it was before, you must have made a mistake. The mass stays the same, whatever happens in time." And so it seems as if the axioms of chemistry are timeless truths about the world—almost like the truths of mathematics. But if we remember now that in some sense even the timeless propositions of mathematics depend on how the world is now, then it is easier for us to see how fruitful might be Wittgenstein's observation in §104 of his *Philosophical Investigations*, according to which in philosophy we "predicate of the thing what lies in the method of representing it" (1953, 46e). And of Lavoisier he writes, in §167 of *On Certainty*,

> It is clear that our empirical propositions do not all have the same status, since one can lay down such a proposition and turn it from an empirical proposition into a norm of description.
>
> Think of chemical investigations. Lavoisier makes experiments with substances in his laboratory and now he concludes that this and that takes place when there is burning. He does not say that it might happen otherwise another time. He has got hold of a definite world-picture—not of course one that he invented: he learned it as a child. I say world-picture and not hypothesis, because it is the matter-of-course foundation for his research and as such also goes unmentioned. (1969, 24e)

Here, I think, lies the connection with Nordmann's first main clause. With regard to this clause, the problem is that in those cases in which it is clear what it means to adjust a language to fit the facts, we are able to compare both language and facts. But what does it mean to adjust a language to fit facts we do not know? There is no possibility of speaking about a "comparison" here in the ordinary way. But that is precisely what is necessary if we want to be justified in agreeing with Lavoisier's words in the way they are written. In the light of what we have just seen, we need not say that what Lavoisier says is wrong. Instead, we can say that Lavoisier is giving a philosophical (or, in Newton's terminology, *metaphysical*) justification for a new way of looking at chemical phenomena and for a new language of the science of chemistry, a justification he takes to be on the same level as that which can be represented with it (the facts). But we have to draw a distinction between a method of presentation—or, as Nordmann puts it, a "space of reasons"—and the matter presented. This new "method of presentation" is no more objective in the sense that it fits better what is presented—the facts—than was the old method. So then is the choice purely arbitrary? No doubt, we do not want to say this either. So what *should* we say?

When *do* we say that something is objective or subjective? There are different cases. One case is particularly interesting for us. We say something is subjective when everyone can say what he or she wants (and defend it) without getting into conflict with the opinions and views of others. Questions that everyone may answer as he or she wishes concern, for instance, certain matters of taste (not all, of course, as Burke taught us). If Priestley says, "Ale is the best substance there is," then Lavoisier has no reason to be troubled; he can answer, quite coolly, "Well, Priestley, ale is the greatest stuff there is *for you; for me* it is of course cognac." But if Priestley says that 2 + 2 = 3, then Lavoisier cannot say: "Well, just for you, not for me." Perhaps he would follow what Gottlob Frege says in a similar case in his *Grundgesetze der Arithmetik* and respond, "Priestley, you must suffer from a new kind of madness" (see Frege 1893, xvi).

Asking ourselves whether Lavoisier's revolution comes closer to the first or the second case, we will undoubtedly say that it is closer to the second. For it is surely not such a matter of taste. In this respect, it makes a lot of sense when Nordmann says that Lavoisier imposes an ahistorical mathematical ideal on the experimental tradition in physical science, whereas Priestley speaks for the experimental or Baconian tradition, because this way of putting the matter emphasizes the *nonarbitrariness*, the nomological or even necessary character, of the theoretical propositions of modern chemistry. But if we use this kind of conceptual framework regarding "subjective" and "objective," then Priestley (and with him the whole Baconian experimental tradition) automatically becomes someone who takes the investigation of, and judgment on, chemical experiments and phenomena to be matters of taste. Although one might make *some* sense out of this classification, this description would go way too far. For here Priestley starts looking like a madman. And of course, as Nordmann has convincingly shown, this does not do justice to Priestley. Here it may be appropriate to start looking at the last of Nordmann's main clauses.

Priestley's Turn

The great Michael Faraday had an assistant in his laboratory, a Mr. Anderson, a former artillery sergeant. He used to say, "I am doing the experiments, and Faraday is doing the talk about them" (translator's remark, Faraday 1859, 26).

There is much more in this aperçu than one might think. As Nordmann, together with Davis Baird, has shown elsewhere, experiments might not only be seen from the point of view of giving answers to theoretical questions (see Baird and Nordmann 1994). Some experiments simply impress us by resulting in "striking phenomena." Nature does not always need to be subjected to an experiment in order to offer us striking phenomena. Earthquakes, firestorms, thunder and lightning, giant squids, obnoxious creatures—we get all this and much more with no delivery charge. These are phenomena that are out of the ordinary. They surprise us and arouse our curiosity. This is true of even the duller members of our own human race. But for the more sensible ones, nature in its entirety is full of "striking phenomena": think of the sounds of the wind or the birds; look at the colors of the flowers in spring and of the leaves in autumn; remember how cold, warm, and hot water feels; and so on. In short: being a "striking phenomenon" is a function with two variables: the phenomenon and the one it is going to strike. Now are there only "striking phenomena" in those cases in which we have a *theory,* and is there only one way to satisfy our curiosity—the way Faraday does it, by searching for a causal explanation within a scientific theory that allows us, so to speak, to systematically reconstruct the phenomenon? No doubt, this is *one* way of satisfying our curiosity. And Lavoisier, Faraday, and many others do this not only in a fascinating manner but in a way that allows us to produce deliberately the phenomena we want. But in this way the phenomena, which are subject to scientific, causal investigations—the facts *as such*—become *uninteresting* in another sense. As Lavoisier puts it: "Among the facts that chemists discover are many that have little significance until they are connected together and assembled into a system" (Lavoisier in a manuscript quoted in Perrin 1986, 665).

Opposed to this is what Priestley, according to Nordmann, demands of the naturalist: "As a modest witness to the operations of nature, Priestley's job is simply to wait and see, to produce striking phenomena." Looking at the facts or phenomena from Lavoisier's point of view, it seems quite reasonable even to "withhold facts of this sort and ponder in the quiet of their studies how to connect them to other facts, announcing them to the public only when they have been pulled together as far as possible"—whereas Priestley takes it as his duty to publicize immediately each and every new experiment and phenomenon he observes. Taken by themselves, both stances are merely consistent. But if we want to say that Lavoisier represents the *new* chemistry, should we not also say that Priestley's ideal of "speedy publication" is *outdated*? Isn't it as if Lavoisier freed chemistry from the status of being a "matter of taste" discipline, designed to satisfy our appetite for "striking phenomena"?

If we regard Priestley merely as someone who preceded Lavoisier, then perhaps we have to say this. But if one tends—as Mr. Anderson, Alfred Nordmann, and others do—to accord at least to some experiments a greater independence vis-à-vis the theory, then one will perhaps be more cautious about Priestley's attitudes. (Some chemists and philosophers of chemistry already are quite cautious, arguing that some developments in chemistry *after* Lavoisier partially rehabilitate some of Priestley's ideas [see Psarros et al. 1996].) There is, it seems to me, a kind of looking at Priestley which puts him in a better light *whatever developments in chemistry brought or will bring about*. This way of looking at Priestley takes him to represent a way of dealing with natural and social phenomena that is in between Lavoisier's and a totally different way. We find such a way in Goethe's works. In his *Maximen und Reflexionen,* we read: "Just don't search for anything behind the phenomena: they themselves are the doctrine. . . . No phenomenon explains itself in and from itself; only many of them, grasped together, ordered methodically, in the end yield something that might count as theory" (Goethe 1998, 432, 434, my translation; see also Schulte 1990; Rehbock 1995; Vendler 1995).

Compare this remark with that of Lavoisier according to which there are many facts among those discovered by chemists that "have little significance until they are connected together and assembled into a system." What Lavoisier means is, of course, that a phenomenon has significance to the extent that it is incorporated into the theory of chemistry. To put the same matter differently, a phenomenon has significance to the extent that its incorporation into chemistry preserves its character as a system of knowledge. In still other words, whatever nature might say, it has only significance to the extent that it is an answer to one of our scientific questions. What would, in contrast, be more natural than to describe Goethe's aim with the words in which Priestley's aim is phrased by Nordmann: the naturalist has to be a "transparent medium of nature" (Nordmann 1994). We have to listen to what nature says in order to find the questions to which what it says might be the answer. The point here is that it is not the theory, or the system of chemistry, that gives each phenomenon its weight, but rather the other way around: a theory—or system of knowledge—serves only as a way of presenting the phenomena.

What is problematic about this method becomes visible if one asks oneself: What is the criterion for the natural scientist's *indeed* having met the requirement of being a transparent medium of nature and not—with his own prejudices, views, and so on—interrupting nature's speaking through his mouth? Now, to follow the famous Wittgensteinian maxim "Don't think, but look!" need not result in the kind of agreement that is typical for mathematics and scientific chemistry. The one who does not see a resemblance need not make a mistake. In this regard, one can call Goethe's and Wittgenstein's method—"the morphological method"—a "subjective method" and label as "objective" a method that is characterized by general agreement. But here we are dealing with a totally different concept of "subjectivity" than the one sketched previously. A "morphological" judgment is no more a judgment of taste than the judgment of those with better knowledge of mankind, even if this is also characterized by less agreement. But contrary to knowledge of mankind, morphology is not designed primarily to allow *explanations* and *prognoses,* but to formulate *internal resemblances.*

Does time come in here anywhere? That there is an internal resemblance between two objects is independent of time insofar as it does not matter at what time the objects exist, if they do exist at all. There is some danger of confusion here. The question of whether there are morphological, or "formal," resemblances between two objects is different from the question of whether there is a certain morphological, "formal" structure in the flow of objects in time. With regard to the first question, it does not matter which object came first in time and which came last. With regard to the second question, we are asking for the *morphé* of time itself. Again, it does not make sense to ask whether this form of time has some features that are due to time, for that would merely be a duplication of the question. Therefore, this method might also be called "ahistorical."

The meaning of this ahistoricism cannot be better expressed than by these words from Søren Kierkegaard's *Journals and Papers:* "There is a most remarkable saying, I know not where, but one which bears the inward stamp of being the kind of utterance which, so to speak, is spoken with the mouth of a whole people. A desperate sinner wakes up in hell and cries out, 'What time is it?' The devil answers, 'Eternity'" (Kierkegaard 1970, 382, entry 831).[1]

For the devil and for God time plays no role. They will never be too late; only we will sometimes be. Our whole life is put on the Great Balance, and the Judgment tells us what its form was. Are you, in the end, a sinner, or is there a way to forgive, since we are all sinners in the first place? Once the word is said, the sense you made out of your life is fixed. Only then can you know what your destiny was. Before this day you might only have guessed, or you might have tried a morphological investigation, looking for the overall form of your life, that which makes up your destiny. What is the point of view according to which all your doings might be presented? What is their meaning? This area might be the *original* place for morphological questions. But they may very well be expanded into the field of nonhuman nature, trying to find out what its sense, or destiny, is. Yet this does not make the difference disappear: "Fate stands in contrast with natural law. One wants to find a foundation for natural law and to use it: not so with fate" (Wittgenstein 1980, 166e). That is, the morphological method is *not* a general alternative to a scientific, theoretical, explanatory method, useful and appropriate for any purpose. For some purposes it is better suited than other methods; for other purposes it is less appropriate.

To illuminate this point with the help of a related subject, I quote a lengthy passage from J. P. Eckermann's notes in *Gespräche mit Goethe in den letzten Jahren seines Lebens* (Talks with Goethe in the Last Years of His Life). For Sunday, June 20, 1831, we read the following:

We talked about some subjects in the natural sciences, especially about the imperfectness and insufficiency of language, by which errors and false views spread out that are later difficult to overcome. [Do you see the connection with Lavoisier?]

"The matter is simply the following," said Goethe. "All languages arise from close human needs, human activities, and general human feelings and views. Now, if a higher human being gains insight into the secret creating and producing of nature, his traditional language won't suffice to express such remote things, which are far away from human affairs. He would have to master the language of ghosts to express his peculiar perceptions. Since this is not so, he has to use human expressions for the perception of unusual natural relations; this way he will almost always fall too short, will pull down or even injure or destroy his subject. [Do you see the difference from Lavoisier?]

Goethe then goes on to give an example:

"Geoffroy de Saint-Hilaire is a man who really has great insight into the spiritual producing and creating of nature; yet his French language forsakes him to the extent that he is forced to use ordinary expressions. And this not only when it comes to secretly spiritual, but also quite visible, purely bodily objects and relations. If he wants to describe single parts of an organic being, he has no other word for them than *materials,* so that, for instance, the bones that make up as equal parts the organic whole of an arm are on *one and the same* level of expression with the stones, beams, and boards used for a house." "Just as improper," Goethe continued, "the French use the expression *composition* when talking about the products of nature. I might very well put together the single parts of a machine built piecemeal and talk about composition with regard to such an object, but not when I have in mind the single parts of an organic whole, lively constructing themselves and pervaded by a common spirit. . . . It is a totally mean word we owe to the French, and we should get rid of it as soon as possible. How can one say that Mozart has *composed* his *Don Juan?*—Composition—as if it were a piece of cake or biscuit mixed out of eggs, wheat, and sugar!—A spiritual creation it is, the single as well as the whole out of *one* spirit and casting, pervaded by the breath of *one* life, with the producer neither trying nor doing it arbitrarily, but being in the grip of his demonic spirit and having to fulfill what he ordered." (Eckermann 1982, 653–55, my translation)

What is remarkable in this passage is not only that it gives us a good impression of what Goethe has in mind when offering the organic, morphological view, but also that it gives us strong hints at his *paradigmatic cases*. They are *works of art*. The whole of nature is for him a work of art, with God as the artist. According to this view, the concept of nature being composed of parts—having an atomic structure, if you like—is not only inessential but also beside the point. What counts is the "spirit" of the creation—its "meaning" or "sense" as we would perhaps put it nowadays, since (normally) works of art express some sense or meaning—which in the case of nature as a whole might appear to us as its, and our, destiny.

Now, instead of taking the difference between Priestley and Lavoisier to be the difference between a historical and an ahistorical interpretation of nature, should we not better say that the difference between these two great men is the difference between two kinds of timeless views on nature, with one of these two views designed to *explain* and *predict* changes in time, and the other designed to tell us what the *sense* of these changes might be, what nature taken as a whole and in its parts might *mean* to us? Is it not the case that nature as a whole as well as nature in many of its parts—take horses and their beauty, think of those sounds called "music" and their irresistible power to make us move in "unnatural" ways, and (how could one forget them) all the paintings we love and that simply consist of shaped colors—is it not indeed the case that what science can tell us is by far not all that counts for us? "Vincent van Gogh loved the color yellow—and certainly not because of its wavelength" (Vendler 1995, 394). Therefore we should honor *both* Lavoisier *and* Priestley, as well as Newton *and* Goethe, although it might seem at first glance that they simply contradict one another. No doubt, they do contradict one another—but that is not the whole story. The fact that Priestley came first and paved the way for Lavoisier's modern chemistry might then be the less interesting thing to be reminded of about these two geniuses.

NOTES

I thank Mario von der Ruhr as well as Stan Theophilou for checking the English translation of this chapter.
1. I owe this reference to Mario von der Ruhr.

REFERENCES

Baird, D., and A. Nordmann. 1994. "Facts-Well-Put." *British Journal for the Philosophy of Science* 45: 37–77.
Eckermann, J. P. 1982. *Gespräche mit Goethe in den letzten Jahren seines Lebens.* Berlin: Aufbau Verlag.
Faraday, M. 1859. *Naturgeschichte einer Kerze.* G. Bugge, intro. and transl. Leipzig: Reclam.
Frege, G. 1893. *Grundgesetze der Arithmetik: Begriffsschriftlich abgeleitet,* vol. I. Jena: Pohle.
Goethe, J. W. 1998. *Maximen und Reflexionen.* Hamburger Ausgabe XII. Munich: DTV.

Kierkegaard, S. 1970. *Søren Kierkegaard's Journals and Papers,* vol. 1. H. V. Hong and E. H. Hong, eds. and transls. Bloomington: Indiana University Press.

Lavoisier, A. 1964. *Elements of Chemistry, in a New Systematic Order, Containing All the Modern Discoveries.* R. Kerr, transl. New York: Dover.

Nordmann, A. 1994. "Der Wissenschaftler als Medium der Natur." In J. Zimmermann and K. Orchard, eds., *Die Erfindung der Natur: Max Ernst, Paul Klee, Wols und das surreale Universum.* Freiburg: Breisgau, 60–66.

Perrin, C. E. 1986. "Lavoisier's Thoughts on Calcination and Combustion." *Isis* 77: 647–66.

Psarros, N., K. Ruthenberg, and J. Schammer, eds. 1996. *Philosophie der Chemie: Bestandsaufnahme und Ausblick.* Würzburg: Königshausen & Neumann.

Raatzsch, R. 1995. "Begriffsbildung und Naturtatsachen." In E. von Savigny and O. R. Scholz, eds., *Wittgenstein über die Seele.* Frankfurt: Suhrkamp, 268–80.

Rehbock, T. 1995. *Goethe und die Rettung der Phänomene: Philosophische Kritik des naturwissenschaftlichen Weltbildes am Beispiel der Farbenlehre.* Konstanz: Verlag am Hockgraben.

Schulte, J. 1990. "Chor und Gesetz: Zur 'morphologischen Methode' bei Goethe und Wittgenstein." In his *Chor und Gesetz: Wittgenstein im Kontext.* Frankfurt: Suhrkamp, 11–42.

Vendler, Z. 1995. "Goethe, Wittgenstein, and the Essence of Color." *The Monist* 78(4): 391–410.

Wittgenstein, L. 1953. *Philosophical Investigations.* G. E. M. Anscombe, transl. New York: Macmillan.

——. 1969. *On Certainty.* G. E. M. Anscombe and G. H. von Wright, eds.; G. E. M. Anscombe and D. Paul, transls. Oxford: Blackwell.

——. 1980. *Remarks on the Philosophy of Psychology,* part I. G. E. M. Anscombe and G. H. von Wright, eds.; G. E. M. Anscombe, transl. Oxford: Blackwell.

12

Metaphors and Theoretical Terms

Problems in Referring to the Mental

Hans Julius Schneider
Department of Philosophy, University of Potsdam

What is it that makes expressions for "mental events" meaningful in cases in which the objects referred to (such as neurophysiological processes or items in the "stream of consciousness") cannot be supplied? It is argued in this chapter that a construal of these expressions as theoretical terms for natural objects is misguided. Instead, they should be taken as metaphors of a special object-constituting kind. For them a cultural-linguistic practice is primary, and the "objects referred to" are secondary. As a consequence, a hermeneutic (as contrasted with a scientific) understanding of the corresponding field of psychology is called for, and in this sense a "limit of science" is revealed.

I begin with a few words about my point of departure. In his paper "On Mind and Matter," Georg Henrik von Wright (1994) discusses the traditionally so-called mind/body problem with the help of a simple example: a sound is heard, and we see how the bodily posture of a person changes in such a way that the body is directed toward the source of the sound. Depending on the circumstances obtaining, we may be confronted (to mention just two clear-cut possibilities) with a causal relation, for instance with a reflex: the sound caused the bodily movement. Or we may be confronted with an action, not just the behavior of a body. In this second case our description of what has happened will include the understanding the person has of her own activity: she was *interpreting* the noise in a certain way, for example as an indication that somebody had arrived whom she was *expecting*.

The turning of her body toward the source of the sound in this case (and under this description) is an intentional action, not just a bodily reflex.

Now the words *to interpret* and *to expect* seem to refer to mental processes or mental events that are taken to be present in the case of action, over and above the bodily movements. Moreover, the acting person is conscious of them. Their presence, so it seems, makes the description of what happened as an action possible.

Following von Wright (1971, 1974; cf. Schneider 1999), I will not apply the term *cause* to a mental event *as mental* (i.e., when it is not identified with a physiological process at the outset, the legitimacy of which step is one of the points at issue here). A satisfying justification of this conceptual and terminological step would have to be as detailed and as sophisticated as von Wright's discussion of these matters, including a discussion of the history of the concept of causality. But the basic idea behind it is that causing the movement of a billiard ball by making a first ball hit it and then explaining the resulting movement of the second one by appealing to the laws of mechanics is a completely different kind of experience (and a different kind of account of it) than asking a person for the time and then making intelligible the fact that one has received a particular answer. This difference remains intact, I claim, regardless of the possibility of genuine statistical causation (such as in quantum theory). The possibility of making the prediction that under certain circumstances the person asked will utter "5:30" with a certain measure of probability is no evidence for the claim that understanding and predicting human action is of the same type as explaining and statistically predicting quantum phenomena.

So the term *cause* (if applied in the human realm) is restricted here to phenomena such as physical stimuli, which can be produced at will. Such stimuli are said to be the "causes" of responses if and only if they elicit these responses. Since mental events such as "x's interpreting y as z" or "x's expecting p" cannot be produced at will and do not in turn produce fixed consequences, they are not causes in the sense adopted here.

Von Wright reminds us that the legitimacy of speaking of a causal relation in this strict sense does not depend on our knowledge of the complete chain of events on the neurophysiological level. According to his well-known "interventionist" theory of causality, it is necessary and at the same time sufficient that we have experimentally assured ourselves of our ability to manipulate the events on the macro level.[1] The legitimacy of calling something an *action* is normally established by giving a convincing explanation for its occurrence in terms of reasons and goals, that is, with reference to the mental realm. If this kind of explanation of a candidate for an action succeeds, we say that what has been observed was *not* a reflex, the sequence was not a *purely* causal chain of events. In doubtful cases, unsuccessful attempts to show by manipulative experiments that what appeared to be an action was "really" just a causal chain can secure the original interpretation as an action. Surely a person might voice the metaphysical belief that any given failure or failures at such an attempt do not disprove that ultimately all phenomena of action *must* yield to a causal description. Such a belief to me seems to be philosophically empty. But fortunately it is not required to make causal science rational: from a purely theoretical point of view (i.e., excluding ethical considerations) it is perfectly reasonable to try to extend the scope of successful causal manipulation to more and more events on the micro level. And this is what science is doing, without being committed to any metaphysical claim.

The subject matter of this chapter is the question of whether (and in what sense) the meaningful use of mental terms *requires* that there be objects referred to. The first candidates I discuss are objects that can be identified either by neurophysiology or by introspection. Is the existence of such objects necessary in all cases to make mental expressions meaningful? Concerning introspection, to be sure, I have not the least doubt that we sometimes dream or that solutions to problems come to our minds, and that we can refer to dreams and ideas with the help of words. So *some* mental terms do refer to happenings in the "stream of consciousness." But my question is the following: Are there *always* "objects referred to" when mental expressions are meaningful? And what do the terms *object* and *to refer to* mean here? In order to pursue these questions I will consider a group of words for which direct referents cannot easily be found, but which are at the same time not so unlike the referring expressions as to render the question of their reference pointless (for example: "to interpret *x* as *y*," "to expect somebody," "to weigh the evidence").[2]

As a possible way of interpreting these words, it has been proposed to regard them as *theoretical terms* in the sense familiar from the philosophy of science. Under this interpretation we assume that there are "natural" or physical entities to which they refer, even if we cannot (or cannot yet) observe these entities directly. Our knowledge in such cases is a knowledge by inference. I will try to cast some doubt on the adequacy of this interpretation.

As an alternative, I propose that we view the terms under discussion as object-constituting *metaphors*. The expression *object-constituting* is meant to indicate that the function of the metaphorical expression is not to add poetic ornamentation to a text that could also be formulated by using a nonmetaphorical expression in such a way that it would refer to the intended object directly and simply. Instead, the function of the metaphor is to open up new realms of objects.[3] How this kind of object-constitution is possible and what particular sense the expression *to refer to an object* acquires by such a step will be discussed with reference to a particular grammatical process that Wittgenstein called "grammatical fiction." It leads to "cultural" (as opposed to "natural") objects, which are not "known by inference" in the same sense as are the objects represented by theoretical terms.

If this is correct, the proper approach to the field of psychology to which the relevant expressions belong would not be that of science but that of hermeneutics, that is, the methodology developed by the *Geisteswissenschaften* to understand meaningful cultural products such as texts or social institutions. In this sense this chapter points to a *limit of science* (in the sense of *Naturwissenschaft*). It strengthens arguments that have been put forward to advocate a "cultural psychology" (Bruner 1990).

Does the Meaning of Mental Terms Depend on the Existence of Corresponding Physiological Processes?

In the history of research on neurophysiological processes we find (as we do in other scientific fields) speculations and models, some of which have turned out in later times to be fictions. These are then called "mere" speculations, mere models; they are false pictures of reality. As an example of such a case, von Wright (1994, 103) mentions that Descartes wrongly believed that nerves are a kind of blood vessel. But contemporary ideas might also turn out to involve fictions. For example, when a neurophysiological theory postulates certain modules or departments of the brain as the places at which certain processes take place that make possible specific achievements, it may turn out that these achievements have to be explained in a quite different way, without recourse to the postulated modules or processes. Something like this might have happened recently when representationalist models of the brain were criticized and supplemented by connectionist models; a postulated "representation module" of a certain kind might have turned out to have been a fiction, corresponding to no empirical reality.

This type of unsuccessful linguistic transfer is what von Wright has in mind when he argues as follows: he imagines that, on the basis of certain expressions we use for mental activities, somebody advances the hypothesis that on the level of neurophysiology there are processes that could be called "sieving" (for example, the sieving of good from bad reasons) or could be described as "something passing through a screen" (von Wright 1994, 104), or (to add an example of my own) as "weighing the evidence" (for instance, for arriving at a conclusion about the desirability of competing goals of action). Hypotheses of this kind are by no means absurd in principle. In advancing them, we take a familiar mental term and try to make a linguistic transfer from the mental to the neurophysiological realm; we attempt to formulate an empirically testable model of those bodily processes we suspect underlie the mental activities or accompany them in a characteristic way.

In spite of this acceptability in principle, it is clear that such attempts can go wrong. It can turn out that a particular model or one of its constituents has no empirical basis. In such a case, the physiological interpretation of words such as *sieving* or *weighing the evidence,* for example, would turn out to be impossible because there are no entities to which they refer. The processes they were meant to describe would have been shown to be "nothing but fictions." Von Wright (1994, 104) uses the word *invention* for what I have here called a fiction: a model of bodily processes that was based on the ordinary mental terms we use for explaining actions turned out to be vacuous; it involved inventions with no corresponding reality behind them.

I now raise the question of what a case such as this means for our original mental use of the terms concerned. Are we still entitled to speak of the "sieving" of good from bad reasons or of "weighing" the evidence in favor or against the desirability of one or another course of action, in spite of the failure of the attempt at a physiological interpretation of these terms? For the moment I merely note that von Wright does not express any doubts in this respect. He discusses cases in which these terms are understood in a purely mental sense, without neurophysiological connotations. This means that the terms involved cannot be understood as forlorn or clumsy attempts to describe processes in the nervous system. Therefore I understand von Wright as claiming that the ordinary mental use of the terms is not undermined by the negative outcome of the attempt to interpret them physiologically. This becomes quite plausible when we remind ourselves of the fact that there is no reason to correct or blame a user of the expression "to know by heart" because from a physiological perspective he referred to the wrong organ—one that, as far as we know, plays no role in the process of memorizing. In its mental use, at least nowadays, the expression "to know by heart" does not refer to a physiological entity at all, so it is not possible to make a mistake in the choice of organ.

The given argument presupposes that a purely mental use of mental terms does indeed exist, and this in turn means that it would be incorrect to maintain that *every* mental term must refer to physiological processes. There seem to be borderline and mixed cases, however, in which physiological components of meaning are indeed present. But in order to do justice to von Wright's argument, we have for the moment to hold on to the presupposition that there are at least *some* genuine or purely mental terms. And I think von Wright is correct here.

It follows from these elucidations that the question of whether or not a mental expression such as "weighing the evidence" makes good sense must be decided independently of the success or failure of attempts to interpret it physiologically. It is indeed possible to be critical about its value in the context of mental talk about actions, and there might be good reasons for such a criticism. For example, one could maintain that the expression "to weigh" suggests a common scale for evaluating totally different things, and that this suggestion is wrong, so that in this respect it is inappropriate to some situations in which we apply the expression "to weigh the evidence." But this internal criticism of an expression, which stays inside the realm of action talk, is independent of the success or failure of giving it a neurophysiological interpretation.

From the point of view of the philosophy and history of language, this is a quite complex case. In a first step, an expression for a physical activity (namely comparing the weights of two material things) is used metaphorically to enter into the mental sphere of deciding between items that do not have any weight in the literal sense. In court, for example, the judge has to weigh the evidence that has been put forward. And then the mental term, which at that point has no neurophysiological connotations, becomes a candidate for a second "metaphorical" step. An attempt is made to transfer the expression to the physical level of neurophysiological processes, and it can turn out (as we had supposed with von Wright) that this attempt fails. It is the failure of this second step that made me speak of the "fictional" character of the term in its neurophysiological interpretation and that led von Wright to call its designatum an "invention." So an expression that on the mental level is unproblematic and respectable can turn out to stand for a neurophysiological fiction without thereby losing its mental legitimacy and significance.

As I have already indicated, the situation can be even more complicated, because one often cannot tell from an expression alone whether or not that expression is meant to have a physiological significance. Many ordinary language expressions are part of a network of metaphorical expressions for a "mental apparatus," and this is enough to make them possible candidates for being interpreted physiologically.[4] The reason is that in principle any apparatus can be proposed as a model for physiological processes. For example, a simple expression such as "to keep in mind" has (like "to weigh the evidence") on the one hand a clear relation to a bodily action (to keep in one's hands, not to let go of something); second, it has an obvious mental use ("to remember"); but third, it can express the idea that a part of the brain is like a container into which things can be put in such a way that they do not "fall out." Whether this last interpretation is intended or whether the expression is meant in a purely mental sense (such as our normal use of "to know by heart") must be decided in the particular context in which it is used.

At this point it is worth mentioning that there are quite a number of psychosomatic expressions, that is, words for mental states and processes, that have a clear and obvious relation to the human body and the bodily experience of the person to whom the mental state is ascribed. Among them are "the words stick in one's throat" and "something goes to the heart" (in German: "einen Kloß im Hals haben," "es verschlägt ihm die Sprache," "ihr liegt etwas im Magen," "er hat Schiß vor der Prüfung"). In many cases, modern medicine agrees that the physical connotations of these expressions indeed have a point, so they can be seen as useful (and in many cases as very old) building blocks of a "folk theory" about the living human body. Whether something parallel can be said about those mental terms that are *not* in this way related to particular bodily experiences remains to be seen. Some authors interpret our ordinary mental talk as a "folk theory" of the mental. To my mind the question here is not whether this is the case, but rather what understanding of "theory" is appropriate in the case at hand and what are the consequences of a decision on this point for seeing a continuity or discontinuity between "folk" and "scientific" theories. I will return to this point.

The "Stream of Consciousness" as a Bridge to Neurophysiological Processes: Psychology as a Science of Entities Hidden from Public View

If it is true that the meaningfulness of purely mental terms does not depend on the possibility of their neurophysiological interpretation, the next question has to be: On what then does it depend? Following traditional conceptions of what it is for a word to be meaningful, one might be tempted to say that mental terms must refer to something that "really exists," that they must not refer to purely fictitious entities. One might be willing to concede that our current ordinary language vocabulary is clumsy and imprecise, or that it is composed of a number of unconnected and possibly even mutually exclusive metaphors and analogies, none of which quite captures what we intend to refer to. But still it seems that there must exist "out there" (or better, inside our bodies) something that we mean, a realm of entities that we aim at with the help of our words, a target we are *trying* to hit. This seems quite obvious when we consider a negative case: most people today will reject an explanation of an action by reference to possession of the agent by a devil; this explanation is a fiction, one would say, because devils do not exist. And their nonexistence might be seen as a sufficient reason for rejecting the explanation.

    This would mean for the positive case—for example, for the expression "to weigh the evidence"—that with its help we try to point to a real, nonfictional process, a kind of inner activity that we in some sense "feel" or are aware of when we are performing it, comparable to the way we feel when we are breathing or when we are holding our breath, or when a forgotten phone number comes to our mind or when it fails to come to mind. And in the same way, so it seems, the expressions mentioned previously, namely "she was *interpreting* the noise in a certain way" and "she was *expecting* somebody" have to refer to felt processes, to inner states or activities. This is of course compatible with the claim that there can be cases of pretending; for example, in order to escape military service a youth might simulate inner conflicts that he does not really experience. But if it were generally the case that there were no entities as the meanings of mental terms, it seems that our practice of explaining actions by reference to them would be empty. It would be a purely linguistic ritual, a case of faith healing, not rooted in reality but floating freely in the realm of fictions.

Whether indeed there must always be "sensations" of this kind to which mental terms refer, something like moments or segments in a "stream of consciousness," I will discuss further in the next section, when I treat Wittgenstein and the meaning of his term "grammatical fiction." Before doing that I mention that there seems to be a strong temptation to say that at bottom or ultimately our mental terms *must* refer to something physical—contrary to what has been said earlier in the discussion of von Wright's paper. And the traditional idea of a stream of consciousness seems to support this: the chain of neuro-physiological happenings that seemingly "must" be what we ulti-mately refer to with our mental terms is envisaged as running parallel to this mental stream. As a mediating step to this claim it is sometimes said that mental events are the "inner view" of the physical events in the body of the person whose mental events they are, and that he alone has this privileged "view from inside."

One is tempted to ask what else could be meant by the mental terms, what could they possibly refer to, how could we explain their nonfic-tional character, if we refuse to accept ghosts and other supernatural entities or processes whose scientific treatment is impossible? What else can a sensation be, if not the awareness of one's own physiological processes, as simple examples such as the experience of the coolness of a drink seem to show? What can the sensation of coolness be, if not the subjective experience of the effect of the drink on the appropriate receptors of the person drinking?

If this were indeed the case for the whole realm of mental expres-sions, even for those we have previously called "purely mental" terms ("to interpret as," "to expect"), then the success or failure of the metaphorical step to a neurophysiological interpretation of mental terms would after all *be* relevant in the context of explaining actions, contrary to what we assumed when we were following von Wright. The reason is the following: even if we tried to defend a rather sophisticated version of the "two sides of one coin" thesis, namely that mental entities are the "inner views" of physiological processes (and accordingly are, under this description, not *simply*, in all respects, identical with neurophysiological processes), even in this case a certain interdependence between the two realms could hardly be denied. If indeed the "inner view" is a particular *view of* its outer (neurophysiological) counterpart, then the correctness of the treatment of the one side must surely have a relation to the correctness of the treatment of the other. This could only be denied, it seems to me, by rejecting the metaphor of the two views right from the beginning. This rejection, by the way, is the course of argument I myself would opt for if time would permit.

However we decide on the legitimacy of the metaphor of the "inside" and the "outside" and on the legitimacy of the conviction that ultimately all mental terms "must" refer to physiological processes, the conception of a "stream of consciousness" suggests that the mental vocabulary of ordinary language is part of a "folk psychology" employing prescientific, possibly mythological, fictions, which are full of self-deceptions. Consequently, it could be seen as a preparatory stage to a scientific psychology, the goal of which would be (among other things) the systematic and experimentally controlled investigation of sensations and inner activities as the items that appear in this "stream." One could concede that folk psychology has always aimed at these real entities, but one would suspect that it does so in a very imperfect manner, one reason being the scientific unreliability of introspective reports. Consequently, a better, more precise, and in the long run the only acceptable treatment of this subject matter is to be expected from the science of psychology, that is, from a systematic study of human experience and behavior.

This well-known formula suggests that the methodology of psychology should be a mixture of so-called introspection and observation. A "motive" is on the one hand a prescientific object of knowledge. Everybody can in some way or other comment on what has "moved" her to a certain action; at least in some cases we know some of our own motives. At the same time, motives are the subject matter of psychological theories of motivation. These theories employ behavior-related models of processes and activities that may be, but need not be, within the conscious grasp of the subject under investigation. We all know the possibility and reality of self-deception. Science, it seems, can disclose the "real motives"; it leaves myths, fictions, and speculations behind and refuses to accept anything but the "real facts." Untested folk theories, accordingly, are like fictions. They are like prescientific fairy tales about a realm in which genuine knowledge should be possible, in the sense in which the status of this knowledge is not inferior to the knowledge of physics. This genuine knowledge can and should be acquired in the usual scientific ways, which will include the formulation of models with the help of theoretical terms, and this procedure should allow us an indirect view of a realm that is (at least in part) hidden from direct observation, not only from our own observation but even more from the observation of other people.

When There Are No Entities in the Stream: Ludwig Wittgenstein and His Concept of a "Grammatical Fiction"

This is precisely the point at which Wittgenstein wades in with some critical questions. His most interesting claim in this context is the thesis that a considerable number of important and perfectly meaningful terms for mental entities just do not stand for sensations, activities, or processes in a "stream of consciousness." I stress once more that we are concerned here with a special class of expressions. Wittgenstein does not deny that we have dreams, that melodies go through our minds, or that we suddenly remember forgotten names or phone numbers. But of interest in this particular discussion are those expressions for which a closer look reveals that there are no "mental entities" correlated with them. Among the examples of mental terms that I have cited, the terms "to interpret as," "to expect," and "to weigh the evidence" belong to this category; they do not, according to Wittgenstein, designate or classify sensations. Still he does not think they are meaningless. They belong to those expressions of our language the meaning of which does not consist of their standing for or classifying an *independent* object. An object is "independent," in the sense intended here, if its existence does not depend on the particular expression (or an equivalent one) in question.

If Wittgenstein is right about this, it follows that (for want of an object) there can be no empirical investigation of such alleged, putative sensations and that our ordinary talk about and understanding of our reasons, expectations, and so forth cannot be improved by an investigation the method of which is inspired by science (in the sense of "natural science" as exemplified by physics, in contradistinction to the humanities, as exemplified by history). This is true regardless of whether the method chosen is designed for directly observable entities, such as billiard balls or pendulums, or for entities that can be studied only indirectly, such as subatomic particles. So the envisaged process of cleansing the folk theories of mythological residues in order to isolate a core of hard facts is not the road to a *science* of psychology. Such a core simply does not exist in this special realm to which Wittgenstein directs our attention.

It has been said that Wittgenstein denied the existence of mental states, that he believed them to be fictions. With respect to the particular range of terms here under discussion, this is a provocative and questionable formulation, but it is not altogether inapt. Wittgenstein *does* indeed use the word *fiction,* but as a part of the complex expression "grammatical fiction," which does not lend itself to an easy and ready understanding (Wittgenstein 1953, I, §307). So we have to ask: Of what kind are the objects that result from "grammatical fictions"? How can referring to them be a meaningful speech act that allows the difference between true and false? Is to make a "grammatical fiction" the same thing as to introduce a theoretical term? I now provide a short account of what I take to be Wittgenstein's point here.[5] It will also help us to get an idea about what psychology could be *instead* of being a ("natural") science.

Wittgenstein's basic claim is that an understanding of the language games in which one engages with the help of mental terms such as "to mean" or "to interpret as" is primary, and that this understanding is a practical ability relating to the use of metaphors and analogies. And it is then a further step to isolate "objects referred to" by constituent expressions present in these ways of talking, a step that is internal to language. To play the language game with the help of complex expressions is primary; to talk about "referring" to "entities" is secondary. To put it simply: it is possible (for example, for a German speaker) to use correctly the expression "to know by heart" without having any knowledge of what bodily organ the expression "heart" refers to. The same can be claimed for the expression "he left her in the lurch" in cases in which it is used by a speaker without special etymological information; he will be unable to answer the question about what object the constituent expression "lurch" refers to, but he can still use

the complex expression meaningfully and with "real-life" consequences. It should be noted, however, that the "lurch" case is different
from the case of mental terms insofar as the latter appear in clusters
(cf., e.g., "mindless," "in mind," "to remind") and can for that reason
not be dismissed as idiomatic singularities without systematic interest.[6]

Correspondingly, Wittgenstein's claim is that in understanding the
functioning of expressions such as "to weigh the evidence" the metaphorical step from physical to mental use can be made independently
of the existence or nonexistence of a special sensation of mental weight
comparison, which the language user would have to recognize as the
referent of the expression. Such a sensation, Wittgenstein tries to convince us, does not exist, and whatever sensation one may have in
conjunction with a particular use of the expression, this sensation is
not what the phrase means or refers to. So the expression "weighing
the evidence" does not refer to a sensation in a manner comparable to
the way in which the expression "itching on my nose" refers to a
sensation or the expression "scratching my nose" refers to a physical
process, or in the sense that the expression "now I remember her
name" refers to an introspectively accessible event.

As the examples "to know by heart" and "to leave somebody in the
lurch" were meant to show, this understanding does not imply that our
mental talk of this type consists of fabulations, without connection to
the "real world." One can speak about one's motives, convictions, and
intentions truthfully or untruthfully; one can lie about knowing a
poem by heart or about having left a friend in the lurch. Accordingly,
Wittgenstein does not use the word *fiction* with the intention of rejecting or devaluing this kind of mental talk; he acknowledges that it is
meaningful and has consequences in "real life." So his view is at right
angles to the alternative that mental objects must be either normal
objects of science (directly or only indirectly observable) or mythical
fictions, separated from reality.

After having seen and comprehended the special semantic character
of the kind of expressions Wittgenstein discusses, one can very well go
on talking about "mental entities," in the way a nominalist can go on
talking about numbers. In both cases only an uncritical or naive way of
understanding this kind of "talking about" is to be eschewed. As we
have seen, it would be wrong in Wittgenstein's eyes to say that mental
entities are like other entities, with the sole difference that they pose
special difficulties for observation or that they can be observed only

indirectly, that is, by taking note of their causal effects. But it would also be misleading to say that the mental act or state of expecting somebody "does not exist," because this can be understood to mean that nobody ever expected anything or anybody, that the language game involved has no application. This is not at all what Wittgenstein wants to say. Instead, his claim is that mental entities are constituted by the process of acquisition of the particular language games involved, in a cultural-linguistic context, and that they are constituted in such a way that it is meaningless to ask what they are outside this cultural context. We can express this by saying that the special semantic process Wittgenstein discusses generates "objects of discourse" or, more broadly speaking, "cultural objects."[7]

What is at issue here can perhaps be clarified by relating it to Rudolf Carnap's (1950) well-known distinction between internal questions ("Is there a prime number between 5 and 9?") and external questions ("Are there numbers?"). Carnap contends that external questions are practical: they are important questions, he admits, but correctly formulated they are not concerned with the existence of entities (such as numbers) but with the usefulness of certain linguistic frameworks. In parallel fashion, the question "Did P act according to intention i_1 or i_2?" should be thought of as an internal question, whereas a discussion of the "language of intentions" (as conducted in this chapter) is concerned with an external question. The objective of this chapter, then, can be described as an attempt to clarify what is involved in different modes of adding new expressions (or new uses of old ones) to a linguistic framework. My claim is that, with respect to the mental terms discussed here, an enlargement of a given language through the introduction of novel metaphorical uses of its expressions differs in kind from enlargements effected by the introduction of theoretical terms.

Returning to Wittgenstein, we can say that at the extreme end of the spectrum of constitutions of the kind he is discussing are cases in which the expression by which the speaker seems to refer to her mental state has no nonlinguistic correlate *at all* that could be isolated as an object independent of the ongoing dialogue. In these cases there is no natural object like a sensation to which the speaker refers. So here the point is not just the acknowledgment that a given form of reference to a given object carries with it a particular "coloring" (as might be seen in the different grammatical gender of the masculine German form *der Mond* as compared with the feminine Italian *la luna*). Instead, it is the acknowledgment that there is no "given object."

Wittgenstein discusses the "act of meaning somebody" as a case in point. Imagine you are beckoning to a person in order to call her to you, and you find you are misunderstood. Then in a second move you say "I meant *x*, not *y*": you say what you meant, you describe the intention you had in the act of beckoning. Wittgenstein wants to make us see that in such a case we are not referring back in time to a "state of mind" of "meaning *x*" or "having the intention to make *x* come to us." The intention is no additional thing or event, that is, additional to the physical gesture of beckoning. But still the "move in the language game," that is, the second step of commenting on the past act of beckoning, is meaningful and normally understood without problems.

Here is the place at which Wittgenstein's expression "grammatical fiction" has its clearest meaning. It designates the linguistic process that culminates in expressions such as "my intention," and it does not mean that these expressions are pointless. This way of talking makes it appear as if there were mental objects on a par with apples and pears, and as if our way of referring to these mental objects is like reaching out for them or pointing to them. But Wittgenstein proposes to see matters the other way around: "Look on the language game as the *primary* thing. And look on the feelings, etc., as you look on a way of regarding the language game, as interpretation" ("Sieh auf das Sprachspiel als das *Primäre*! Und auf die Gefühle, etc. als auf eine Betrachtungsweise, eine Deutung, des Sprachspiels!") (Wittgenstein 1953, I, §656). Generalizing, he says:

The paradox [i.e., that he seems at the same time to deny and not to deny the existence of mental states] disappears only if we make a radical break with the idea that language always functions in *one* way, always serves the same purpose: to convey thoughts—which may be about houses, pains, good and evil, or anything else you please. (Das Paradox verschwindet nur dann, wenn wir radikal mit der Idee brechen, die Sprache funktioniere immer auf *eine* Weise, diene immer dem gleichen Zweck: Gedanken zu übertragen—seien diese nun Gedanken über Häuser, Schmerzen, Gut und Böse, oder was immer.) (Wittgenstein 1953, I, §304)

That our way of expressing ourselves, our "form of representation," suggests one uniform way of "referring to something" is responsible for the fictional character of some of the contents we express with this form. The word *fiction* here does not mean that we can take for granted the usual way of "referring to something" and treat as the only peculiarity of the mental case under discussion that the act of reference miscarries—does not hit its target—because such a target (as in "Little Red Riding Hood") does not exist. Mental states are not fictions in the fairy tale sense, but they are *grammatical* fictions, fictions produced by our grammar, not by a fabulist. To quote Wittgenstein once more:

"Are you not really a behaviorist in disguise? Aren't you at bottom really saying that everything except human behavior is a fiction?" If I do speak of a fiction, then it is of a *grammatical* fiction. ("Bist du nicht doch ein verkappter Behaviourist? Sagst du nicht doch, im Grunde, daß alles Fiktion ist, außer dem menschlichen Benehmen?" Wenn ich von einer Fiktion rede, dann von einer *grammatischen* Fiktion.) (Wittgenstein 1953, I, §307)

Folk Psychology, Theoretical Terms, and the Limits of Science

Summing up what has been argued so far, we can formulate the following propositions:

1. When we adopt Wittgenstein's view that not even an introspectively observable "mental state" has to exist in order that a mental term be meaningful, we are able to agree with von Wright that a negative result of an attempt to interpret a mental term neurophysiologically does not militate against the mental use of the term. When even the alleged "inside view" of a neurophysiological state is unnecessary to give a meaning to such a term, a fortiori the "outside view" is not necessary either.

2. Instead, mental terms have their place in meaningful language games, even if (in the extreme cases here under discussion) they do not denote entities that exist apart from or outside these particular language games, either "in introspection" or "in the brain."

3. One can use the phrase "grammatical fiction" in a positive sense: it points to the constitutive character and at the same time to the meaningful use of the expressions concerned. They in a way "create" mental entities, and their use is not arbitrary; they are inside the realm of truth and deception.

In this section I consider what sort or type of psychology would be hospitable to the philosophical arguments developed in my discussion of the views of Wittgenstein and von Wright. I offer first a short comment on what I view as a positive development in psychology itself, and then a longer, negative comment about what I take to be a clearly formulated step in the direction I have argued against.

The positive development in psychology is the following. For a number of years now Jerome Bruner (1990)—one of the fathers of the so-called cognitive turn, which sought to overcome behaviorism and to reintroduce the mental into psychology—has proposed in some detail another step toward an enrichment of his field of study.[8] His contention is that for the psychologist "cognition" must mean more than "information processing," and this claim has led him to sketch an outline of what he has called "cultural psychology." One short passage from Bruner will show just how close he comes to the ideas of Wittgenstein discussed in this chapter: "The fact of the matter is that we do not have much of an idea of what thought *is*, either as a 'state of mind' or as a process. . . . It may be simply one of those 'oeuvres' that we create after the fact" (Bruner 1996, 108).

In order to do justice to these cultural acts of creation, Bruner tries to sketch the outlines of a hermeneutic psychology. Hermeneutic psychology would not be modeled after science, that is, it would not try to develop theories about hidden objects that can be studied only indirectly by building models and seeing whether the derived observable consequences do in fact materialize as predicted. So Bruner does not treat mental entities in the same way as physicists treat subatomic particles and other objects that can be studied only by examining their causal effects. Instead, the objects of cultural psychology would be (in the area discussed here) mental states *as constituted by a specific culture*, among other things by its ways of talking about intentions, attitudes, and dreams. Bruner's ideas serve as a reminder that certain developments in psychology itself point in the direction advocated in this chapter.

To make clearer what I understand Bruner to argue *against*, I now consider some aspects of the naturalistic, science-oriented understanding of psychology recently formulated by Carrier (1998) in a beautifully clear paper (see Carrier and Mittelstrass 1989).[9] I offer two interpretations of his vision of psychology: the first is more or less the polar opposite of what I have been advocating here, and the second amounts to the question of whether Carrier's view should not after all be construed as an unnecessarily restrictive *version* of cultural psychology.

The ground is so well prepared that I can be brief. Here then is the first interpretation, according to which Carrier would turn out to be my diametrical opponent. A psychological theory such as motivation theory has as its subject matter human behavior and the largely unknown causes or influences that bring it about. For example, it talks about "stimulating" and "activating" motives, and about "measuring their relative strength" (Carrier 1998, 222). That these influences are "largely unknown" means that they are not directly observable, and since we know of the possibility of self-deception, their general nature as well as (very often) their presence in particular cases seem to be unknown.

The ideal of reliable knowledge for Carrier is modern physics, and since the physics of subatomic particles shows that we can construct theories about nondirectly observable entities he finds here both a reason for hope and a goal toward which the psychologist can strive. So physics is taken as the model for investigating the mental lives of other people (i.e., of people other than the investigator). These mental states or events are treated as unobservables—as entities about which the psychologist can obtain knowledge only indirectly.

The resulting psychological theory traffics in hidden "inner" forces and their overt consequences. The terms used in the theory refer to items that have their primary existence as elements of a theoretical model, but a secondary existence by virtue of the success of the model, notably in predicting certain aspects of the behavior of experimental subjects. Carrier deems it unnecessary to make any claims about the exact nature of the "forces" under investigation. In this respect the theory is functional or instrumentalist. As long as input/output relations can be stated with some success, the question of what might be hidden in the "black box" of the mind is taken to have no point: hidden entities are hidden, and that is it. And even if it would turn out one day that some of them (or even all of them) do not exist at all (or not nearly in the way the theory represents them), this would not really matter for our investigations today. We must take the best explanation we can get in order to cope with the world, to handle things successfully, even if this explanation cannot answer *all* our questions.

A less agnostic version that seems to be in the backs of the minds of many proponents of such a conception of psychology (and Carrier explicitly confesses to it in his paper) claims that the "real" entities of which the model is a model are neurophysiological processes and states. Carrier explains that we must refer to such entities, for example, when we try to formulate psychological ceteris paribus laws: "Since it is difficult to imagine how such intentional states could produce other mental states and behavior, they are thought to be realized at some more basic (probably neurophysiological) level, and their transition is supposed to be governed by laws applying to these realizations" (1998, 220). Under such an interpretation the mental states mentioned in the psychological model are indeed less mysterious in their character: they are neurophysiological states. Yet a huge gap opens between assertions and conclusions in the model on the one hand and current knowledge in neurophysiology on the other. It is clear that to say of a person that she is in a state of "learned helplessness" (Carrier's example) cannot be taken as even a rough approximation for characterizing a neurophysiological state of a particular nature, one differentiated from other forms of depression. Such a state may well be a fiction or invention in the sense discussed by von Wright.

On the other hand the expression "learned helplessness" can be readily used to explain somebody's chronic failures. It can even be introduced as a technical term in psychotherapy by relating episodes from the lives of real people that we understand in our familiar way of understanding biography. And only after the successful introduction of a technical term in this *hermeneutic* fashion can we investigate whether there are physiological processes correlated with the mental phenomenon, so that we might even find, in severe cases, a medication to control it, for example, a drug to prevent suicide. The availability of the hermeneutic understanding contradicts the claim that what we are "really" talking about in psychology are processes about which we have hardly any knowledge.

In my view, then, Carrier's scientific construal of psychology has the following serious disadvantages:

1. It makes hardly any use of the knowledge of the mental image that every one of us has about him- or herself and that has been set forth in world literature (or better, in the literatures of the different cultures of the world), not to speak of art and other cultural products.

2. It does not bridge the gap between first-person knowledge of my own motives and the third-person knowledge one can have of other

people's inner lives. Indeed it misrepresents (or makes unintelligible) the way in which, in the realm of the mental, "learning about oneself" and "learning about others" go hand in hand.[10]

3. It misrepresents the knowledge a culture has accumulated about its own mental life: "folk psychology" did not *arise* as scientific model-building, nor does it have the shape that one would expect of such a model. Seemingly local similarities (between folk psychology and scientific psychology) must be accounted for in reverse fashion: scientific and technical thinking has entered into our self-descriptions as a source of metaphors.

4. The talk of "hidden entities" evades the question of the special character of the terms concerned; it misses all the semantic points I have tried to make in this chapter, following Wittgenstein and von Wright.

5. The application of the notion of the "best explanation" and the functionalist interpretation of psychological theory restricts the meaning of mental terms to the context of their application to others. It favors the objective of influencing other people over and against the goal of coming to terms with one's own mental life.

6. The nonagnostic version (that we "really" mean neurophysiological states by mental terms) is either simply false (in the sense that I cannot mean what I do not know; see the previous discussion) or else useless at the present stage of inquiry because we are unable to bridge the gap between our mental terms and their currently available putative "neurophysiological counterparts."

So much for the first of my two interpretations.

My second interpretation of Carrier's conception of psychology is as merely a special version of cultural psychology, but an unnecessarily restrictive one. According to this interpretation, model-building as envisaged by Carrier is a language game of the same type or status as the ones discussed by Wittgenstein. It employs a network of metaphors and thereby generates "grammatical fictions." It can even, as Sigmund Freud (1969, 529) said about his theory of drives, be called a "mythology." As explained earlier, however, according to this interpretation there is no analogy between psychology and sciences that study "hidden entities" by examining their effects. Talking about "motives" is more like talking about the Greek god Eros than like talking about subatomic particles. Someone who views psychology in this way ac-knowledges that it is neither his reference to a stream of consciousness nor his reference to neurophysiological states that makes mental terms meaningful. Nor are they "theoretical terms" familiar from the phi-losophy of science. Rather it is the use we make of them in the context of nonscientific language games that makes these terms meaningful. These language games have their customary application in speaking about oneself *and* about others. They are elements of our cultural tradition, a tradition that is much older and much broader in its means of articulating the mental than the young science of psychology.

Surely it is legitimate to criticize parts of this cultural tradition, to propose other ways of talking, as I have explained previously for "to weigh the evidence." And surely it is legitimate to refuse recourse to the devil (or to Eros, for that matter) as part of an explanation of certain acts. But to do so and to give reasons for doing so is nothing else than to participate in the culture, to go on with the tradition of transmitting and modifying the language games of the mental. Once the idea that a motive is a hidden or unobservable natural entity is given up, and once it is acknowledged that it is a *cultural* entity instead, there is no reason to introduce the above-mentioned restrictions that characterize a functionalist understanding of the mental—an under-standing modeled after theory-building in the physics of subatomic particles. The hiddenness of motives differs in kind from the hidden-ness of particles, and so too do the kinds of steps that must be taken to get them into better view.

Summary

Beginning with von Wright's observation that some meaningful mental terms might be found to designate "inventions" if an attempt were made to apply them directly to alleged neurophysiological processes, I have raised the question of what it is instead that makes them mean-ingful. Following Wittgenstein, I tried to show that in many cases not even introspectively observable "inner events" are available as "ob-jects referred to" to make the corresponding mental expressions mean-ingful. Instead, in these cases the cultural practice of engaging in "lan-guage games" must be seen as primary, and the "objects referred to" as secondary, that is, as constituted by such practices, without having an independent "natural" existence. As a consequence, a hermeneutic (and in this sense "nonscientific") approach to these practices is needed in order to do justice to the ontological status of the "objects" under investigation. In particular, the ability to understand the workings of metaphors in the constitution of its objects will be a methodologically central ingredient of such a "cultural psychology" (as proposed by Bruner).

This hermeneutic (*geisteswissenschaftliche*) methodology stands in contrast to a science-oriented (*naturwissenschaftliche*) methodology, in which a comparably privileged position is reserved for the introduction of theoretical terms that denote unobservable entities (as has been recently worked out by Carrier). According to the view advocated here, psychological "folk theories" share the same footing with cultural psychology and so cannot be viewed as ineffectual or misguided precursors of scientific psychological theories. Insofar as the hermeneutic approach is the more adequate one for (certain areas of) psychology, the case presented here has uncovered a "limit of science," namely a deep-seated limitation of scientific psychology.

NOTES

I thank my commentator Prof. Robert Brandom from the University of Pittsburgh for his useful comments generally. I also thank Profs. Gerald Massey and Martin Carrier for valuable comments and Prof. Massey and Timothy Doyle for the time they devoted to improving my language.

1. A warning against a common misunderstanding: Von Wright's interventionist concept of causality by no means denies that lightning can cause a barn to burn down. It only says that to describe an event of lightning as a cause means to treat it as being of the same kind as events produced by our scientific experiments. The same holds for events of a cosmological dimension, and I take this way of treating macro and micro events as belonging to the same *kind* as one of the characteristics of Newtonian science. For a more detailed discussion of some other aspects of this concept of causality, see Schneider (1999).

2. This means that I do not treat expressions such as logical constants or numerals as theoretical terms in the sense under discussion here, because they are of a different kind altogether.

3. See Schneider (1997b) for a more detailed discussion.

4. For a rich fund of examples see Johnson (1987); for a critical discussion of Johnson's conclusions cf. Schneider (1995a).

5. For a more detailed discussion see Schneider (1997a).

6. I thank Prof. Robert Brandom from the University of Pittsburgh for pointing out the necessity of mentioning this difference.

7. For a discussion of concepts and numbers as "objects of discourse" see Schneider (1995b).
8. For a controversial discussion of a similar approach see Straub and Werbik (1999).
9. I thank Martin Carrier for letting me work with his (at the time) unpublished paper.
10. For a detailed account of this point see Kerr (1986), despite its misleading title.

REFERENCES

Bruner, J. 1990. *Acts of Meaning.* Cambridge, Mass.: Harvard University Press.
———. 1996. *The Culture of Education.* Cambridge, Mass.: Harvard University Press.
Carnap, R. 1950. "Empiricism, Semantics, and Ontology." *Revue Internationale de Philosophie* 4: 20–40. [Reprinted in Carnap, R. 1956. *Meaning and Necessity.* Chicago: University of Chicago Press, 205–21.]
Carrier, M. 1998. "In Defense of Psychological Laws." *International Studies in the Philosophy of Science* 12(3): 217–32.
Carrier, M., and J. Mittelstrass. 1989. *Mind, Brain, Behavior: The Mind-Body Problem and the Philosophy of Psychology.* New York: De Gruyter.
Freud, S. 1969. *Neue Folge der Vorlesungen zur Einführung in die Psychoanalyse. 32. Vorlesung: Angst und Triebleben,* Studienausgabe, vol. 1. A. Mitscherlich, A. Richards, and J. Strachey, eds. Frankfurt: S. Fischer.
Johnson, M. 1987. *The Body in the Mind: The Bodily Basis of Meaning, Imagination, and Reason.* Chicago: University of Chicago Press.
Kerr, F. 1986. *Theology after Wittgenstein.* Oxford: Blackwell.
Schneider, H. J. 1995a. "Die Leibbezogenheit des Sprechens: Zu den Ansätzen von Mark Johnson und Eugene T. Gendlin." *Synthesis Philosophica* 19-20 [10(1-2)]: 81–85.
———. 1995b. "Begriffe als Gegenstände der Rede." In I. Max and W. Stelzner, eds., *Logik und Mathematik: Frege-Kolloquium Jena 1993.* Berlin: De Gruyter, 165–79.
———. 1997a. " 'Den Zustand meiner Seele beschreiben': Bericht oder Diskurs?" In W. R. Köhler, ed., *Davidsons Philosophie des Mentalen.* Paderborn: Schöningh, 33–51.
———. 1997b. "Metaphorically Created Objects: 'Real' or 'Only Linguistic' "? In B. Debatin, T. R. Jackson, and D. Steuer, eds., *Metaphor and Rational Discourse.* Tübingen: Max Niemeyer, 91–100.
———. 1999. "Mind, Matter, and Our Longing for the 'One World.' " In G. Meggle, ed., *Actions, Norms, Values: Discussions with Georg Henrik von Wright.* Berlin: De Gruyter, 123–37.
Straub, J., and H. Werbik. 1999. "Handlung, Verhalten, Prozess: Skisse eines integrierten Ansatzes." In their *Handlungstheorie: Begriff und Erklärung des Handelns im interdisziplinären Diskurs.* Frankfurt: Campus, 27–48.

Von Wright, G. H. 1971. *Explanation and Understanding.* Ithaca, N.Y.: Cornell University Press.

———. 1974. *Causality and Determinism.* New York: Columbia University Press.

———. 1994. "On Mind and Matter." *Journal of Theoretical Biology* 171: 101–10.

Wittgenstein, L. 1953. *Philosophische Untersuchungen/Philosophical Investigations.* New York: Macmillan.

13

Unity and the Limits of Science

Margaret Morrison

Department of Philosophy, University of Toronto

The theme of this volume, the limits of science, is interesting for several reasons. On the one hand science, as a human activity, sometimes serves to limit itself. Technological advances determine the kind of experimental findings that are possible in, for example, the area of high-energy physics. As a result some theoretical predictions about the fundamental constituents of matter cannot be verified until experimental physics can surpass the limits currently in place, that is, until accelerators can be built that are capable of producing energies high enough to create conditions under which certain kinds of particles can be detected. But these are in some sense practical limitations—ones that can, in principle, be overcome.

There are, however, other kinds of limitations for which the boundaries are less fluid; they concern not only the long-term goals science sets for itself but also our philosophical picture of what science has achieved and what it is capable of achieving. For example, in scientific contexts there is some debate about whether it is even possible to arrive at a physically meaningful theory that can unify the four fundamental forces; yet few would doubt that science has produced some degree of unity and uncovered at least some of the fundamental constituents of matter. Philosophically, if one adopts an antirealist or instrumentalist interpretation of science even these claims may become contentious. Proponents of such a view suggest that the very idea that we could say, with any certainty, that we know what nature is like is to pass beyond the limits of what science can reasonably tell us. Hence we have an in-principle limitation imposed on science that is grounded in a philosophical assessment of the powers of human cognition.

If, on the other hand, one is a realist, the possibilities increase; given the success of science we have no good reasons to doubt its claims or its ability to extend our knowledge beyond its present limits. In other words, we have no reason to doubt the veracity of what science has claimed to discover, and the possible limits of science are the result of contingent features of technological or theoretical development. These limits can and supposedly will be overcome in the future; current knowledge is cited as providing reasons for such belief.

Although this latter view seems largely uncritical, the antirealist position seems prima facie too conservative. At some point along the continuum between realism and antirealism, there must be room for philosophical analyses that can characterize and evaluate the achievements of science and the implications of those successes for human knowledge, and do so without unnecessary skepticism or unbridled enthusiasm. This kind of philosophical reflection and appraisal is necessary not only for developing a coherent epistemology of science but also because of the lack of consensus in the scientific community regarding both actual and potential achievements. Although it is not my intention to develop such an epistemology in this short chapter, I do try to outline, using examples, what I take to be the right sort of approach for assessing scientific achievements and limitations.

Specifically I address the issue of unification within science as a particular instance of the limits of science. Whether this takes the form of ontological reduction or integration of different phenomena within the framework of a single theory, it can be assessed in terms of the accomplishments, goals, and hence limits of science. In what follows I look briefly at the claims made on behalf of one aspect of scientific unity: theory unification in physics. Several questions come to mind when thinking about the limits of science from this perspective: How should we understand the unity science has achieved? Has it in fact achieved unity? Are there philosophical problems associated with this unity? What are the limits of this unity?

These kinds of questions involve both a philosophical and an empirical dimension; in other words, we must begin by telling an empirical story about unity that is grounded in historical documentation before we can draw philosophical conclusions. The specific conclusion for which I argue is that it is a mistake to structure the unity/disunity debate in metaphysical terms; that is, one should not view success in constructing unified theories as evidence for a global metaphysics of unity or disunity. Whether the physical world is unified or disunified is an empirical question, the answer to which is both *Yes* and *No* depending on the kind of evidence we have and the type of phenomena with which we are dealing. In some contexts there is evidence for unity whereas in others there is not. But nothing about that evidence warrants any broad, sweeping statements about nature "in general" or the world as it "really" is, that is, how its ultimate constituents are assembled. Constructing unified theories for those phenomena is a practical problem for which we only sometimes have the appropriate resources—the phenomena themselves may simply not be amenable to a unified treatment. Seeing unity in metaphysical terms shifts the focus away from practical issues relevant to localized contexts toward a more global problem—erecting a unified physics or science—one that in principle we may not be able to solve because nature may not wish to cooperate. Hence to inquire about the limits of science with respect to unification is first and foremost an empirical matter that is context specific.

But that need not preclude one from drawing philosophical conclusions, as I intend to do. There *are* interesting philosophical issues that arise from examining cases of theory unification and examples of unity within science. For instance, there are different ways in which unified theories are constructed (i.e., the different ways that unity manifests itself as a synthesis, a reduction, or a combination of the two), each of which has different implications for claims about the unity of specific phenomena governed by a theory and what the limits of that unity might be. There are also distinct ways in which the idea of unity is present in the practice of science: a unity of method, a practical kind of unity that results from a culture of instrumentation; and the extent to which theory and experiment can be "unified" within a particular discipline. These have broader implications for a general thesis about the unity of science, a claim that I will not address here. The important point to remember is that eliminating metaphysics from the unity/disunity debate need not entail a corresponding dismissal of philosophical analysis.

In what follows I discuss some examples of unity and disunity, particularly at the level of theory construction, in the hope of showing that the debate, conceived as a metaphysical issue, is misguided, not only for the reasons mentioned earlier but also, and perhaps more importantly, because neither category in and of itself provides a coherent framework capable of describing the natural world. Debates about the limits of science become fruitful when they take as their starting point an analysis of the accomplishments of empirical research. Science has undoubtedly achieved a level of unity, but in order to understand its limits we must know its nature. Uncovering the character of unity and disunity is a philosophical task, one that contributes to an understanding of the limits of science.

Unity, Explanation, and Mathematics

In many ways the unity/disunity debate in the 1990s became for the philosophy of science what the realism debate was in the 1980s. In fact the two are rather closely connected; it has been suggested that one should consider a unified theory as more likely to be true or better confirmed and that particular entities or structures that unify phenomena should be considered real (see, e.g., Friedman 1983). This connection between unity, theory confirmation, and realism often involves scientific explanation. By virtue of its unifying power a theory is said to be capable of explaining more phenomena and of explaining how they are connected with each other. In fact, explanatory power is often described in terms of unity: the best explanation is the one that unifies the most phenomena. The question then arises: How could a theory achieve this unity if it did not latch on to some fundamental elements of the way the world is configured? This brings me to the first philosophical question, namely whether unity should be characterized in terms of explanatory power. In other words, is there more to unification than simply the theory's ability to explain diverse phenomena? This question has important implications for the limits of science and the question of whether explanatory power alone provides empirical evidence for an ontological unity in nature.

The difficulty of answering this question is partly due to the many different ways in which one can think about both unity and explanation. Let me begin with the case of Maxwellian electrodynamics. In earlier work (Morrison 1990) I argued that Maxwell's unification of electromagnetism and optics was successful in large part due to his use of the Lagrangian formulation of mechanics—a very general mathematical structure that is applicable to a number of different kinds of systems.[1] Maxwell initially formulated the field equations with the aid of a mechanical ether model that he considered fictitious. His later work, in "A Dynamical Theory of the Electromagnetic Field" (Maxwell 1965c [1865]) and in the *Treatise on Electricity and Magnetism* (Maxwell 1954 [1873]), was an attempt to divorce the field equations from the ether model and to use Newton's method of "deduction from phenomena" to establish their independent validity. Although he was not exactly successful in this deduction, the important issue is the way the Lagrangian formalism figured in the attempt (see Morrison 1992). Maxwell's stated goal for the *Treatise* was to examine the consequences of the assumption that electric currents are simply moving systems whose motion is communicated to each of the parts by forces, the nature of which "we do not even attempt to define, because we can eliminate them from the equations of motion" (Maxwell 1954 [1873], §552) by the method of Lagrange. The power of this formal approach provided Maxwell the benefits of a mathematically precise theory without commitment to a specific mechanical account of how electromagnetic forces were propagated. Gone were the mechanical hypotheses prominent in his earlier models. Instead, one could now conceive of a moving system connected by means of an "imaginary mechanics used merely to assist the imagination in ascribing position, velocity, and momentum to pure algebraic quantities" (Maxwell 1954 [1873], §555).

Although there can be no doubt that Maxwell provided a unified account of electromagnetism and optics, to what extent is the theory explanatory? In other words, is there a connection between the unifying power of the theory and its ability to explain a variety of physical phenomena? Again the answer is *Yes* and *No*. Maxwell showed that electromagnetic waves and light waves obeyed the same laws and traveled at the same velocity. In the earlier work (1965a [1856], 1965b [1861–62]), in which he made extensive use of mechanical models, he expressed, as a tentative conclusion, that the electromagnetic and luminiferous ethers were one and the same thing. This more substantial conclusion was absent from the later Lagrangian formulations of the work because none of it depended on a particular hypothesis about the specifics or even the material existence of an ether. To that extent we can see that there is a type of reductive explanation, via the equations, of the behavior of electromagnetic and optical phenomena, but clearly no causal or substantial theoretical explanation of the field theoretic nature of these phenomena or of how they were propagated in space. In that sense then the unity was achieved at the *expense* of explanatory power. Maxwell was able to provide a comprehensive theory through his use of the Lagrangian formalism, a technique that enabled him to ignore details about the phenomena that were necessary for a full theoretical understanding.

Indeed, this lack of a theoretical picture can be seen as one of the reasons for the controversy that surrounded the theory long after its development. People were confused about the relationship between matter and the field, about Maxwell's notion of charge, and about the origin and function of the displacement current, the mechanism that formed the foundation for Maxwell's field theoretic approach. The unity resulted from the powerful mathematical techniques—techniques that are applicable to a variety of phenomena regardless of their specific nature. The generality of the Lagrangian approach is at once both its strength and its weakness—capable of producing unity at the expense of a deeper theoretical explanation of the phenomena. In that sense the limits placed on theory construction and unification were a function of the availability of certain kinds of mathematical structures as well as the limitations resulting from a lack of knowledge of the physical systems involved. Maxwell was simply unable to explain how the electromagnetic waves hypothesized by his new field theory were propagated through space at the speed of light.[2]

Despite this, one would be hard pressed to deny that Maxwell had successfully unified electromagnetics and optics. In fact the kind of unity that he achieved could be classified as reductive or ontological. Although electric and magnetic forces remain distinct within the theory, he did succeed in showing that light waves and electromagnetic waves were one and the same; that is, they obeyed the same laws and were manifestations of the same processes. An interesting historical dimension of this case is that empirical evidence for the existence of electromagnetic waves came fifteen years after the publication of the *Treatise* and twenty-five years after the first formulation of a unified theory. During that period Maxwell's theory was embraced by some of his British colleagues but received little attention on the continent. Interestingly enough, even after Hertz's experimental production of these waves in 1888, some, including Kelvin, still opposed Maxwell's formulation of field theory on the basis of what they saw as theoretical inconsistencies at its core and the lack of a suitable explanatory foundation. To that extent the "unity" that the theory produced had little if any impact on its acceptance.

To return to the main point, how pervasive is this connection between unity and mathematical structures? Another interesting example of how mathematics rather than physical phenomenology can motivate unity is the case of the electroweak theory. Although the unification of the electromagnetic and weak forces represents more of a "unity through synthesis" than Maxwell's reductive approach, it nevertheless shares with electrodynamics the property of being largely mathematical in nature. To see how this is so we must look at the role played by gauge theory in developing the theoretical framework for the electroweak theory.

From Mathematics to Physics

If one considers simply the phenomenology of the physics, there initially seemed to be no way in which the electromagnetic and weak forces could ever be unified, if for none other than the simple reason that the particles that carry these forces are, in the former case, massless and electrically neutral while the weak carrier is massive. In addition, weak processes like beta decay involve the exchange of electric charge. Indeed, those responsible for the initial formulation of the theory, especially Sheldon Glashow and Steven Weinberg (private correspondence), claim that they were trying to understand weak interactions (in particular beta decay) rather than attempting to unify the weak force with electromagnetism. But, rather than building from the ground up, the process of theory construction in this case involved using symmetry principles and gauge theory to develop a framework for fundamental interactions. The power of gauge theory and symmetries was well known from successes in quantum electrodynamics. There it was possible to show that from the conservation of electric charge one could, on the basis of Noether's theorem, assume the existence of a symmetry; the requirement that it be local forces one to introduce a gauge field that turns out to be just the electromagnetic field. So, from the fact that the Lagrangian must be locally gauge invariant it is possible to predict the existence of the electromagnetic field, since symmetry can be preserved only by introducing this new gauge field. Hence, the mathematics can, in a sense, be seen to dictate the physics; alternatively, one might claim that the two simply become inseparable.

What is significant about these symmetry groups is that they are more than simply mathematizations of certain kinds of transformations; they are in fact responsible for generating the physical dynamics. In electrodynamics local symmetry requirements are capable of determining the structure of the gauge field, which in turn dictates, almost uniquely, the form of the interaction, that is, the precise form of the forces on the charged particle and the way in which the electric charge current density serves as a source for the gauge field.[3]

What was needed in the electroweak case was a symmetry principle that would relate the forms of the weak and electromagnetic couplings. The initial model involved a symmetry group SU(2) × U(1), a combination of the U(1) symmetry of electrodynamics and the SU(2) group associated with weak interactions. However, this mathematical structure provided no natural way of introducing the particle masses associated with the weak force into the theory. Although they could be added by hand, this destroyed the invariance of the Lagrangian since gauge theory requires the introduction of only massless particles. Hence, one could not reconcile the *physical* demands of the weak force with the *formal* demands of gauge theory. The solution to the problem came with "spontaneous symmetry breaking." The idea is that when a local symmetry is spontaneously broken the vector particles acquire a mass through a phenomenon known as the Higgs mechanism. The equations of motion of the theory have an exact symmetry that is broken in the physical system by the Higgs mechanism. The particles that carry the weak force acquire a mass and are governed via the structure of the Higgs field; hence the masses do not affect the basic gauge invariance of the theory.

So the Higgs phenomenon plays two related roles in the electroweak theory. First, it explains the discrepancy between the photon and boson masses: the photon remains massless because it corresponds to the unbroken symmetry subgroup U(1) associated with the conservation of charge; the bosons have masses because they correspond to the SU(2) symmetries that are spontaneously broken. Second, the avoidance of an explicit mass term in the Lagrangian allows for the possibility of renormalizibility. Once this mechanism was in place, the weak and electromagnetic interactions could be successfully unified under the SU(2) × U(1) gauge symmetry group.

One of the most interesting features of the electroweak theory that emerges from the interaction between the mathematics and the physics is the *kind* of unity that is achieved. Previously we saw that in Maxwell's electrodynamics the Lagrangian formalism provided the theoretical structure for electromagnetism and optics without any physical details of the system itself; the SU(2) × U(1) gauge theory, taken by itself, furnishes a similar kind of framework. It specifies the form of the interactions between the weak and electromagnetic forces but provides no theoretical explanation of why the fields must be united. And, unlike the role of gauge theory in quantum electrodynamics, in the electroweak case gauge theory alone is incapable of generating a unified theory directly from basic principles; it requires, in addition, the introduction of symmetry breaking and the Higgs mechanism.

Moreover, there is a strong sense in which the SU(2) × U(1) gauge theory allows only for the *subsuming* of the weak and electromagnetic forces under a larger framework, rather than the kind of reductive unity initially provided by Maxwell's electrodynamics. This is due to the fact that the weak and electromagnetic forces remain essentially distinct; the unity that is supposedly achieved results from the unique way in which the forces interact, that is, the mixing of the fields. The theory retains two distinct coupling constants: one (q) for the electromagnetic field and another (g) for the SU(2) gauge field. In order to calculate the masses of the particles, the coupling constants are combined into a single parameter θ known as the Weinberg angle, which fixes the ratio of the couplings. In order for the theory to be unified θ must have the same value for all processes. But the theory does not provide direct values for this parameter; as a result it cannot furnish a full account of how the fields are mixed. The mixing is not the result of constraints imposed by gauge theory itself; rather, it follows from an assumption imposed at the outset, namely that leptons can be classified as particles possessing isospin and are governed by the SU(2) symmetry group.

The crucial feature that facilitates the field interactions is the non-Abelian (noncommutative) structure of the symmetry group rather than something derivable from the phenomenology of the physics.[4] In any gauge theory the conserved quantity serves as the source of the gauge field in the same way that electric charge serves as the source of the electromagnetic field. In Abelian gauge theories (such as electromagnetism) the field itself is not charged, but in non-Abelian theories the field is charged with the same conserved quantity that generates it. In other words, the non-Abelian gauge field generates itself, and as a result these fields interact with each other, unlike those in electrodynamics. The form of the interactions (also determined by the structural constraints of the symmetry group) leads to cancellations between different contributions to high-energy amplitudes. Hence, both the unity and renormalizibility of the electroweak theory emerged as a result of the very powerful structural features imposed by the mathematical framework of the non-Abelian gauge group. So, although the Higgs mechanism allows for the possibility of producing a unified theory (there was no way to add the boson masses otherwise), the actual unity is the result of the constraints imposed by the isospin SU(2) group and the non-Abelian structure of the field.

What is interesting about this case is that the particles that act as the sources of the fields, and of course the forces themselves, remain distinct, but the theory provides a unified formal structure that enables one to describe how these separate entities can interact and combine. To that extent the theory exhibits a unity at the level of theoretical structure while at the same time retaining a measure of disunity with respect to its substantive aspects, namely the particles that carry the forces.[5]

As we saw previously, Maxwell's equations facilitated a reduction that was explanatory in only a very limited way; most importantly, the physical processes that lay at the foundation of the theory were left unexplained. Although the theoretical structure was capable of describing the behavior of optical and electromagnetic phenomena, no explanation of the theory's physical assumptions—for example, the nature of the field and the propagation of waves—was provided. Similarly in the electroweak case, the Higgs mechanism facilitates the unification but does not explain the mixing of the fields; the Weinberg angle *represents* the mixing, yet no value for θ is given from within the theory. In both of these cases the unification results from specific properties of a mathematical structure, and neither structure provides a comprehensive explanatory story about the phenomena themselves. Consequently unity and explanation must be differentiated; the mechanisms that produce the unification are not mechanisms that enable us to explain why or how the phenomena behave as they do.

The moral of the story is twofold. When theories are unified it can sometimes be traced to the power of the mathematical structures used to *represent* the phenomena rather than to a well-worked-out qualitative account of how the phenomena/processes relate to one another. In addition we can see—even limiting ourselves just to examples from physics—that theoretical unity is not characterizable in a monolithic way. It can take on different forms, and, perhaps most importantly, we see that unity is something that can exist at one level of theoretical structure without any substantive implications for a unified metaphysics or ontology. In that sense, the idea of theoretical unification neither presupposes as an initial premise nor implies as a conclusion ontological reduction. Sometimes the two may coincide, but this is by no means a reason for seeing them as part of the same undertaking.

What then are the implications of this conclusion for our topic, the limits of science? Given that the notion of unity in and among the sciences is frequently cited as a primary goal, the extent to which it can and has been achieved speaks directly to the limitations of scientific theorizing. What we have seen here is that the unity achieved in localized contexts of theory construction cannot be taken as evidence for global claims about the ability of science to provide a unified ontological description of the empirical world. The unification produced by physical theories is not only diverse in nature but also limited in scope. Consequently, the limits associated with the project of constructing a unified theory or science arise from the multifaceted character of unity itself. These limits are determined in part by the nature of physical systems themselves but also by the way in which we define unity in science. Consequently any claims regarding the ability of science to produce a "grand unified theory" based on current success must be interpreted in light of an analysis of the kind of unity that has thus far been achieved. Taking seriously the current evidence, one might reasonably conclude that a unity conceived of as ontological reduction (as opposed to an integrative synthesis) is considerably more likely to be subjected to limits that cannot "in principle" be overcome. However, the seventeen or so free parameters associated with the unification of the weak, electromagnetic, and strong forces suggest that even hopes for an integrative or synthetic unity might be exceeding the possible limits of what one typically takes to be "unification."

Disunity and the Limits of Science

Thus far I have mainly discussed how different notions of unity impose limits for the project of providing a unified theory in physics. There are other ways in which the limits of unity are evident, specifically in the various explicit examples of disunity in particular sciences. Perhaps the most pervasive kind of disunity exists in cases in which the theoretical treatment of a specific phenomenon involves the use of several different models, some of which many be inconsistent with others. The nucleus is a case in point. A variety of models are used to describe the structure of the nucleus, depending on the kind of phenomenon we want to understand. Nuclear fission makes use of the liquid drop model, which relies strictly on a classical description. Other kinds of nuclear data require the shell model, which incorporates quantum properties, such as spin statistics, that cannot be included in the liquid drop model. Additional experimental results, such as those of some scattering experiments, are explained using the compound nucleus model, which builds in quantum mechanical properties that cannot be added to the shell model. Still other phenomena, such as high-energy neutron scattering, require the optical model. Although quantum chromodynamics (QCD) is the theory governing nuclear forces, because of its structure it is applicable only to high-energy domains; hence we require models when dealing with nuclear phenomena or processes that take place at low energies. Nor does QCD tell us how to construct nuclear models; instead the models are developed in a piecemeal way for the purpose of classifying and explaining experimental results.

The way models function in science provides good evidence for a kind of disunity that points to two concerns: first, the epistemological fact that in some cases we are simply unable to construct the kind of comprehensive theories that allow us to treat certain kinds of physical systems in a unified way, and second, the ontological claim that disunity is a reflection of the diversity present in the physical world. In the case of nuclear physics, models are not just approximations to theory or derived from theoretical structure in a way that allows them to be part of a single theoretical framework. Instead they stand on their own as ways of representing and explaining the physical world. The fact that models possess this kind of autonomy is evidence for a disunity inherent in many aspects of science, not just between disciplines but within theoretical domains themselves. Another example that illustrates the way in which models point to disunity via their depiction of physical systems comes from hydrodynamics. Despite the existence of well-established theories in the field, prior to 1904 these theories were inapplicable to certain kinds of fluid flows. That is, the theories themselves could not yield models or suggest approximations that could be employed in treating empirical systems. What was needed was a way of reconciling theory with experimental results, something that was finally achieved by using two different theoretical pictures to treat a fluid flowing past a solid boundary. In other words, the same homogenous fluid required two separate theoretical treatments to explain the long-standing discrepancy between the theory of classical hydrodynamics and the empirical science of hydraulics.

The problem was that classical hydrodynamics evolved from Euler's equations describing a frictionless, nonviscous fluid; but this theory could not account for the treatment of viscous fluids nor for the flows of fluids of low viscosity (air and water) past a solid boundary. Although the Navier-Stokes equations dealt with frictional forces, their mathematical intractability made it impossible to develop a systematic theoretical treatment for these nonideal fluids. And simply ignoring viscosities that were extremely small led to glaring conflicts between theoretical predictions and experimental outcomes. The difficulty was solved in 1904 by Ludwig Prandtl (1961 [1904]), who built a small water tunnel that would recreate fluid flows and furnish a visualization of the flows in different regions of the fluid. From this physical model he was able to construct a mathematical model that could treat the flow problem in an accurate way.

The obstacle was that the theory of perfect fluids assumes there are no tangential forces on the boundary between the fluid and a solid wall (resulting in a "slip" caused by a difference in relative tangential velocities). In real fluids there are intermolecular attractions that cause fluid to adhere to a solid wall, thus giving rise to shearing stresses. However, even in cases of very low viscosity (air and water) the condition of "no slip" prevails near a solid boundary. What the water tunnel revealed was that flow around a solid body could be divided into two parts: the thin layer in the neighborhood of the body, where friction plays an essential role (the boundary layer), and the region outside this layer, where it can be neglected. The reason viscous effects become important in the boundary layer is that the velocity gradients are much larger than they are in the main part of the flow, owing to the fact that a substantial change in velocity is taking place across a very thin layer (Landau and Lifshitz 1959).

If one assumes that this boundary layer is very thin, the Navier-Stokes equations can be simplified into a solvable form; as a result an analytical continuation of the solution in the boundary layer to the region of ideal flow can be given (Gitterman and Halpern 1981). Hence one divides the fluid conceptually into two regions, the boundary layer, where the Navier-Stokes equations apply, and the rest of the fluid, where a different set of solvable equations (those of classical hydrodynamics) applies. Of course there are no strict "boundaries" for the part of the fluid that constitutes the boundary layer; rather, there is a constant flow of fluid particles in and out owing to Brownian motion. But this presents neither a conceptual nor a mathematical problem. The particularly interesting thing about this case is that it is the phenomenology of the *physics* that dictates the use of different approximations for the same homogenous fluid. Although the solutions ultimately depend on the structural constraints provided by classical hydrodynamics and the Navier-Stokes equations, the kind of approximations used in the solutions come not from a direct simplification of the mathematics but from the phenomenology of the fluid flow as represented by Prandtl's water tunnel (Morrison 1999). In other words, a resolution to the problem could not have been attained simply by mathematical manipulation of the equations of hydrodynamics or the Navier-Stokes equations; it was impossible to reconcile the physical differences between real and ideal fluids simply by retaining the "no-slip" condition in a mathematical analysis of inviscid fluids.

What exactly does this example tell us about disunity? Primarily it tells us something about the relationship between models and theories in dealing with concrete phenomena. There are some cases, such as Newtonian physics, in which we need models to apply the theory; for example, we use the pendulum model as a way of applying the second law in the description of harmonic motion. In situations such as this the models one uses are models of the theory in the sense that they are generated from the basic framework of Newtonian mechanics. In that sense the models do not really indicate a departure from the fundamental theoretical structure in the treatment of phenomena. However, many cases of modeling are not like this. QCD is typically characterized as a unified theory—an element of the standard model that unifies the weak, strong, and electromagnetic interactions. Yet, as we have seen, when one is interested in nuclear phenomena the theory is of little help in either accounting for or suggesting models of how one might explain these processes. Instead phenomenological models of nuclear structure are developed piecemeal, taking aspects of classical and quantum mechanics to account for different kinds of events, such as fission or scattering. Hence, unlike the example of the pendulum, a kind of disunity exists between the theory and the models, as well as between the different models themselves.

To some extent the case of the boundary layer resembles the nuclear case in that the model was completely unmotivated by the structure of either theory. The approximations that were subsequently added to the Navier-Stokes equations were dependent on the way in which one conceptualized the fluid layer surrounding a solid body. In that sense the model is not just a way to apply theory; rather it provides an entirely different conceptualization of the phenomena than the theory itself provides. Once this new picture was in place, the theory could be mathematically manipulated to generate appropriate solutions. What this reveals is that the presence of a well-established background theory sometimes requires the existence of separate phenomenologically based models to account for phenomena considered to be within the theoretical domain.

My point is not one of simply discovering new knowledge and extending current theories. Instead, I am making a methodological claim about how science is carried out and how that practice involves models as an essential ingredient. These models enable us to use the theories we have and represent physical systems in a variety of ways. Yet they stand apart from the kind of fundamental theory that we typically see as the key ingredient for unifying nature. The existence of models and their function in these contexts point to a kind of theoretical disunity that is an integral part of physical science, a part that also stands side by side with a significant measure of unity.

At the beginning of this chapter I claimed that one of my goals was to argue that the unity/disunity debate as a metaphysics of nature rests on a false dichotomy. Inasmuch as this debate should be seen as a local matter with empirical foundations, so too should questions concerning the limits of science. Philosophical methods and concepts are important for analyzing the structure and implications of the evidence for both unity and disunity, and hence for the boundaries of scientific inquiry. They should not be used as a means for extending the domain of that evidence to claims about scientific practices and goals that are merely promissory.

Taken as a metaphysical thesis, unity and disunity as a particular instance of the limits of science take on a structure resembling a

Kantian antinomy. The resolution consists of showing that both thesis and antithesis (in this case unity and disunity) are mutually compatible, since neither refers to some ultimate way in which the world is constructed. As an empirical thesis about the physical world both are right, but as a metaphysical thesis both are wrong, simply because the evidence, by its very nature, is inconclusive. With respect to the broader question—does science have limits?—again, understood empirically, the answer is both *Yes* and *No*. Insofar as it is possible to imagine current limits being overcome, we can think of science as limitless in certain respects. Yet there is also a sense in which science will never be free of certain limits; as a human activity its practice will always be subject to constraints.

It is perhaps trivial to claim that the limits of science are constantly changing. Yet, if we take that statement seriously we begin to see the futility of framing it as a metaphysical issue. One need not adopt a metaphysical perspective to engage in philosophical analysis; and given the progress and methods of empirical science, there is perhaps nowhere where metaphysics is less helpful than in that domain.

NOTES

1. The difference between the Lagrangian formulation of mechanics and the Newtonian one is that in the former the velocities, momenta, and forces related to the coordinates in the equations of motion need not be interpreted literally. The method consists of expressing the elementary dynamical relations in terms of the corresponding relations of pure algebraic quantities, which facilitates the deduction of the equations of motion. Hence the quantities expressing relations between parts of a mechanical system that appear in the equations of motion are eliminated by reducing them to "generalized coordinates."

2. The point here is not that explanatory power requires a mechanical explanation of the phenomena (although this was certainly the view in nineteenth-century field theory). Rather, it is that some type of theoretical understanding about the origin and existence of the displacement current—the very foundation of Maxwell's field theoretic approach—is required to prevent the theory from being regarded as little more than phenomenological.

3. A local symmetry is one that can vary from point to point, as opposed to a global symmetry, in which transformations are not affected by position in space and time (e.g., rotation about an axis).

4. Non-Abelian gauge field theory was discovered by Yang and Mills in 1954.

5. For more on this see Morrison (2000).

REFERENCES

Friedman, M. 1983. *Foundations of Space-Time Theories*. Princeton, N.J.: Princeton University Press.

Gitterman, M., and V. Halpern. 1981. *Qualitative Analysis of Physical Problems*. New York: Academic Press.

Landau, L. D., and E. M. Lifshitz. 1959. *Fluid Dynamics*. Oxford: Pergamon Press.

Maxwell, J. C. 1954 [1873]. *A Treatise on Electricity and Magnetism*. New York: Dover.

———. 1965a [1856]. "On Faraday's Lines of Force." In Niven 1965, 1:155–229.

———. 1965b [1861–62]. "On Physical Lines of Force." In Niven 1965, 1:451–513.

———. 1965c [1865]. "A Dynamical Theory of the Electromagnetic Field." In Niven 1965, 1:526–97.

Morrison, M. 1990. "A Study in Theory Unification: The Case of Maxwell's Electrodynamics." *Studies in the History and Philosophy of Science* 23: 103–45.

———. 1992. "Two Kinds of Experimental Evidence." *PSA* 1: 49–62.

———. 1999. "Models as Autonomous Agents." In M. S. Morgan and M. Morrison, eds., *Models as Mediators*. Cambridge: Cambridge University Press, 38–65.

———. 2000. *Unifying Scientific Theories: Physical Concepts and Mathematical Structures*. New York: Cambridge University Press.

Niven, W. D., ed. 1965. *The Scientific Papers of James Clerk Maxwell*. 2 vols. New York: Dover.

Prandtl, L. 1961 [1904]. "Über Flüssigkeitsbewegung bei sehr kleiner Reibung." Reprinted in *Gesammelte Abhandlungen*. 3 vols. Berlin: Springer-Verlag, 3:574–83, 752–77.

Yang, C. N., and R. Mills. 1954. "Conservation of Isotropic Spin and Isotropic Gauge Invariance." *Physical Review* 96: 191–95.

Prospects for the Special Sciences

14

Limits and the Future of Quantum Theory

Gordon N. Fleming

Department of Physics, Pennsylvania State University

> For the sake of persons of . . . different types, scientific truth should be presented in different forms.
>
> —Maxwell (1965 [1890], 2:220)

Quantum theory may be the preeminent example of a scientific theory the history and current status of which are indicative of what many regard as our present confrontation with the limits of science. It is a framework theory (Shimony 1987, 401) more broadly applied and extensively empirically corroborated than any other in the history of science. It is widely believed to be, in principle, universally applicable to the physical world in essentially its present form (with serious qualifications regarding only the general relativistic, early universe, or quantum gravity contexts). And yet, after seventy years of intense analysis and argument, it is still not accompanied by an interpretation that is widely convincing to those of us who care about such matters. Most of us who embrace one or another of the many interpretations or approaches to quantum theory do so as an *approach* to understanding which seems promising rather than as a declaration of where understanding lies as a fait accompli. To claim that one already understands quantum theory, as regards its implications for the structure of the world and beyond the merely technical rules for its application, is to risk becoming the leader or a follower in a small cult!

It is fashionable today to claim that the majority of professional physicists actually belong to the large cult of the unreconstructed Copenhagen interpretation, but that is not my experience. Most physicists are still unaware of the strenuous activity that goes on to find genuinely adequate interpretations or completions of quantum theory. They may pay lip service to the Copenhagen story they imbibed from their quantum theory textbooks, and a few of them have concocted their own versions, which live pretty much in happy private vacuums. But when pressed on the issues concerning the conceptual foundations of quantum theory, they often as not will resign themselves to the notion that, beyond a point we may have already reached, nature, at the quantum level, is not to be understood but merely accepted. Our struggles and differences sometimes seem pointless to them.

Nor have the leading lights of the field offered much more. Thus, commenting on the problem of understanding quantum theory, Richard Feynman (1967, 129) wrote: "the difficulty really is psychological and exists in the perpetual torment that results from your saying to yourself, 'but how can it be like that?' which is a reflection of uncontrolled but utterly vain desire to see it in terms of something familiar. . . . I think I can safely say that nobody understands quantum mechanics. . . . Nobody knows how it can be like that." In the frustrating but fascinating book *The End of Science* (1997), John Horgan records observations by Steven Weinberg on the paradoxes of quantum theory ("I tend to think these are just puzzles in the way we talk about quantum mechanics") and on adopting the many worlds interpretation as a way to eliminate these puzzles ("I sort of hope that whole problem will go away, but it may not. That may just be the way the world is. . . . I think we'll be stuck with quantum mechanics. So in that sense the development of quantum mechanics may be more revolutionary than anything before or after") (75–76). Horgan also quotes Murray Gell-Mann on the strangeness of quantum theory: "I don't think there's anything strange about it! It's just quantum mechanics! Acting like quantum mechanics! That's all it does!" (215).

In his own book, *Dreams of a Final Theory* (1992), Weinberg offers a bit more direction. In the midst of a long list of the injuries done to science by the influence of positivism, he writes, "The positivist concentration on observables like particle positions and momenta has stood in the way of a 'realist' interpretation of quantum mechanics, in which the wave function is the representation of physical reality" (181).

And what do we who expend serious effort investigating the conceptual foundations of quantum theory have to offer the professional physicist in the street (where many of the new ones are these days!)? What can I, for example, present to the hungry, as yet unjaundiced, eyes of the advanced physics undergraduates and first-year graduate students who sign up for my course on the competing approaches to quantum theory?

Well, I try to find time to introduce them to Bohmian interpretations, quantum logic interpretations, Everett relative state or many worlds interpretations, many minds interpretation(s), statistical or ensemble interpretations, modal interpretations, decoherence interpretations, consistent histories interpretations, potentiality or propensity interpretations, relational or information theoretic approaches, and dynamical state reduction completions. (Have I forgotten any?) With the passage of time some of the divisions between these approaches have become blurred. Most of the approaches I have listed represent camps that display alternatives or even factionalism within their ranks. Arthur Fine (1996) surveyed the possible interpretations of Bohmian mechanics itself, defined as the formalism David Bohm developed for mounting his interpretation of standard quantum theory. In the same volume Jeffrey Bub (1996) showed how both Copenhagen quantum theory and Bohmian mechanics can fall under the umbrella of modal interpretations as very special cases.

What are we to make of this situation? Have we reached the limits of achievable communitywide agreement regarding the interpretation or completion of this seemingly fundamental theory? Is it to remain forever the case that we will each have to choose a particular approach to quantum theory from among this (no doubt growing) spectrum of options on the basis of personal philosophical and psychological predilections and aesthetic tastes, essentially unassisted by really *compelling* philosophical argument or scientific considerations? Are we in this sense, irreducibly, as Maxwell put it in my opening quote, "persons of . . . different types"? Have we degenerated (as I occasionally find myself gloomily thinking) into scholastics debating how many interpretations can dance on the head of a formalism?

Some perspective may be gained by recalling briefly that interpretational disputes over theories are not at all new to physics. Since much of the disagreement between champions of different approaches to quantum theory is over what really exists in this quantum world, we are reminded of the waxing and waning fortunes of the concepts of force and energy as fundamental constituents of reality during the eighteenth and nineteenth centuries. This evolution was punctuated by Hamilton and Jacobi's development (Dugas 1988, 390–408) of the calculationally powerful transformation theory based on energy concepts and variational principles and, ultimately abortively, by Hertz's (1956 [1899]) effort to eliminate the force concept altogether from mechanics in favor of geometrical constraints in a higher-dimensional configuration space. Although force is still with us, energy has clearly achieved ontological priority in this century by virtue of its association with symmetry principles, its relativistic amalgamation with mass, and the central role the energy operator plays in quantum theory.

Another case in point is the long evolution of the controversy between energy conservation as the all-encompassing law of nature and atomism (Scott 1970). This disagreement, which may be seen as an instance of the clash between ontologies of continuity and discreteness, became entangled with issues of positivism in the late nineteenth century (Gillespie 1960, 495–99; Meyerson 1960 [1930], 346–51). With increasing access to indirect empirical evidence of the existence of atoms, such as Brownian motion, atomism won that battle. And today, in our quantum theory, continuity and discreteness coexist well enough in the eigenvalue spectra of self-adjoint operators, but not quite so well in the dualism of the continuous unitary evolution and the discrete stochastic state reduction of the state vector. The desire to eliminate state reduction helps to drive most of the approaches to quantum theory I have listed. And a new chapter in the continuity versus discreteness controversy may now be emerging with the current ideas on the possible discrete structure of space-time itself coming from the quantum gravity research community (e.g., Baez 1994).

A close analogue to some of the interpretation issues in quantum theory is the controversy between the champions of a mechanical foundation for the electromagnetic field and those claiming autonomous reality for the field. Here Hertz rode the winning bandwagon. His masterful book summarizing his research on electromagnetic waves (Hertz 1962 [1893])—along with the writings of Oliver Heaviside (1950 [1893–1912]), who strongly influenced Hertz (Hunt 1991, 180–82) and finally Einstein's kinematical derivation of the Lorentz-Poincaré transformation equations—took the wind out of the sails of the mechanists. Yet the sails were becoming slack anyway, as Einstein

(Schilpp 1949, 1:27) later recalled: "One got used to operating with these fields as independent substances, without finding it necessary to give oneself an account of their mechanical nature; thus mechanics as the basis of physics was being abandoned, almost unnoticeably, because its adaptability to the facts presented itself finally as hopeless." Heaviside (1899), who believed in the mechanical ether, exemplified this process when he wrote to Hertz: "My experience of so-called 'models' is that they are harder to understand than the equations of motion!"

I see a close analogy between this example and some of the interpretation problems of contemporary quantum theory because just as ether models were driven by the desire to understand the new electromagnetic physics in familiar mechanical terms, so I agree with Feynman as quoted previously in seeing the Bohmian interpretations—and, to a lesser degree, the modal interpretations—as driven by the desire to understand the new quantum physics (except it is not very new any longer) in familiar classical terms.

A considerable assist to the resolution of each of these historical controversies was provided by the continual acquisition, over the years, of new data. Of this one would always ask: Which of the disputed approaches handles it better, or sheds more light on it, or even suggested it? We, however, in our quantum morass, reflect the sentiments of David Mermin's (1990) fictional professor Mozart: "All particle physics has taught us about the central mystery is that quantum mechanics still works. Perfectly, as far as anyone can tell. What a letdown!" The same could be said concerning high-energy astrophysics, the experimental study of the micro-macro transition, and, of course, the nonlocal correlations and delayed choice experiments that corroborate quantum theory as they violate the Bell inequalities. Will we ever find empirical evidence of the limits of applicability of quantum theory in its present form?

I think we will, but do not ask me for a time schedule! But why should I think we will at all? First of all, the dynamical reduction models that treat state reduction as a real physical process already deviate from standard quantum theory. In fact in the relativistic domain it is difficult to ensure that they do not deviate so much that we must already discard them! I look forward to the experimental tests that will emerge to distinguish them from standard quantum theory. Furthermore, with the exception of the statistical or ensemble interpretations, in which the state vector for a quantum system encodes *only* epistemic probability data reflecting our ignorance of reality, all of the competing interpretations attribute some degree of physical reality to the state vector itself, or one of its representations. Now as a consequence of the history of the ascendance of the energy concept, which I briefly considered, all genuine constituents of the physical world have come to be expected to contribute to or carry some of the world's energy. If this is true of the state vector (and why not?), the simplest way would be for the state vector to appear in the Hamiltonian, thus rendering the theory nonlinear and the famous-cum-notorious quantum superposition principle approximate! Nonlinear modifications of Schrödinger's equation have been considered by several workers, and the sizes of the nonlinear modifications have, in some instances, been experimentally determined to be less than very small upper bounds (see Shimony 1993, 2:48–54, for a useful assessment). But that was before decoherence theory was developed, and one wonders if decoherence processes may mask the nonlinearity.

Finally, on behalf of nonlinear modifications of quantum theory, it has many times been conjectured that just as the successful formulation of a theory of quantum gravity may require that general relativity, which is nonlinear, become probabilistic, so it may require that quantum theory, which is probabilistic, become nonlinear. We cannot, unfortunately, really be sure, since superstring theorists think quantum gravity will be found within the framework of their scheme (Davies and Brown 1988), and superstring theory sits squarely within linear homogeneous quantum theory. Of course, if we can get nonlinear modifications to quantum theory only through quantum gravity, the way in which that would help our present problems (which seem to have nothing to do with gravity) is hard to see. But then Roger Penrose (1989, 1994) may come to our aid! I do not mean to suggest that nonlinearity and the breakdown of the superposition principle would be a panacea. But it would surely be a profound conceptual difference, and one would hope that not all of the presently competing approaches to quantum theory would survive the innovation! I really do think that the discovery of a breakdown of the superposition principle—along energy theoretic lines, and possibly supplanting but also possibly augmenting the current dynamical reduction theories—is a likely development. But again, do not ask me for a time schedule!

Nor is it *impossible* that some innovations in the empirically testable content of quantum theory itself will emerge from the camps of the nonstandard *interpretations*. For example, the liberal wing of the Bohmian camp (Valentini 1996) is exploring plausible ways of altering Bohmian mechanics to deviate from standard quantum theory under exotic circumstances. Research into the family of modal interpretations frequently leads to theorems concerning the mathematical structure of the quantum theory formalism (e.g., Bub and Clifton 1996). These theorems, quite apart from the interpretations that motivate them, may provoke experimental searches into obscure nooks and crannies of the quantum world, where we may find the limitations of quantum theory.

Another possibility, more fashionable in character than my mundane energy considerations, is that a breakdown of the superposition principle may emerge in the domain of complexity, that is, of complex systems. Since one definition of a complex system is a system containing many fundamental constituents with a significant (but not too large) number of connections between them (Kauffman 1995), macroscopic bodies can belong. But macroscopic bodies are precisely the most mysterious in a quantum world, for they are the ones that persistently display quasi-classical behavior. The dynamical reduction completions of quantum theory (Pearle 1996) aim to account for just this by having macroscopic bodies automatically subject to continual rapid-fire state reductions of the desired sort. The experimental exploration of systems on the fringes of complexity (mesoscopic systems?) would seem to be a natural way to look for evidence of these nonlinear processes.

At the same time it is desirable to continue experimentally pressing the limits of the circumstances under which we *can* observe superposition. For example, macroscopic quantum interference devices, the neutron interferometer, the monitoring of individual quantum particles in Penning traps, and the observations of early deviations from exponential decay of unstable states continue to provide striking manifestations of superposition. How far can this be pushed? Already the classic instances of single photons, electrons, protons, and neutrons interfering with themselves have been greatly surpassed by recent experiments in which such massive entities as whole beryllium atoms and molecules have been manipulated into superpositions of mesoscopically distinct states and have displayed self-interference (Monroe et al. 1996). Some workers are developing interferometric techniques that have them anticipating de Broglie interference of small rocks and live viruses! Although it is extremely unlikely that we will ever see in the laboratory the entangled state of Schrödinger's cat, we may one day see the entangled state of Schrödinger's retrovirus, or maybe even Schrödinger's biological cell. Perhaps it is in these directions that we will find the limits of quantum theory.

But perhaps not! What if the worst prevails? What if we never discover any breakdown of quantum theory in essentially its present linear homogeneous form? What then?

It could be that most of the existing interpretations will continue to attract adherents indefinitely. Insofar as an interpretation goes beyond quantum theory in ways that, as a matter of principle, cannot be subject to empirical inquiry (which is the case for the conservative wing of the Bohmian camp, for the modal interpretations, for the many worlds interpretation, and for some others), choosing among them would not involve scientific considerations in any significant way. The choices would be overwhelmingly driven by philosophical, psychological, and aesthetic considerations.

On the other hand, it could happen—and this would be my personal hope—that these interpretations will come to be seen less as opportunities for untestable ontological commitments and more as provocative formal approaches to quantum theory that assist in suggesting novel experiments or solving novel problems. In this way they would resemble choosing a gauge in electromagnetism. A wise or clever choice can be very helpful in "seeing" the problem at hand, but when the work is done one remembers that a gauge choice is just a useful formal freedom, not a metaphysics to be embraced! If and when such an attitude would prevail, a natural impulse would then be to reformulate quantum theory in a "gauge-independent" manner in this sense. Such a reformulation would not provide the most potent approach to individual problems. Choosing a gauge would remain the way to go for that purpose. But the gauge-independent formulation would help to delineate the aspects of the theory one could safely take seriously, the aspects one could tentatively invest with ontological content.

The other approaches (decoherence, consistent histories, potentiality, dynamical reduction, information theoretic)—which were motivated from the start by concerns for conceptual tightening of standard quantum theory and/or interest in various empirical research programs (and which do not go beyond quantum theory in ways that, in principle, cannot be subject to empirical inquiry)—would continue playing those roles and reciprocally be streamlined and amalgamated with one another as a result of progress in those research programs. My guess is that the resulting amalgamation would be synonymous with the gauge-independent reformulation of the previous paragraph.

In these ways we would gradually get used to quantum theory and come to accept a description of it and the world in which it works in terms of ontological concepts as closely and usefully tied to the empirically corroborable aspects of the formalism as possible. A healthy balance between qualified realism and qualified positivism would prevail. That, at least, is my vision.

REFERENCES

The references are not always to original sources. They have been chosen as representative examples of the literature and for ease of access.

Baez, J., ed. 1994. *Knots and Quantum Gravity.* Oxford: Oxford University Press.
Bub, J. 1996. "Modal Interpretations and Bohmian Mechanics." In Cushing et al. 1996, 331–42.
Bub, J., and R. Clifton. 1996. "A Uniqueness Theorem for Interpretations of Quantum Mechanics." *Studies in History and Philosophy of Modern Physics* 27B: 181–220.
Cushing, J. T., A. Fine, and S. Goldstein, eds. 1996. *Bohmian Mechanics and Quantum Theory: An Appraisal.* Dordrecht: Kluwer.
Davies, P. C. W., and J. Brown. 1988. *Superstrings: A Theory of Everything.* Cambridge: Cambridge University Press.
Dugas, R. 1988. *A History of Mechanics.* New York: Dover.
Feynman, R. 1967. *The Character of Physical Law.* Cambridge, Mass.: MIT Press.
Fine, A. 1996. "On the Interpretation of Bohmian Mechanics." In Cushing et al. 1996, 231–50.
Gillespie, C. C. 1960. *The Edge of Objectivity.* Princeton, N.J.: Princeton University Press.
Heaviside, O. 1899. Letter to Hertz, August 14. Heinrich Hertz Deutsches Museum.
———. 1950 [1893–1912]. *Electromagnetic Theory.* 3 vols. New York: Dover.
Hertz, H. 1956 [1899]. *The Principles of Mechanics Presented in a New Form.* New York: Dover.
———. 1962 [1893]. *Electric Waves.* Trans. D. E. Jones. New York: Dover.

Horgan, J. 1997. *The End of Science*. New York: Broadway.

Hunt, B. J. 1991. *The Maxwellians*. Ithaca, N.Y.: Cornell University Press.

Kauffman, S. 1995. *At Home in the Universe*. Oxford: Oxford University Press.

Maxwell, J. C. 1965 [1890]. "Address to the Mathematical and Physical Sections of the British Association." In W. D. Niven, ed., *The Scientific Papers of James Clerk Maxwell*. 2 vols. Cambridge: Cambridge University Press, 2:215–29.

Mermin, D. 1990. "What's Wrong with Those Epochs?" *Physics Today*, November, 9–11.

Meyerson, E. 1960 [1930]. *Identity and Reality*. New York: Dover.

Monroe, C., D. M. Meekhof, B. E. King, and D. J. Wineland. 1996. "A 'Schroedinger Cat' Superposition State of an Atom." *Science*, May 24, 1131–35.

Pearle, P. 1996. "Wavefunction Collapse Models with Nonwhite Noise." In R. Clifton, ed., *Perspectives on Quantum Reality*. Dordrecht: Kluwer, 93–110.

Penrose, R. 1989. *The Emperor's New Mind*. Oxford: Oxford University Press.

——. 1994. *Shadows of the Mind*. Oxford: Oxford University Press.

Schilpp, P. A. 1949. *Albert Einstein: Philosopher-Scientist*. 2 vols. Evanston, Ill.: Tudor.

Scott, W. L. 1970. *The Conflict Between Atomism and Conservation Theory 1644–1860*. London: Macdonald Elsevier.

Shimony, A. 1987. "The Methodology of Synthesis: Parts and Wholes in Low-Energy Physics." In R. Kargon and P. Achinstein, eds., *Kelvin's Baltimore Lectures and Modern Theoretical Physics*. Cambridge, Mass.: MIT Press, 399–424.

——. 1993. *Search for a Naturalistic Worldview*. 2 vols. Cambridge: Cambridge University Press.

Valentini, A. 1996. "Pilot-Wave Theory of Fields, Gravitation and Cosmology." In Cushing et al. 1996, 45–66.

Weinberg, S. 1992. *Dreams of a Final Theory*. New York: Pantheon.

15

Limits to Biological Knowledge

Alex Rosenberg
Department of Philosophy, University of Georgia

In this chapter I argue that one particular science faces limits that do not confront other sciences, and that these limits reflect a combination of facts about the world and facts about the cognitive and computational limitations of the scientists whose business it is to advance the frontiers of this science. The science is biology, and the limitations I claim it faces are those of explanatory and predictive power. In the first part of this chapter I advance a contingent, factual argument about the process of natural selection that consigns the biology in which we humans can take an interest to a kind of explanatory and predictive weakness absent in our physical science. I then show how these limitations are reflected in at least two of the ruling orthodoxies in the philosophy of biology: the commitments to the semantic approach to theories and to physicalist antireductionism.

If I am correct about the limits to biological knowledge, we must face some serious issues in our conception of what scientific adequacy and explanatory understanding consist of. To see why, consider what Nicholas Rescher writes in *Predicting the Future:*

Scientific adequacy . . . involves a complex negotiation in which both prediction and explanation play a symbiotic and mutually supportive role:

(1) To qualify as well established, our explanatory theories must have a track record of contributing to predictive success.

(2) To qualify as credible, our predictions must be based upon theories that militate for these predictions over and against other possibilities.

(3) Our explanatory theories should be embedded in a wider explanatory framework that makes it possible to understand why they enjoy their predictive successes.

The ideal is a closed cycle of theory-driven improvements in predictive performance that are themselves explicable in a wider context of theoretical understanding. (1998, 168)

This ideal can be approached only through the provision of scientific laws of increasing power and generality. For neither scientific explanation nor prediction can proceed without nomological generalizations.

Of course the closed cycle of theory-driven improvements in predictive performance of which Rescher speaks is an ideal, one that is nowhere fully attained. My claim is that biology is far more limited in its ultimate degree of attainment of this ideal than are the physical sciences, because the only generalizations of which biology is capable will not provide for the cycle of coordinated improvement in explanation and prediction that characterizes increasingly adequate science. This fact about biology reflects as much on the biologist as it does on the phenomena the biologist seeks to explain and predict. Were we much smarter, physics and chemistry would remain very much as they are, but biology would look much different. The history of chemistry and physics is a history of an increasingly asymptotic approach to the closed cycle of theory-driven improvement in prediction that makes us confident that we are uncovering the fundamental regularities that govern the universe. If we were smarter—if we could calculate faster and hold more facts in short-term memory—quantum mechanics, electromagnetism, thermodynamics, and physical and organic chemistry would still be characterized by physical theories recognizably similar to the ones we know and love.

But if we were smarter, then either the generalizations of biology as we know it would give way to others, couched in kind-terms different from those that now characterize biology, or they would give way to the laws and theories of physical science, aided only by a principle of natural selection. Were this to happen, the limitations on attaining the closed cycle of theory-driven predictive improvements about the phenomena we now call biological could be abridged. But the resulting science would not be recognizably biology. And besides, we are not going to transcend our natural cognitive and computational abilities any time soon.

The reason for the predictive weakness and consequent explanatory limits of biology is of course to be found in the generalizations of which it is capable. To see why biology is capable only of weak generalizations, we need merely reflect on two considerations: the mechanism of natural selection and the inevitability for us of individuating biological kinds by their causal roles.

Natural selection "chooses" variants by *some of their effects,* those that fortuitously enhance survival and reproduction. When natural selection encourages variants to become packaged together into larger units, the adaptations become functions. Selection for adaptation and function kicks in at a relatively low level in the organization of matter. As soon as molecules develop the disposition chemically, thermodynamically, or catalytically to encourage the production of more tokens of their own kind, selection gets enough of a toehold to take hold. Among such duplicating molecules, at apparently every level above the polynucleotide, there are frequently to be found multiple *physically distinct* structures with some (nearly) identical rates of duplication, different combinations of different types of atoms and molecules, that are close enough to being equally likely to foster the appearance of more tokens of the types they instantiate. Thus, so far as adaptation is concerned, from the lowest level of organization onward there are frequently *ties* between structurally different molecules for first place in the race to be selected. And, as with many contests, in case of a tie, duplicate prizes are awarded. For the prizes are increased representation of the selected types in the next "reproductive generation." This will be true up the chain of chemical being all the way to the organelle, cell, organ, organism, kin group, and so on.

It is the nature of any mechanism that selects for effects that *it cannot discriminate between differing structures with identical effects.* Functional equivalence combined with structural difference will always increase as physical combinations become larger and more physically differentiated from one another. Moreover, perfect functional *equivalence* is not necessary. Mere functional similarity will do. So long as two or more physically different structures have packages of effects each of which has roughly the same repercussions for duplication in the same environment, selection will not be able to discriminate between them unless rates of duplication are low and the environment remains very constant for long periods of time. Note that natural selection makes functional equivalence–cum–structural diversity *the rule and not the exception* at every level of organization above the molecular. In purely physical or chemical processes, where there is no opportunity for nature to select by effects, structural differences with equivalent effects are the exception, if they obtain at all.

Now if selection for function is blind to differences in structure, then there will be nothing even close to a strict law in any science that individuates kinds by selected effects, that is, by functions. This will include biology and all the special sciences that humans have elaborated. From molecular biology through neuroscience, psychology, sociology, economics, and so on, individuation is functional. That cognitive agents of our perceptual powers individuate functionally should come as no surprise. Cognitive agents seek laws relating natural kinds. Observations by those with perceptual apparatus like ours reveal few immediately obvious regularities. If explanations require generalizations, we have to theorize. We need labels for the objects of our theorizing even when they cannot be detected, because they are too big or too small or mental. We cannot individuate electrons, genes, ids, expectations about inflation, or social classes structurally because we cannot detect their physical features. But we can identify their presumptive effects. This makes most theoretical vocabulary "causal role" description.

It is easy to show that there will be no strict exceptionless generalizations incorporating functional kinds. Suppose we seek a generalization about all *F*s, where *F* is a functional term, like *gene* or *wing* or *belief* or *clock* or *prison* or *money* or *subsistence farming*. We seek a generalization of the form $(x)[Fx \rightarrow Gx]$. In effect our search for a law about *F*s requires us to frame another predicate, Gx, and determine whether it is true of all items in the extension of Fx. This new predicate, Gx, will itself be either a structural predicate or a functional one. Either it will pick out Gs by making mention of some physical attribute common to them, or it will pick out Gs by descriptions of one or another of the effects (or just possibly the causes) that everything in the extension of Gx has. Now there is no point in seeking a structural, physical feature that all members in the extension of Fx bear: the class of Fxs is physically heterogeneous just because they have all been selected for their effects. It is true that we may find some structural feature shared by most or even all of the members of F. But it will be a property shared with many other things—like mass, or electrical resistance, properties that have little or no explanatory role with respect to the behavior of members of the extension of Fx. For example, the exceptionless generalization that "all mammals weigh more than 0.0001 gram" does relate a structural property (weight) to a functional one (mammality), but this is not an interesting biological law.

Could there be a distinct functional property different from F shared by all items in the extension of the functional predicate Fx? The answer must be that the existence of such a distinct functional property is highly improbable. If Fx is a functional kind, then the members of the extension of Fx are overwhelmingly likely to be physically diverse, owing to the blindness of selection to structure. Since they are physically different, any two Fs have nonidentical sets of effects. If there is no item common to all these nonidentical sets of effects, selection for effects has nothing to work with. It cannot uniformly select all members of F for some further adaptation. Without such a common adaptation, there is no further function all Fs share.

Whether functional or structural, there will be no predicate Gx that is linked in a strict law to Fx. We may conclude that any science in which kinds are individuated by causal role will have few if any exceptionless laws. So long as biology and the special sciences continue so to individuate their kinds, the absence of strict laws will constitute a limitation on science. How serious a limitation will this be?

Not very, some philosophers of science will say. There is a fairly widespread consensus that many ceteris paribus generalizations have explanatory power, and that they do so because they bear nomological force—for all their exceptions and exclusions. This thesis is the linchpin of most accounts of the integrity of what Fodor has called "the special sciences." Indeed, according to this view, my argument simply reveals that biology is one such "special science"; what is more, since all the other special sciences individuate functionally, we now have an explanation of why none embodies exceptionless generalizations. A defense of the nomological status and explanatory power of exception-ridden generalization may even be extended, as Nancy Cartwright (1983) has argued, to the claim that many generalizations of the physical sciences are themselves bedecked with ceteris paribus clauses; accordingly, the ubiquity of such generalizations in biology is no special limitation on its scientific adequacy.

Whether or not there are nonstrict laws in physics and chemistry, there is a good argument for thinking that the exception-riven generalizations—the "nonstrict laws" of biology—will not be laws at all. For the existence of nonstrict laws in a discipline requires strict ones in that discipline or elsewhere to underwrite them. If there are no strict laws in a discipline, there can be no nonstrict ones in it either.

The trouble with inexact ceteris paribus laws is that it is too easy to acquire them. Because their ceteris paribus clauses excuse disconfirming instances, we cannot easily discriminate ceteris paribus laws with nomological force from statements without empirical content maintained come what may and without explanatory force or predictive power. Now the difference between legitimate ceteris paribus laws and illegitimate ones must turn on how they are excused from disconfirmation by their exceptions. Legitimate ceteris paribus laws are distinguished from vacuous generalizations because the former are protected from disconfirmation through apparent exceptions by the existence of *independent* interfering factors. An interfering factor is independent when it explains phenomena distinct from those the ceteris paribus law is invoked to explain. This notion of an independent interferer has been explicated by Pietroski and Rey (1995) in the following terms:

A law of the form,

$$ceteris\ paribus, (x)(Fx \rightarrow Gx) \tag{1}$$

is non-vacuous, if three conditions are filled: Fx and Gx don't name particular times and places, since no general law can do this; there is an interfering factor, Hx, which is distinct from Fx and when Gx doesn't co-occur with Fx, Hx explains why; and Hx also explains other occurrences which do not involve the original law $(x)(Fx \rightarrow Gx)$.

More formally, and adding some needed qualifications,

(i) the predicates Fx and Gx are nomologically permissible.

(ii) $(x)(Fx \rightarrow Gx$ or (EH) (H is distinct from F and independent of F, and H explains not-Gx) or H together with $(x)(Fx \rightarrow Gx)$ explains not-Gx).

(iii) $(x)(Fx \rightarrow Gx)$ does sometimes explain actual occurrences [i.e., the interfering factors, H, are not always present], and H sometimes explains actual occurrences [i.e., H is not invoked only when there are apparent exceptions to (1)]. (89–90)

So ceteris paribus laws are implicitly more general than they appear. Every one of them includes a commitment to the existence of independent interferers.

This conception of ceteris paribus laws captures intuitions about such laws that have been expressed repeatedly in the past. But note that in this account of nonstrict laws there will have to be some laws in some discipline that are strict after all. If all the laws in a discipline are ceteris paribus laws, then the most fundamental laws in that discipline will be ceteris paribus laws as well. But then there are no more fundamental laws in the discipline to explain away its independent interferers. Unless there are strict laws somewhere or other to explain the interferers in the nonstrict laws of the discipline, its nonstrict statements will not after all qualify as laws at all. Without such explainers, the nonstrict laws will be vulnerable to the charge of illegitimacy if not irretrievable vagueness, explanatory weakness, and predictive imprecision. Or they will be singular statements describing finite sets of casual sequences. They will fail the test of Pietroski and Rey's definition of nonvacuous ceteris paribus law.

Of course, if the fundamental laws in a discipline are not ceteris paribus, but are strict laws, this problem will not arise. This I would argue is indeed the case in physics and atomic chemistry, for that matter. The Schrödinger wave equation does not obtain ceteris paribus, nor does light travel at constant velocity ceteris paribus. But as we have seen, in biology functional individuation precludes strict laws, and there are no nonstrict ones either, unless of course physical science can provide the strict laws to explain away the independent interferers of the nonstrict laws of biology. Alas, as we shall see, this is not an outcome that is in the cards. But apparent generalizations that are neither strict nor nonstrict laws do not have the nomological force that scientific explanation requires. They are not generalizations at all. If the biological explanations in which these nongeneralizations figure do explain, they do so on a basis different from that of the rest of natural science.

Of course biology could circumvent limits on the discovery of laws governing the processes it treats, if it were to forgo functional individuation. By identifying the kinds about which we theorize and predict structurally, it could avoid the multiple-realization problem bequeathed by the conjunction of functional individuation and natural selection.

But this conclusion hardly constitutes methodological advice anyone is likely to follow in biology or elsewhere beyond physical science. The reason is that forgoing functional individuation is too high a price to pay for laws: the laws about structural kinds that creatures like us might uncover will be of little use in real-time prediction or intelligible explanation of phenomena under descriptions of interest to us.

What would it mean to give up functional individuation? If you think about it, most nouns in ordinary language are functional; in part this preponderance is revealed by the fact that you can make a verb out of almost any noun these days. And for reasons already canvassed, most terms that refer to unobservables are functional as well, or at least pick out their referents by their observable effects. What is more to the point, the preponderance of functional vocabulary reflects a very heavy dose of anthropomorphism, or at least human interests. It is not just effects, or even selected effects, that our vocabulary reflects, but selected effects important to us either because we can detect them unaided, or because we can make use of them to aid our survival, or both. We cannot forgo functional language and still do much biology of phenomena above the level of the gene and protein. *Plant, animal, heart, valve, cell, membrane, vacuole*—these are all functional notions. Indeed, *gene* is a functional notion. To surrender functional individuation is to surrender biology altogether in favor of organic chemistry.

Let us revisit Rescher's strictures on the relation of explanation and prediction to scientific adequacy. First there is the requirement that to be well established our explanatory theories must have a track record of contributing to predictive success. It is a fact about the history of biology that contributions to predictive success—particularly the prediction of new phenomena, as opposed merely to prediction of particular events—are almost entirely unknown outside the most fundamental compartments of molecular biology, compartments where functional individuation is limited and nature has less scope to accomplish the same end by two or more different means. The startling predictions of new phenomena that have secured theoretical credibility—for example, the prediction that DNA bases code for proteins—emerge from a structural theory par excellence—Watson and Crick's account of the chemical structure of DNA. Even in cases in which relatively well-developed theory has inspired the prediction of hitherto undetected phenomena, these predictions, as already noted, have characteristically been disconfirmed beyond a limited range. Beyond molecular biology, the explanatory achievements of biological theory are not accompanied by a track record of contributions to predictive success—with one exception.

The generalizations in biology that have consistently led to the discovery of new phenomena are those embodied in the theory of natural selection itself. Though sometimes stigmatized as Panglossian in its commitment to adaptation, evolutionary theory's insistence on the universal relentlessness of selection in shaping effects into adaptations has led repeatedly to the discovery of remarkable and unexpected phenomena. This should come as no surprise, since the theory of natural selection embodies the only set of laws—strict or nonstrict—to be discovered in biology. It is because they are laws that they figure in my empirical explanation of why there are no (other) laws in the discipline.

Continuing to credit biological theory with explanatory power in the absence of predictive power is of course tantamount to surrendering Rescher's strictures on scientific adequacy. For we must surrender the ideal of "a closed cycle of theory-driven improvements in predictive performance that are themselves explicable in a wider context of theoretical understanding" (Rescher 1998, 169). By and large, philosophers of biology have accepted that generalizations connecting functional kinds are not laws of the sort we are familiar with from physical science. But instead of going on to rethink the nature of biology, this conclusion has led them to try to redefine the concept of scientific law to accommodate the sort of non-nomological generalizations to which biological explanations do in fact appeal. But philosophers who have done this have not recognized the collateral obligation to provide an entirely new account of the nature of explanation and its relation to prediction required by this redefinition of scientific law.

Elliot Sober (1993) exemplifies this approach to redefining scientific law with the greatest candor. He writes:

Are there general laws in biology? Although some philosophers have said no, I want to point out that there are many interesting if/then generalizations afoot in evolutionary theory.

   Biologists don't usually call them laws; models is the preferred term. When biologists specify a model of a given kind of process, they describe the rules by which a system of a given kind changes. Models have the characteristic if/then format we associate with scientific laws . . . they do not say when or where or how often those conditions are satisfied. (15)

Sober provides an example:

R. A. Fisher described a set of assumptions that entail that the sex ratio in a population should evolve to 1:1 and stay there. . . . Fisher's elegant model is mathematically correct. If there is life in distant galaxies that satisfies his starting assumptions, then a 1:1 sex ratio must evolve. Like Newton's universal law of gravitation, Fisher's model is not limited in its application to any particular place or time. (16)

True enough, but unlike Newton's inverse square law, Fisher's model is a mathematical truth, as Sober himself recognizes:

Are these statements [the general if/then statements] that models of evolutionary processes provide empirical? In physics, general laws such as Newton's law of gravitation and the Special Theory of Relativity are empirical. In contrast, many of the general laws in evolutionary biology (the if/then statements provided by mathematical models) seem to be nonempirical. That is, *once an evolutionary model is stated carefully, it often turns out to be a (nonempirical) mathematical truth.* I argued this point with respect to Fishers' sex ratio argument. (71)

If the generalizations of biology are limited to mathematical truths, then there are indeed few laws in this science. Sober recognizes this fact:

If we use the word tautology loosely (so that it encompasses mathematical truths), then many of the generalizations in evolutionary theory are tautologies. What is more we have found a difference between biology and physics. Physical laws are often empirical, but general models in evolutionary theory typically are not. (72)

As Abraham Lincoln once said, calling a dog's tail a leg does not make it one. Sober provides little reason to identify as laws the models that characterize much of theoretical biology. But he is right to highlight their importance. Many other philosophers of biology have dwelt on the central role that models play in the explanations biologists offer and in the limited number of predictions they make. Indeed, in the philosophy of biology the semantic approach to theories—which treats theories as nothing more than sets of models of the very sort Sober has described—is almost a universal orthodoxy (see, e.g., Beatty 1981; Thompson 1988; Lloyd 1993). Most exponents of this approach to the nature of biological theorizing admit openly that based on this conception, the biologist is not out to uncover laws of nature. Thus Beatty (1981) writes, "On the semantic view, a theory is not comprised of laws of nature. Rather a theory is just the specification of a kind of system—more a definition than an empirical claim" (410). Models do not state empirical regularities, do not describe the behavior of phenomena; rather they define a system. Here Beatty follows Richard Lewontin (1980): in biology, "theory should not be an attempt to say how the world is. Rather, it is an attempt to construct the logical relations that arise from various assumptions about the world. It is an 'as if' set of conditional statements" (28). This is a view that has been expressed before, by instrumentalist philosophers of science. Despite its known limitations as a general account of scientific theorizing, we may now have reason to believe that it is more adequate a conception of biological theorizing than of theorizing in physical science. At least we can see why in biology it may be impossible to say how the world is nomologically speaking, and it may be important to seek a second-best alternative to this end. (I argue for this view more extensively in Rosenberg 1994.)

In physical science models are presumed to be waystations on the road to physical truths about the way the world works; yet there can be no such expectation in biology. Models of varying degrees of accuracy for limited ranges of phenomena are the most we can hope for. A sequence of models cannot expect to move toward complete predictive accuracy, or even explanatory unification.

Consider the set of models that characterize population biology—beginning with a simple two-locus model that reflects Mendel's "laws" of independent assortment and segregation. The sequence of models must be continually complicated, because Mendel's two laws are so riddled with exceptions that it is not worth revising them to accommodate their exceptions. These models introduce more and more loci, probabilities, recombination rates, mutation rates, population sizes, and so on. Some of these models are extremely general, providing a handle on a wide range of different empirical cases, whereas some are highly specific, enabling us to deal with only a very specific organism in one ecosystem. But what could the theory that underlies and systematizes these Mendelian models be like? First of all, suppose that since the models' predicates are all functional, the theory will be expressed in functional terms as well. But we know already that any theory so expressed will itself not provide the kind of exceptionless generalizations that would systematize the models in question—that would explain when they obtain and when they do not obtain. For this theory about functionally described objects will itself be exception-ridden for the same reasons the models of functionally described objects are.

Could a theory expressed in nonfunctional vocabulary systematize these models, explain when they work and when they do not? Among contemporary philosophers of biology and biologists, surprisingly enough, the answer to this question is that no such theory is possible. Yet these philosophers and biologists do not recognize that this answer commits them to grave limits on biological knowledge, limits imposed by our cognitive powers and interests. That is, if, as is widely held, there are no more fundamental but nonbiological explanations for biological processes, then biological explanations employing functional generalizations and mathematical models really do have autonomous explanatory force. But this can only be because biology is far more a reflection of our interests and powers than physical science.

The widely held autonomy of biological explanations rests on a thesis of *physicalist* antireductionism that has become almost consensus in the philosophy of biology. The explanations that functional biology provides are not reducible to physical explanations, even though biological systems are nothing but physical systems. The consensus physicalist antireductionism relies on two principles:

1. The principle of autonomy of biological kinds: the kinds identified in biology are real and irreducible because they reflect the existence of explanations that are autonomous from theories in physical science.

2. The principle of explanatory ultimacy: at least sometimes processes at the biological or functional level provide the best explanation for other biological processes, and in particular better explanations than any physical explanation could provide.

These two principles require that the biological systems have *distinctive* causal powers—powers with biological effects different from the effects of the structural assemblages that implement them. Why? Without distinctive causal powers, there is no basis to accord biological kinds autonomy from structural kinds or biological explanations with adequacy and autonomy. But these principles must be reconciled with physicalism—the mereological dependence of biological systems on their constituent structural implementations. Principles (1) and (2) derive autonomy from a causal/explanatory role. If according biological kinds distinctive causal powers is incompatible with physicalism, the autonomy of biology is purchased at the price of an unexplainable metaphysical emergentism little different from a form of vitalism with which no one is willing to be saddled.

The antireductionist must block the shift of causal powers from the biological down to its structural implementation. The only way to do this in a way that is compatible with physicalism is to embrace the thesis that causation is conceptually dependent on explanation and the claim that explanation is a heavily pragmatic notion. Explanation is pragmatic roughly when it is viewed not as a relation just between propositions, but as a relation between questions and answers offered in the context of an interlocutor's beliefs. These beliefs reflect presuppositions of interlocutors that together determine the correctness or goodness or explanatory worth of various answers.

Here is how pragmatism about explanation combined with its conceptual priority to causation can save physicalist antireductionism. Suppose that some biological system a has some functional property F and a's having the functional property F is implemented by a having some complex physical property M.

So a has F because a is M, and the way F is realized in nature is through the physical mechanism M.

Now how can

$$a \text{ has } F$$

explain

$$a \text{ has } G$$

(where G is some effect of being an F) without

$$a \text{ has } M$$

(also) explaining why

$$a \text{ has } G?$$

After all, a's being an F is "nothing but" a's being an M.

The only way this can happen is when *explains* is a relation not just between matters of fact, but between them and cognitive agents with

varying degrees of knowledge. For someone who does not know that
a's Fing is realized by a's Ming, the substitution of M for F in the
explanation will not work. If we do not know about M, or understand
how it does F, then a's having M does not explain to us what a's having
F explains to us.

This means that when explanation is treated as "subjective" or
"pragmatic," its direction can move downward from the functional to
the structural. And if causation is just explanation (or depends on it), it
too can move "downward" from the biological to the molecular, following the direction of explanation.

At this late date in the operation of selection, almost all the complex
disjunctive structural properties on which biological ones supervene
will be too complex for cognitive agents like us to uncover, to state, or
to employ in real-time explanation, prediction, or both. Under these
circumstances, the biological property may provide the best, or the
only, explanation (for us) of the phenomenon in question. According
to this view ultimate biological explanations of the sort principle (2)
envisions may be possible: it may turn out that sometimes the instantiation of biological properties autonomously and indispensably
explains other biological processes, without the functional property's
instantiation being explainable (to us) by appeal to the instantiation of
structural kinds we could uncover and use in real time.

Downward causation of the sort principles (1) and (2) require will
thus be compatible with physicalism after all. Biological explanations
will be ultimate because there will be contexts in which correct structural responses to explanatory demands fail to explain to us. In these
causes the direction of causation will be downward from the functional to the structural if it follows the direction of explanation.

But notice that by hinging the explanatory power of biological theory on the interests and limitations of interlocutors and inquirers,
physicalist antireductionism commits itself to limiting the character of
biological science in a way that physics and chemistry are not limited.
The character of these two disciplines seems to be entirely independent
of any controversy about the nature of explanation. Whether we adopt
a pragmatic or a causal or even a syntactic account of explanation will
have no impact on the likelihood that these disciplines can uncover
explanatory generalizations, laws, or theories. If anything, the nomological achievements of the physical sciences suggest that in these
disciplines we have been able to uncover truths with explanatory

power for any and all inquirers, no matter how powerful their intellects. The same cannot be said for biology. Its character will be contingent on what counts as explanatory for us.

Most philosophers of biology will be unhappy with this relativization of biology to our cognitive limits and interests. They will argue that this conclusion simply reflects a fundamental misconception of biology according to the inappropriate model of physical science. According to some of these philosophers biology differs from chemistry and physics not because of its limits but because of its aims. Unlike physics or chemistry, biology is a historical discipline, one that neither embodies laws nor aims at predictions any more than human history does. This approach stems in part from Darwin's own conception of the theory of natural selection as in large measure a historical account of events on this planet, and therefore not just a body of universal generalizations that, if true, are true everywhere and always. Contemporary versions of this approach surrender Darwinian pretensions to nomological status for the theory of natural selection and defend its epistemic integrity as a historical account of the diversity and adaptedness of contemporary flora and fauna.

Again, interestingly, the outstanding exponent of this view is Elliot Sober. Following Thomas Goudge (1961), Sober (1993) insists that Darwin's theory is not a body of general laws but a claim about events on and in the vicinity of the earth:

The two main propositions in Darwin's theory of evolution are both *historical hypotheses*. . . . The ideas that all life is related and that natural selection is the principal cause of life's diversity are claims about a particular object (terrestrial life) and about how it came to exhibit its present characteristics. (7)

Moreover, Sober is committed to the claim, originally advanced by Dobzhansky (1973), that "nothing in biology makes sense except in the light of evolution." These two commitments generate a thoroughly historical conception of all of biology:

Evolutionary theory is related to the rest of biology in the way the study of history is related to much of the social sciences. Economists and sociologists are interested in describing how a given society currently works. For example, they might study the post World War II United States. Social scientists will show how causes and effects are related within the society. But certain facts about that society—for instance its configuration right after World War II—will be taken as given. The historian focuses on these elements and traces them further into the past.

Different social sciences often describe their objects on different scales. Individual psychology connects causes and effects that exist within an organism's own life span. Sociology and economics encompass longer reaches of time. And history often works in an even larger time frame. This intellectual division of labor is not entirely dissimilar to that found among physiology, ecology, and evolutionary theory. (7)

So evolutionary theory is to the rest of biology as history is to the social sciences. History is required for complete understanding in biology because biological theories can only provide an account of processes within time periods of varying lengths, and not across several or all time periods. Why might this be true? This thesis will be true in cases in which the fundamental generalizations of a discipline are restricted in their force to a limited time period. History links these limited-time disciplines by tracing out the conditions that make for changes in the generalizations operative at each period. Some "historicist" philosophers of history argue, like Marx or Spengler, for time limits on explanatory generalizations by appeal to an ineluctable succession of historical epochs, each of which is causally important for the epochs that follow, no matter how long afterward they occur. Others have held that the best we can do is uncover epoch-limited generalizations and identify the historical incidents that bring them into and withdraw their force.

If biology is viewed as a historical science, the absence of exceptionless generalizations in it will be no more surprising than their absence in history. For such generalizations seek to transcend temporal limits, and it is only within these limits, apparently, that historical generalizations are possible.

Leaving aside the broad and controversial questions about this sort of historicism in general, it is clear that many social scientists, especially sociological and economic theorists, will deny that the fundamental theories of their disciplines are related to history in the way Sober describes. Consider economic theory. Though economic historians are interested in describing and explaining particular economic phenomena in the past, the economic theory they employ begins with generalizations about rational choice, which they treat as truths about human action everywhere and always. The theory is supposed to obtain in postwar America as well as in seventh-century Java. Similarly, sociological theory seeks to identify universal social forces that explain commonalities among societies owing to adaptation, and differences between them owing to adaptation to historically different local geographical, meteorological, agricultural, and other conditions.

For these sociological or economic theorists, the bearing of history is the reverse of Sober's picture. These disciplines claim, like physics, to identify fundamental explanatory laws; history *applies* them to explain particular events that may test these theories.

Unlike Sober's view of the matter, according to this view social science is more fundamental and history is derivative. And if there is an analogy to biology it is that evolutionary theory is to the rest of biology as fundamental social theory is to history. This means that Sober is right about the rest of biology, but for the wrong reasons. Biology is a historical discipline, but not exactly because evolutionary theory is world-historical and the rest of biology is temporally limited history. This is *almost* right. The rest of biology *is* temporally limited history. But the main principles of Darwin's theory are not historical hypotheses. They are the only transtemporally exceptionless laws of biology. And it is the application of these laws to initial conditions that generates the functional kinds that make *the rest* of biology implicitly historical.

Evolutionary theory describes a mechanism—blind variation and natural selection—that operates everywhere and always throughout the universe. For evolution occurs whenever tokens of matter have become complex enough to foster their own replication so that selection for effects can take hold. Its force even dictates the functional individuations that terrestrial biologists employ and which thereby limit the explanatory and predictive power and thus the scientific adequacy of their discipline.

In our little corner of the universe the ubiquitous process of selection for effects presumably began with hydrocarbons and nucleic and amino acids. That local fact explains the character of most of the subdisciplines of biology. Their explanations are "historically" limited by the initial distribution of matter on the earth and the levels of organization into which it has assembled itself. So their generalizations are increasingly riddled with exceptions as evolution proceeds through time.

Thus biology is a historical discipline because its detailed content is driven by variation and selection operating on initial conditions provided in the history of this planet, combined with our interest in functional individuation. This explains why biology cannot approach very

closely Rescher's standards for scientific adequacy. Its apparent generalizations are really spatio-temporally restricted statements about trends
and the co-occurrence of finite sets of events, states, and processes.
There are no other generalizations about biological systems to be uncovered, at least none to be had that connect kinds under biological—
that is, functional—descriptions. This explains why biology falls so far
short of Rescher's standards. Its lowest-level laws will be the laws of the
theory of natural selection. But these laws are too general, too abstract,
to fulfill our technological, agricultural, medical, and other biological
needs. What will fill our needs are not laws but useful instruments
couched in the functional language that suits these interests and our
abilities to deal with complexity. The puzzle with which these limits on
biology leave us is this: with what sort of understanding can these
useful instruments—couched in the language of functional biology—
provide us?

REFERENCES

Beatty, J. 1981. "What's Wrong with the Received View of Evolutionary Theory?"
In P. Asquith and R. Giere, eds., *PSA 1980*, vol. 2. East Lansing, Mich.: Phi-
losophy of Science Association, 397–439.
Cartwright, N. 1983. *How the Laws of Physics Lie*. Oxford: Oxford University
Press.
Dobzhansky, T. 1973. "Nothing in Biology Makes Sense Except in the Light of
Evolution." *American Biology Teacher* 35: 125–29.
Goudge, T. 1961. *The Ascent of Life*. Toronto: University of Toronto Press.
Lewontin, R. 1980. "Theoretical Population Genetics in the Evolutionary Syn-
thesis." In E. Mayr and W. Provine, eds., *The Evolutionary Synthesis*. Cam-
bridge, Mass.: Harvard University Press, 19–31.
Lloyd, E. 1993. *The Structure and Confirmation of Evolutionary Theory*. Prince-
ton, N.J.: Princeton University Press.
Pietroski, P., and G. Rey. 1995. "When Other Things Aren't Equal: Saving *Ceteris
Paribus* Laws from Vacuity." *British Journal for the Philosophy of Science* 46:
81–110.
Rescher, N. 1998. *Predicting the Future*. Pittsburgh: University of Pittsburgh
Press.
Rosenberg, A. 1994. *Instrumental Biology or the Disunity of Science*. Chicago:
University of Chicago Press.
Sober, E. 1993. *Philosophy of Biology*. Boulder, Colo.: Westview.
Thompson, P. 1988. *The Structure of Biological Theories*. Albany: State Univer-
sity of New York Press.

16

A Physicist's Comment to Rosenberg

Michael Stöltzner
Vienna Circle Institute, Vienna, Austria

In *Instrumental Biology or the Disunity of Science,* Alex Rosenberg
advocates the former in order to prevent the latter. If we intend to
maintain an empiricist methodology and a physicalist metaphysics *for*
the whole of science, we must admit that—due to our limited cognitive
faculties—"biology is *relatively* more instrumental a science than the
physical sciences" (Rosenberg 1994, 1–2). Rosenberg's relative instrumental disunity is rooted in a thoroughly realistic interpretation of the
theory of natural selection—the only biological theory that, by his
account, contains laws in the same sense as do physics and chemistry.
Thus the axioms of natural selection are analogous to Newton's axioms. Because of the immense complexity of biological kinds and their
interrelations, however, this basis does not suffice for scientifically
relevant generalizations. This stark defect in explanatory power constitutes a principal limitation on biological science.

It is Rosenberg's reliance on such analogies or disanalogies between
physics and biology that allows a philosopher of physics to enter the
debate. My comments are meant to narrow the ditch Rosenberg digs
between realistic physics and instrumental biology *from the side of*
physics. There are, to my mind, several examples in contemporary
physics in which those features of complexity attributed exclusively to
biology do already occur. Accordingly Rosenberg's account of present
physical theory turns out to be too classical. On the other hand, once it
is admitted that contemporary biology is far from being reducible to a

collection of a few simple axioms, it comes closer to the actual physics of the Newtonian age than to our present highly abstracted and mathematically refined state of affairs. Then there existed many laws of nature much humbler than "grand unified theories." Striving for "Newton's ideal of empirical success" (Harper 1997) seems to be more reasonable for contemporary biologists than prematurely choosing between reductionism or disunity on the basis of an inadequate physical analogy. At bottom, I will argue, it is Rosenberg's erroneous identification of "unity of science" with ontological reductionism that enforces such drastic conclusions.

What Unity of Science—If Any?

Rosenberg's book tries to reconcile an appropriate estimation of contemporary biology with the unity-of-science thesis. Those who subscribe to "epistemological pluralism," as does John Dupré (1993), view biology as "soft science" whose kinds do not enjoy the same right as those of physical science. Instead, "the claim that biology is a relatively instrumental science can provide the required reconciliation and so vindicate a reasonably strong version of the unity of science" (Rosenberg 1994,14).

How strong then is the bearing of Rosenberg's "unity of science" thesis? In my view, it is already strong enough to enforce both major theses of his chapter in this volume (Chapter 15): the existence of a priori limits on explanation in biology and the denial of lawfulness to well-established biological generalizations. This high price obtains because he equates unity of science with physicalist reductionism—both taken as *metaphysical* positions. "The claim is that nature is at bottom both simple and regular . . . that is, physical theory deals with the most fundamental and ubiquitous forces and substances. . . . This doctrine of the substantive unity of science is perhaps better known under the rubric of *reductionism*" (Rosenberg 1994, 9). The present comment intends to show that this metaphysical stance biases the empirical arguments listed in Rosenberg's chapter.

Already in his 1994 book, Rosenberg stresses that his "thesis about the cognitive status of biology does not require that we take sides on . . . this dispute between instrumentalists and scientific realists" (Rosenberg 1994, 4). This, however, does not revive logical empiricists' metaphysical neutrality, because Rosenberg's metaphysically reductionist conception of the unity of science turns his "merely supposed" realism about physical entities once again into a metaphysical position because metaphysical unity of science blocks the relativization of ontology to a conceptual framework in a Kantian or Carnapian sense. Had, on the other hand, Rosenberg opted for instrumentalism, it is not clear to me how in this case he could at all maintain *his* conception of unity of science.

To logical empiricists, unity of science was a *methodological,* not an ontological, ideal, toward which scientists work by introducing appropriate conceptual frameworks for their theories and striving for laws *before* attempting to link or unify them at best by using a physicalist language. Otto Neurath's project of an *International Encyclopedia of Unified Science* expressed this ideal more than anything else. George Reisch (1997) convincingly argues that Neurath's antireductionist "unity of science" is quite distinct from Dupré's ontological pluralism. Hence Neurath's encyclopedism might serve as a viable alternative to Rosenberg's ontological way of avoiding disunity at the price of rejecting theory reduction *tout court.* Neurath's concept of an "orchestration of the sciences" does not preclude theory reduction, but it leaves stepwise and piecemeal reductions open to scientific practice.

No Cheer for Ontological Reductionism in Physics

Is physics really so simple that ontological reductionism holds true, as Rosenberg claims?[1] To my mind, there are some good reasons why this position becomes more dubious the closer we come to the most fundamental physical theories. Layer-cake reductionists, such as Steven Weinberg, believe in a pyramid of theories that is crowned by the alleged theory of everything (TOE), to which all "arrows of explanation" converge. Although TOE advocates readily admit that not all natural phenomena can actually be calculated from the topmost principles, they insist that *objective reductionism* "is simply true" (Weinberg 1993, 42). The levels in the pyramid (figure 16.1) are divided in three respects:

1. The main division is made according to energy and length scale.

2. Big-bang cosmology makes it possible to associate them with a sequence of cosmological epochs.

3. Theories of higher levels are mathematically more general than lower ones.

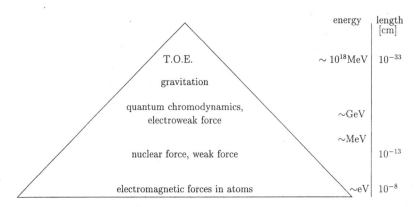

Figure 16.1 The reductionist's pyramid of physical laws.

A closer look at proposed candidates for ontological reduction often tells a different story inasmuch as the transitions between different levels are quite intricate. This has led Walter Thirring to take a different stand[2]:

1. The laws of a lower level are not completely determined by the laws of the higher level, even though they do not contradict the latter. What seems to be a fundamental fact on one level may seem entirely accidental if seen from a higher level.

2. The facts of lower levels depend more on the particular circumstances than on the laws of higher levels. Laws of the higher level, however, may be necessary to resolve internal ambiguities on a lower level.

3. The hierarchy of laws has evolved in the course of the evolution of the universe. Newly created laws initially did not exist as laws, merely as possibilities.

For example, the explanation of why the matter in and around us is stable against clustering into a superdense plasma hinges not only on quantum field theory but also on the fact that the lightest particle in the atomic world, the electron, is a fermion (see Thirring 1990 for details). Yet this may simply be the result of a symmetry-breaking phase transition in the early universe. Obviously so important a contribution to explanation has more weight than a mere measured parameter. So in this case scientific explanation consists of a successful *partial* reduction of a macroscopic fact to quantum theory *and* a hitherto irreducible fact of atomic physics. Contrary to the dreams of TOE-reductionists, one must grant a certain ontological autonomy to the lower theoretical levels. This does not preclude that cosmology might be able to specify the conditions under which symmetry breakings unavoidably occur, although their outcome, the mass and nature of the lightest particle, will probably remain an irreducible fact. Not aiming at complete reduction may even permit one to climb higher in the pyramid. Long-time instabilities of our planetary system, for instance, arise from resonances in the perturbations of a planet's orbital motion. It can be shown that orbital stability depends more crucially on number theory than on the particular form of the gravitational force.

Granting the respective levels a weak ontological autonomy that does not preclude partial theory reduction, there are, to my mind, three ways to proceed honestly on a particular level, if a successful reduction is not in sight:

1. We can demand that the theory be fundamentally valid for all scales even if we know that this is not the case. This reflects our belief that it will ultimately come out as a well-defined limit of a higher-level theory.

2. We can explicitly cut the connection to the higher level by renormalization or by introduction of a fundamental length. Then we obtain a *phenomenological theory* that contains *effective* parameters that represent the excluded microstructure and that are determined by experimental observation or fitted by numerical approximation.

3. We can hope that *mathematical* structures, axiomatic frameworks, might be common to some levels and that this unification would also specify where the irreducible facts enter that characterize a particular *physical* level.

I will use these alternatives later as a pattern for how one could possibly treat weak biological generalizations in a way analogous to those of contemporary physics. But first I use them to illustrate the capacities required of an old acquaintance that Rosenberg revitalizes to back up his metaphysical unity-of-science thesis and the resulting relative instrumentalism: Laplace's demon.

Were We Much Smarter?

Were we much smarter, "if we could calculate faster and hold more facts in short-term memory" (Rosenberg, Chapter 15), physics and chemistry would more quickly approach the fundamental regularities of the real world, but they would still be couched in largely the same terms and expressed by a small set of simple laws similar to those we know today. If we did not believe that "we humans have been able increasingly to discern these [absolute] truths [about the way things in the world are disposed,] . . . our amazing technological advances would be a transcending mystery" (Rosenberg 1994, 4). So much for ontological reductionism in physics.

If, on the other hand, we were much smarter about the phenomena of life, then "the resulting science would not be recognizably biology" (Rosenberg, Chapter 15). Instead, we would obtain an entirely different science or physics supplemented with the principle of natural selection. Could there exist some Laplacian demon who accomplishes this reduction that is unmanageable by human beings like us and cashes in the metaphysical unity of science? Do mere computational powers suffice? Here I disagree with Rosenberg. Taking seriously the levels of physical theories we find in nature, the demon has to accomplish much more than just to evaluate a TOE. He must explain why the levels that are actually present in nature (down to the organic) have emerged in cosmological evolution. Hence the question about the ontological status of theoretical kinds whose reduction is too complex or even impossible on scientific grounds does not concern biology exclusively, but physics as well. Biological kinds, nevertheless, face further ontological problems that are atypical for physical objects.

Functional Equivalence–cum–Structural Diversity

Already at the molecular level, natural selection "*cannot discriminate between differing structures with identical* [or sufficiently similar] *effects*" (Rosenberg, Chapter 15). Moreover, from a very low level of organization on, functional equivalence–cum–structural diversity will be the rule rather than the exception. That "biological kinds" can be defined and individuated only functionally poses a priori limits on biological science—or so Rosenberg claims.

Physico-chemical structures and biological functions are related by *many-to-many* relations. Thus all laws about such functional kinds will inevitably contain exceptions and ceteris paribus clauses. On the other hand, it follows naturally from Rosenberg's physicalism that for every physico-chemical structure one can find an environment in which it is more adaptive than its fellow structures within the functional kind (see Rosenberg 1994, 98). Combining these two aspects has, to my mind, consequences for the stability of functional kinds. They are by definition stable to changes of environmental boundary conditions that concern the defining function. But a functional kind is unstable to other changes if it lives on a saddle point in the mountains of selective values. Hence together with some functional or structural property other than the defining function, the functional kind can be divided into subclasses. Combining many functional relations into a suitable network, one might be able to individuate a nontrivial kind as a knot within a network of biological laws. Such a strategy also reflects the fact that scientists usually do not appraise isolated laws of the form "All swans are white" that could be falsified by a single observation (see in this respect Lakatos 1978, 19). I admit that such a network and the selective kinds so individuated are very likely to be too complex for us. But the knots of this network are not necessarily selective tokens or individuals instead of types, which are commonly required to figure in reasonable laws.

My main point here is that if one does not restrict oneself to *isolated* laws, the complexity argument is not so straightforward as Rosenberg believes. After all, there is a lower complexity bound involved. "For all their complexity, these type identities [in which evolutionary biology is expressed] are just manageable enough to permit something approaching reduction to molecular mechanisms of some species-specific instances of evolutionary and genetic regularities. . . . This is how evolutionary biology links up to the rest of science" (Rosenberg 1994, 170). Hence, if evolutionary biology's complexity ranges *above* the maximum of our cognitive capacities but *below* the connectibility to molecular biology, I do not see any basis for Rosenberg's claim that complexity will *forever* exceed our cognitive faculties, whatever scientific revolution occurs in biology.

How to Measure Relative Instrumentalism

In this section I draw a parallel between the problem of functional kinds in biology and the strategies of identifying non–directly observable physical objects through their effects by constructing suitable networks of lawlike relations. A particle detector in high-energy physics, for instance, typically generates a group of coincidence and anticoincidence measurements that identify the particle. Of course, today's physicists act within a rather limited (at least by biologists' standards) set of possibilities that is given by the standard model of elementary particle physics and certain theoretically posited extensions, such as supersymmetry. History teaches, however, that physicists usually applied rather similar measurement and identification procedures before such theoretical models were at hand. The atomic spectra of almost all elements known at the beginning of this century, for instance, were measured and classified before there was a theoretical explanation of the object these discrete spectral lines represented. Hence physicists searched for *laws* although the indirect (or functional) identification did not yield a well-defined structural kind.

Philosophers who consider objects prior to lawlike relations are committed to "realistic" (or "ontological") interpretations of quantum mechanics, which claim that a particle has a definite position at every instant of time. But this ontology cannot be cashed out as predictions, with the result that these interpretations contain an in-principle limit to the empirical accessibility of particle positions, quantum potentials, parallel universes, and the like. Hence to the realist's mind, even physics suffers from a limitation in Rosenberg's sense.[3] Working physicists are then forced into instrumental talk about measurement results in the style of the Copenhagen interpretation. This instrumental part of physics, however, is of a nature starkly different from that of biology because the limitations concern the most simple quantum systems. The infamous double-slit experiment is not a matter of complexity that can be easily linked to human cognitive capacities. Neither can the quantum mechanical measurement problem be solved by increasing computational powers at will.

But if complexity is not the only way to establish instrumentalism, then it cannot serve as the *unique* measure to judge which part of science is more instrumental than another. Dropping a metaphysically realistic unity of science and ontological reductionism—which Rosenberg assumes for methodological purposes only—does not win the case because determinate quantum mechanical instrumentalists emphasize the irreducible role of the macroscopic observer in a way similar to that by which Rosenberg brings into play our human cognitive faculties.

On the Economy of Ceteris Paribus Laws

Are the generalizations of biology, from molecular genetics on up, fated to remain ceteris paribus laws without end? Such "exception-ridden generalizations" can acquire explanatory power if they are given nomological force. Then "they are excused from disconfirmation by their exceptions." To avoid turning them into vacuous tautologies, one needs *independent interferers* that explain "phenomena distinct from those the ceteris paribus law is invoked to explain" (Rosenberg, Chapter 15). To this end, Rosenberg cites a sufficient condition for nonvacuity that has been proposed by Pietroski and Rey (1995, 92) and maintains the following: "Unless there are strict laws somewhere or other to explain the interferers in the nonstrict laws of the discipline, its nonstrict statements will not after all qualify as laws at all." Because of the problem of individuating biological kinds, "the theory of natural selection embodies the only set of laws—strict or nonstrict—to be discovered in biology" (Rosenberg, Chapter 15). Moreover, complexity prevents reduction of the interferers to the strict fundamental laws of physics and chemistry.

The most startling feature of Rosenberg's argument is that Pietroski and Rey designed their sufficient condition only to exempt ceteris paribus laws from an explicit statement of the ceteris paribus conditions or an explicit reduction of the independent interferers to strict laws, for "ceteris paribus clauses are needed in science precisely when it is *not* clear what the 'other things' are" (Pietroski and Rey 1995, 87). According to them,

[ceteris paribus] clauses are "checks" written on the banks of independent theories, their substance and warrant deriving from the substance and warrant of those theories which determine whether the check can be cashed. . . .

However, on our view, the details of this [explanatory] commitment [assumed with the check] *do not need to be spelled out* [by listing the *cetera* and why they are *paria*]: the second disjunct [principle (2) in Rosenberg, Chapter 15] involves only *an existential quantification over interfering factors, and not a citation of the factors themselves.* The check need not actually be cashed in the statement of the law, which can thereby retain its ceteris paribus status for ever. (Pietroski and Rey 1995, 82, 93)

Accordingly, to my mind, the check metaphor naturally extends to the relational networks mentioned previously. Banknotes are checks on which, for example, the chief cashier of the Bank of England "promise[s] to pay the bearer on demand the sum of £20." Yet even in the old days, when currencies were tied to actual gold or silver deposits in the state treasury, few banknotes were actually cashed in. Trusting the warranty of the national banks is a necessary precondition to the development of a modern monetary system and a market economy. If, on the contrary, a considerable portion of a country's population cashes in banknotes (for other currencies or goods), an economic collapse is looming. In an analogous manner, well-established sciences such as biology operate by cross-checking different ceteris paribus laws, thus spreading trust and warranty through the nomic network of the discipline. If, however, the foundations of the science were to lose trust, a paradigm change would be just as likely as a currency failure would be if banknotes lost trust.[4]

Pietroski and Rey also contemplate the massive—possibly infinite—complexity of the actual world. Contrary to Rosenberg, for whom strict and finitely statable basic laws are needed in order to have any laws in a science, they do not hold that "a true ceteris paribus law is just a strict law 'in disguise,'" such that it could be made strict by explicitly listing the interferers or enabling conditions. Moreover, they "think that laws need not be—and if current science is an indication, are not—strict." Instead, the prevalence of ceteris paribus laws in science shows the importance of "partial explanations of phenomena." Indeed, as stated earlier, in physics this strategy is more promising than pinning vague hopes on the (necessarily strict) TOE to come. In a footnote, Pietroski and Rey offer a large dose of optimism: "It may well turn out that, on an adequate understanding of laws, there are far more laws in the world than philosophical tradition has invited us to suppose. For example, it may well be that good recipes . . . are chains of little ceteris paribus laws. . . . Psychology might turn out to be filled with such 'small' laws that might prove immensely explanatory of (say) human reasonings and cognitive failures" (Pietroski and Rey 1995, 103, 96, 99). Although I am rather skeptical about cooking recipes, searching for the possibility of small and humble laws seems to be a more promising road than Rosenberg's precocious rejection. In the following section I argue that this view enjoys considerable historical support.

Classical and Modern Laws of Nature: On the Proper Physical Analogies for Contemporary Biology

If there is only one law in all biology, how many are there in physics? Matthias Schramm begins a paper on "Roger Bacon's Concept of a

Law of Nature" thus: "In contemporary science the term does not matter any longer; it is still used only when scientists get philosophical. I cannot imagine that a physicist or a chemist would call one of his discoveries a new law of nature" (Schramm 1981, 197, my translation). This apparently provocative statement stresses that physicists of our century were mainly after symmetry principles or the equations expressing them. For example, Maxwell's equations unified four laws of nature—among them Coulomb's law and Ampère's law—that were classical laws pertaining to definite experiments within a limited context. Because of its almost universal validity, Newton's law of gravitation of course occupies the most prominent place within this class. Most classical laws, however, were of humbler aspirations. It happened in those days that the curator of the Royal Society, Robert Hooke, had a contract according to which "at every session of the Society . . . he had to demonstrate three or four experiments proving the new laws of nature" (Arnol'd 1990, 11).

The law carrying Hooke's name is a paradigmatic case for the Newtonian epoch. It lies within the framework of Newton's three axioms—which Newton himself also called laws. For small elongations it describes very well the oscillations of arbitrary springs. Unlike gravitation, the constant in Hooke's law is not universal, but characterizes each particular spring; that is, it is a *phenomenological* constant subject to experimental determination. If the elongation of the spring surpasses a certain limit, then Hooke's law fails and the elongations are larger than those proportional to the force. Beyond the limit of elasticity, deformations become irreversible (in the region of plasticity) until the spring breaks.

Once this difference between classical physics and modern physics is made clear, whether one calls Einstein's equations or the Schrödinger equation a law—as most philosophers of science do—is mainly a question of terminology. In Rosenberg's discussion of biological laws, both notions of *law* are intermingled. This can already be seen in his two examples: the Schrödinger wave equation and the fact that light travels at constant velocity (Rosenberg, Chapter 15). The former is indeed the fundamental building block of quantum theory, although it requires an interpretation in order to make predictions. The latter is of lower systematic order because one can almost completely deduce special relativity from its symmetry group.

The theory of natural selection—as Rosenberg formulates it—is a law in the modern sense. In fact, he even provides an axiomatization and praises the latter's abstractness as its main virtue. Moreover, the law is independent of the hereditary mechanism. "At best, requiring the theory of natural selection to advert to genes is like requiring Newtonian mechanics to make mention of the continents" (Rosenberg 1994, 110). Rosenberg restates the axioms of natural selection of Mary B. Williams (1970) in the following way:

1. There is an upper limit to the number of organisms in any generation of a line of biological descent.
   2. For each organism there is a positive real number that describes its fitness in a particular environment.
   3. If there are hereditary variations in a line of descent that is better adapted to the environment—that is, has a higher fitness number—then the proportionate size of this line of descent will increase over the long run.
   4. So long as a line of descent is not on the verge of extinction, there is always a set of relatively better adapted lines of descent whose proportion of the population will increase over the long term from generation to generation. (Rosenberg 1994, 106–7)

So if Rosenberg continually emphasizes the analogy of the theory of natural selection to Newtonian mechanics, this refers to the latter's axiomatic framework, which could also be fulfilled by forces other than gravity, such as spring tension. Similarly, the Williams-Rosenberg axiomatic theory of evolution allows for various hereditary mechanisms.

"Undoubtedly there are other ways to express the central laws of the theory of natural selection." Rosenberg accordingly reformulates Dawkins and Hull's theory of natural evolution. Whereas in the Williams-Rosenberg account the whole framework represents the law, in the Dawkins-Hull formulation—after adjusting the definitions of *replicator, interactor,* and *selection*—"apparently the statement that selection occurs qualifies as such a law" (Rosenberg 1994, 117, 116). Such a conventionalist element is quite natural for axioms, but it causes problems for Rosenberg's metaphysically realistic reading of laws (in the modern sense).

General symmetry principles (modern laws) usually need additional specifications beyond mere initial data or boundary conditions—which suffice in the case of classical dynamical laws. Our universe is one solution to Einstein's equations. Elementary particles are particular representations of the fundamental groups. If the fundamental law is statistical, unpredictable phase transitions might occur whose result could be essential for the shape of entire theories, such as atomic physics. It is not misguided to call these additional parts of physical theories laws (in the classical sense) as well, because they have categorical validity, although they might just contain a fact about our universe that is not reducible to a universal theory.

Contrary to Rosenberg, it seems to me possible that weak biological generalizations fulfill reasonable criteria for this classical concept of a law. After all, the fact that our evolution is Darwinian should be termed a law, even if some of evolution's traits are partially reducible to the axioms. Moreover, in its formal development biology is a rather young discipline that seems to be closer to the physics of the centuries before the large unifications took place (from Maxwell onward) than to present-day physics. The decade-long debates about what "theoretical biology" might be like are telling in this respect.

Instrumental Biology and Physical Phenomenology

Laws of nature contain quantities whose mode of measurement has to be fixed without circularity:

> Newtonian mechanics . . . provides no way of measuring force except through its effects on accelerations. . . . By appeal to Hooke's law, which ultimately depends on Newton's laws, . . . we can measure force independently of accelerations and so free $F = ma$ from the charge of vacuity. . . . Plainly, however, the axioms of Newtonian mechanics ought not to advert to compressed springs. By the same token, the theory of natural selection should not advert to the multifarious sources of hereditary variation in fitness. (Rosenberg 1994, 119–20)

Once again, Rosenberg confuses two notions of *law*. Hooke's law is a completely acceptable way of measuring mechanical forces of any kind, but as a single law instantiating Newton's axioms it cannot serve as a universal justification of the concept of force as independent of accelerations, or—what amounts to the same—as an independent definition of mass.

Basically there are two strategies to accomplish such a definition: First, the mechanical mass of a body can be defined as a quantity invariant under certain operations, such as translations, divisions and recombinations, and melting. Second, an atomistic theory explains mechanical mass in terms of atomic masses and these masses in terms of elementary particles and gauge fields. Although the last step is incomplete to date, macroscopic masses are reducible at least one micro-level upward, such that nothing prevents a realistic reading of mass although one knows that mass ultimately "supervenes" on compli-cated processes between such particles as quarks, gluons, leptons, and photons. In effect, this knowledge even justifies the neglect of micro-scopic sublevels in applying classical or semiclassical mechanics as an approximative theory.

Rosenberg considers fitness "as a primitive undefined term in the theory of natural selection, exactly on a par with the role of force or mass as a primitive term in Newtonian mechanics." Fitness supervenes on the probabilistic propensities that reflect all "organismal-cum-environmental" factors that causally contribute to reproductive suc-cess. "Supervenience is not a sort of definitial reduction; it is the denial that such reduction is possible" (Rosenberg 1994, 120, 126)—chiefly because of our limited cognitive capacities and without admitting any objectively irremovable indeterminacy in evolution.

Rosenberg claims that a realistic interpretation of the theory of natural selection is guaranteed by treating fitness as a primitive con-cept. But such a realism owing to cutting off microscopic levels yields what physicists call a phenomenological theory. For Rosenberg cannot hope for a definition of fitness in terms of fundamental entities because a realistic position would involve the omniscient viewpoint, which is based on entirely different concepts that appertain to a deterministic theory of those evolutionary processes that to human scientists appear as drift.

More generally, instrumentalism about biology explicitly breaks re-ducibility by identifying those biological kinds of explanatory interest to human creatures with limited cognitive faculties. Such a strategy has its counterpart in phenomenological or effective approaches that are applied in physics mainly in order to overcome computational diffi-culties. For instance, one introduces by hand a fundamental length in order to run calculations on a lattice. The influence of all other micro-scopic levels is absorbed in parameters that are empirically determined or freely adjusted in order to fit the data. High-energy physics and solid-state physics contain plenty of such models and the correspond-ing calculation recipes. Nevertheless, in order not to count as barely ad hoc, physical phenomenology requires some lawlike background (such as the standard model or the mechanism of Cooper pairs) or some partial reduction already achieved (as in the case of semiclassical ap-proaches, in which microscopic quantum effects are approximated by classical potentials). Thus it seems to me that in biology instrumental-ist pragmatics also requires at least a certain dose of realism for some laws (in the classical sense) beyond natural selection in order to back a model. Of course, from the viewpoint of Rosenberg's metaphysical realism such an empirical realism appears to be "as if" at best. But it reflects the fact that there are microscopic levels, in terms of which reductions (at least piecemeal) of mechanisms, such as Cooper pairing, should not a priori be excluded.

Many laws of the classical type have by now acquired phenomeno-logical status. Hooke's humble law still serves the best-paid watch-makers. It should be borne in mind, however, that these laws were historically also a first step in explanation. If biology as a whole is still in this early stage of theoretization and formalization, then it might be advisable to seek a methodology that is closer to the Newtonian epoch than to Hilbert's program of axiomatization of physics.

Newton's Ideal of Empirical Success

From his standpoint, Rosenberg may read the present comment as an attempt "to redefine the concept of scientific law" and accordingly oblige me "to provide an entirely new account of the nature of expla-nation and its relation to prediction required by this redefinition" (Rosenberg, Chapter 15). Well, here is one that is patterned after Newton's methodology. According to Bill Harper and George Smith (1995), a "theory explains a phenomenon when it delivers equiva-lences that make the phenomenon measure a parameter of the theory which specifies its cause. On this view, what counts as empirical suc-cess in a theory is to have its parameters be accurately measured by the phenomena which they purport to explain" (147). If such parameters are specified, theoretical questions can be made decidable by experi-ment, and the phenomena provide an open-ended body of data to increase accuracy in the measurement of the parameter. Perturbation theory since Clairaut brought the final breakthrough of Newton's celestial mechanics based on universal gravitation because the relative masses of the bodies in the solar system could be measured with a perfection that later—in the case of Mercury's perihelion—raised doubts even about Newtonian theory itself. The impressive success of perturbation theory let Newtonian mechanics become an instance of hypothetico-deductive reasoning.

There are three reasons why Newton's ideal of empirical success could be advantageous to modern biology. First, according to Newton's view, "the phenomena from which the theory was deduced did not have to be exact. The deduction allowed approximative reasoning" that yielded eventually the correct power $1/r^2$ for the gravitational force. Similarly, "approximative reasoning" allows one to cope with the many ceteris paribus clauses that characterize the identification of functional kinds in biological laws. "We are characterizing the crucial feature separating Newton's idea of induction as *rendering general propositions inferred from phenomena* from *mere piling up of instances of a generalization* as the provision of equivalence that allows all the phenomena to be construed as agreeing measurements of the same theoretical parameter" (Harper and Smith 1995, 144, 146). That the theory of natural selection can already be cast in various axiomatic frameworks of the sort that Newton's methodology at the time still had to establish should not tempt one to premature deductivism.

Second, it is important to note that the parameters measured are typically not the fundamental constants mentioned earlier, but humbler invariants, such as the gravitational acceleration g. Thus Fisher's sex ratio, which is often mentioned as a good candidate for a biological law, could be established as a lawlike invariant that, admittedly, does have exceptions. As in the case of Kepler's area law, increasingly precise measurement of the parameter establishes the reality of the lawful generalization characterized by it. Of course, there is always a certain contingency or factual constraint in the choice of the parameters measured, whether in *Drosophila* or our solar system.

Third, empirical success is, of course, strongly fostered by reliable mathematical models taken over from physics. For instance, the hypercycle model of Eigen and Schuster (1979) can be interpreted as a two-dimensional Ising model. Even the simplicity of an axiomatic framework can be counted as empirical success. Newton's ideal allows a certain methodological continuity between physics and biology. Moreover, by this account, those who claim that most alleged biological laws (such as Fisher's) are nothing but mathematical truths cannot deny that these models have empirical content. This is an important prerequisite for appreciating successful partial reductions to mathematically simple axioms. Today mechanical mass, for instance, is simply the nontrivial center of the Galilei group.

Biology as a Historical Discipline

In the last part of his chapter, Rosenberg criticizes the view that biology is a historical discipline in which "the fundamental generalizations

. . . are restricted in their force to a limited time period." Whether or not this view is actually held by Elliot Sober—who has recently objected to this reading by Rosenberg (see Rosenberg 1997 and Sober 1997)—I agree with Rosenberg's maintaining that the theory of evolution should "identify fundamental explanatory laws; history *applies* them" (Rosenberg, Chapter 15).

The mechanism of blind variation and natural selection operates universally on all clumps of matter sufficiently complex to foster their own replication. What if some day physicists find that there are only a few such replicative complex structures that were consistent with a very delicate fine tuning of the physical constants of nature? Cosmological possibilities are so vast that a selection by historical facts, taken as final boundary conditions, is unavoidable. This type of reasoning is unluckily termed the "weak anthropic principle" (see Earman 1987 for a critical account). It appears to me as a necessary counterpart to causal models or laws, which—owing to symmetry breakings or the fundamentally statistical nature of the underlying theory—can only provide a class of possible outcomes. Such a finding suggests restricting the axiomatic framework containing the laws or the allowed models further in order to accommodate more tightly the only possible facts. Would not that be at least a weak form of Rescher's cycle (as given in Chapter 15) between an axiomatic framework and historical data of evolution?

Interestingly, Pietroski and Rey's account of ceteris paribus laws explicitly includes cases for "which the *cetera* have been *paria* only *once* in the history of the world . . . since it is at least *possible* that this true causal claim is an instance of a general law" (Pietroski and Rey 1995, 99). Cosmological inflation, once made reasonably precise, seems to be a candidate for such a singular law. Pretending epistemological normality by postulating parallel or pulsating universes would open the door to arbitrary absurd *ceteris*. Moreover, Pietroski and Rey can easily accommodate as ceteris paribus laws irreducible results of phase transitions and the fact that one within a large class of solutions or representations is realized. Previously I argued that they come quite close to the classical concept of a law that supplements the axiomatic (modern) law.

In the end, I should stress the general merit of Rosenberg's approach. In spite of the caveats listed here, his account of biological science seems to me much more appealing than either those that want to separate biology entirely from physics or those that want to make it a simple branch of physics. Biology and physics should not attempt philosophically to force one into the Procrustean bed of the other. But they can learn from each other by appropriate analogies. So let me close with a final one. Finality claims in physics—whether exalted hopes for a TOE or those in the tradition of Du Bois-Reymond's *Ignorabimus*—have usually been overcome and have led to a revision of the epistemological standards for the theory.

NOTES

I thank Martin Carrier, Bill Harper, Jim Lennox, and George Reisch for encouraging discussions.

1. I refrain from discussing the case of chemistry here. The growing literature in the philosophy of chemistry suggests that the complexity of chemical systems is not far from that of biological ones. This casts doubts on Rosenberg's equating chemistry with physics without further ado.

2. This position, first proposed in Thirring (1995), is elaborated in Stöltzner and Thirring (1994) and Stöltzner (1995).

3. Opponents of the realistic interpretations of quantum mechanics will argue that this limitation of quantum physics is merely a product of enforcing metaphysical realism at any cost (see Stöltzner 1998 for further references). Rosenberg remarks that the Copenhagen interpretation "by itself will not allow for subjectivity at the level of biological organization" (Rosenberg 1994, 15). But I do not see how Rosenberg could neatly separate both levels, since phenomena of the atomic scale are likely to be present in evolutionary drift processes—if one applies a reasonably rich notion of drift. Roberta L. Millstein's (1996) criticism of Rosenberg is to the point here, at least to a physicist's mind.

4. Here the reader might wonder which body of scientists is analogous to the central bank. But note that the United States had a national currency without a central bank from 1862 until 1913; therefore my analogy does not require simple authoritarianism in science.

REFERENCES

Arnol'd, V. I. 1990. *Huygens & Barrow, Newton & Hooke.* Basel: Birkhäuser.
Dupré, J. 1993. *The Disorder of Things: Metaphysical Foundations of Science.* Cambridge, Mass.: Harvard University Press.
Earman, J. 1987. "The SAP Also Rises: A Critical Examination of the Anthropic Principle." *American Philosophical Quarterly* 24: 307–17.
Eigen, M., and P. Schuster. 1979. *The Hypercycle—A Principle of Natural Self-Organization.* Berlin: Springer-Verlag.
Harper, W. 1997. "Isaac Newton on Empirical Success and Scientific Method." In J. Earman and J. D. Norton, eds., *The Cosmos of Science: Essays of Exploration.* Pittsburgh: University of Pittsburgh Press, 55–86.

Harper, W., and G. E. Smith. 1995. "Newton's New Way of Inquiry." In J. Leplin, ed., *The Creation of Ideas in Physics: Studies for a Methodology of Theory Construction.* Dordrecht: Kluwer, 113–66.

Lakatos, I. 1978. "Falsification and the Methodology of Scientific Research Programmes." In J. Worrall and G. Currie, eds., *The Methodology of Scientific Research Programmes: Philosophical Papers,* vol. 1. Cambridge, Mass.: Cambridge University Press, 8–101.

Millstein, R. A. 1996. "Random Drift and the Omniscient Viewpoint." *Philosophy of Science* 63: S10–S18.

Pietroski, P., and G. Rey. 1995. "When Other Things Aren't Equal: Saving *Ceteris Paribus* Laws from Vacuity." *British Journal for the Philosophy of Science* 46: 81–110.

Reisch, G. A., Jr. 1997. "How Postmodern Was Neurath's Idea of Unity of Science?" *Studies in the History and Philosophy of Science* 28: 439–51.

Rosenberg, A. 1994. *Instrumental Biology or the Disunity of Science.* Chicago: University of Chicago Press.

———. 1997. "Critical Review: Sober's *Philosophy of Biology* and His Philosophy of Biology." *Philosophy of Science* 63: 452–64.

Schramm, M. 1981. "Roger Bacons Begriff vom Naturgesetz." In P. Weimar, ed., *Die Renaissance der Wissenschaften im 12. Jahrhundert.* Zürich: Artemis, 197–209.

Sober, E. 1997. "Some Comments on Rosenberg's Review." *Philosophy of Science* 63: 465–69.

Stöltzner, M. 1995. "Levels of Physical Theories." In W. DePauli-Schimanovich, E. Köhler, and F. Stadler, eds. *The Foundational Debate: Complexity and Constructivity in Mathematics and Physics.* Dordrecht: Kluwer, 47–64.

———. 1998. "On Various Realisms in Quantum Theory." In A. Pagnini and M. C. Galavotti, eds., *Experience Reality and Scientific Explanation.* Dordrecht: Kluwer, 161–84.

Stöltzner, M., and W. Thirring. 1994. "Entstehen neuer Gesetze in der Evolution der Welt." *Naturwissenschaften* 81: 243–49.

Thirring, W. 1990. "The Stability of Matter." *Foundations of Physics* 20: 1103–10.

———. 1995. "Do the Laws of Nature Evolve?" In M. P. Murphy and L. A. J. O'Neill, eds., *What Is Life? The Next Fifty Years: Speculations on the Future of Biology.* Cambridge: Cambridge University Press, 131–36.

Weinberg, S. 1993. *Dreams of a Final Theory.* London: Vintage.

Williams, M. B. 1970. "Deducing the Consequences of Evolution." *Journal of Theoretical Biology* 29: 343–85.

17

The Limits of Experimental Method

Experimenting on an Entangled System:
The Case of Biophysics

Giora Hon
Department of Philosophy, University of Haifa

> It is interesting to contemplate an entangled bank.
> —Charles Darwin (1976 [1859], 459)

> All is dovetailed together.
> —Charles Sherrington (1953, 89)

Setting limits to knowledge is a worthwhile enterprise of no less importance than the unwearied endeavor to move the boundaries of knowledge a little further. Indeed, the establishment of such limits may provide a cornerstone for the construction of a new theory or a profound insight into scientific practices. This century has seen two such contributions in science: the uncertainty principle and the incompleteness theorem. Undoubtedly, these results of Heisenberg and Gödel constitute essential elements of our current knowledge of, respectively, the physical and the formal—physics and mathematics. Both results exhibit a fundamental proof of limitation: a proof of the impossibility of simultaneously measuring accurately certain physical properties of a system and of proving certain propositions within a given system.

A good example of such a contribution in philosophy is Wittgenstein's *Tractatus* (1966 [1921]); indeed it aims not at removing boundaries but rather at placing a limit. Wittgenstein's objective was explicitly to draw a limit, a limit to thought, or rather, as he corrected himself in the preface, not to thought, but to the expression of thoughts. It should be stressed that Wittgenstein included in the preface to the *Tractatus* the very last and by now famous sentence of

the *Tractatus*, namely the limiting conclusion: "wovon man nicht re-
den kann, darüber muß man schweigen" (1984 [1921], 9).
   The idea of the limit is prevalent in the *Tractatus*. Philosophy
displays the world to us as having limits (Anscombe 1967, 169). The
world has a totality; it can be divided; it should be viewed as a limited
whole (Wittgenstein 1966 [1921], §§1–1.2, 6.45). Parallel to this limited whole, the *Tractatus* establishes a limit to thinking: "[Philosophy]
must set limits to what can be thought; and, in doing so, to what
cannot be thought" (§4.114). As Stenius pointed out, in §6 of the
Tractatus, where Wittgenstein sets down the general form of a meaningful sentence, every proposition bars a way "on which one might feel
tempted to try and overstep this boundary" (Stenius 1960, 13).
   Stenius interpreted the *Tractatus* in the Kantian tradition of limitations. He observed Wittgenstein as seeking to make deductions concerning the limits of theoretical discourse. The investigation of this
limit is the investigation of the "logic" of language (Stenius 1960,
218). From this perspective the *Tractatus* may be called, as Geach
suggested, a "critique of pure language" (Stenius 1960, 220n2).
   By making its way outward, from what can be thought to what
cannot be thought (§4.114)—that is, by setting limits—philosophy,
according to Wittgenstein, accomplishes its sole objective: the logical
clarification of thoughts (§4.112). The *Tractatus* is a shining example
of the fruitful results of the heuristic that seeks boundaries and establishes limits.
   However, philosophy, as Wittgenstein sees it, is not part of the
natural sciences. The place of philosophy, he says, is above or below
the natural sciences, but not beside them (§4.111). Propositions of
natural science therefore have nothing to do with philosophy (§6.53).
Wittgenstein presents Darwin's theory as an example: this theory, he
states, "has no more to do with philosophy than any other hypothesis
in natural science" (§4.1122). Natural science is indeed the sphere of
the empirically discoverable from which philosophy, according to the
Tractatus, is excluded. After all, "logical propositions cannot be
confirmed by experience any more than they can be refuted by it"
(§6.1222).

The Problem

A question arises as to the nature of the "empirically discoverable": Is
it limited? Are there limits to what can be known by the senses or, for

that matter, by their instrumental extension? In the final chapter of *The Origin of Species,* in the recapitulation and the conclusion, Darwin complained bitterly about the hostile reactions that he had received. To be sure, he did not expect to convince experimental naturalists who had hidden their ignorance under such expressions as the "plan of creation," and the "unity of design" (Darwin 1976 [1859], 453). Darwin pinned his hopes on the future, when the reception accorded *The Origin of Species* would be seen, he confidently wrote, as "a curious illustration of the blindness of preconceived opinion." With poignant irony and a tinge of cynicism he commented that "Nature may be said to have taken pains to reveal, by rudimentary organs and by homologous structures, her scheme of modification, which it seems that we wilfully [*sic*] will not understand" (454, 452). This is an expression well worth repeating: *that we willfully will not understand.*

I argue that nature may indeed be said to take pains to reveal her schemes, but, contrary to Darwin, we may not be able to know and understand her. That is, we might not be able to see what nature tries to reveal. My concern is neither with the well-known Baconian charge of prejudices and preconceived ideas, nor with the physiological limitations of the senses, nor for that matter with physical constraints. My concern is chiefly methodological: the possible limitations that the very experimental method may impose. No "expiations and purgations of the mind," to use Bacon's expression, would do to facilitate the entering of "the true way of interpreting Nature" (Bacon 1859 [1620], Book 1, lxix, 51).

What then is an experiment? And what is the problem with it when biological issues are at stake? In the spirit of Wittgenstein's search for a limit, I demonstrate that experiments, as they are commonly conceived of, are methodologically limited by the requirement that they represent propositionally the presupposed state of the system under examination. I then argue that, in contrast to inanimate systems, living systems do not possess states; hence the study of living systems by, say, biophysical experiments rests on a false presupposition. The researcher is therefore inevitably led astray. This limitation casts a long shadow over the possibility of discovering laws or rules of biological systems using the current methodology of experimentation. I proceed to suggest a different way of looking at the living unit, the cell, and end on a Wittgensteinian note that it is better to be silent on matters that are not accessible to knowledge: if experiments on the cell rest on a false presupposition, then silence is a better investment, and the time has come to take a break and go back to first principles.

What Is an Experiment?

There is by now a proliferation of opinions regarding just what an experiment is (see, e.g., Buchwald 1995; Galison 1997; Heidelberger and Steinle 1998). However, this is not the place to discuss critically the different views concerning the nature of experiment; therefore I shall sketch my own position (Hon 1998) and proceed to address the problem.

An experiment is a procedure, a physical process, that can be cast into an argument of a formal nature. A crucial characteristic of an experiment is that its result constitutes a claim to physical knowledge, and it is this claim to physical knowledge that distinguishes a mere procedure from an experiment. Being a claim to knowledge, the conclusion of an experiment may be seen as the result of a chain of reasoning concerning the behavior of some material system. The conclusion is then connected with a certain mode of reasoning. An experiment can be made to exhibit an inference from premises to a conclusion—the argument of the experiment.

An argument is a sequence of propositions of a special form: each proposition is either an assumption, a premise, or a conclusion that arises through inference schema of propositions earlier in the sequence. An experiment, I claim, can be cast into a formal argument whose propositions instantiate partly states of affairs of material systems and partly inference schema and some lawful, causal connections. In other words, an experiment implies an argument the premises of which are assumed to correspond to the states of the physical systems involved, for example, the initial conditions of some material systems and their evolution in time. These premises warrant the argument's conclusion.

This is a fundamental point in characterizing the implied argument of a physical experiment: ultimately, the experimenter aims to secure premises that correspond *correctly* to the actual physical situation of the experiment. The requirement of correct correspondence between physical state and proposition sets this kind of argument apart from any other argument in which the premises may be conditional, hypothetical, or counterfactual. An experiment is therefore a procedure, a physical process, that has a logical facet of a rhetorical force. Indeed, an experiment is often used as an instrument of persuasion for reaching agreement on certain situations.

The general idea of an experiment, to follow Hans Radder, is that "some information about the object can be transferred to the apparatus by means of a suitable *interaction*. That is, the interaction should produce an (ideally complete) correlation between some property of the object and some property of the apparatus. From this it follows that the theoretical description of object and apparatus should also 'interact': they need to have at least some area of intersection" (Radder 1995, 58; emphasis in the original). In other words, the theoretical result of the experiment is inferred from some theoretical descriptions. There should then be in the experimental procedure two basic stages.

Experimenters indeed perform two different tasks: they prepare a system and then they test it (Peres 1990, 317–19). A *preparation* is a procedure that should be completely specified or at least reproducible. This is crucial for the completion of the experiment, since on the basis of this knowledge rests the belief that the premises of the implied argument correspond correctly to the physical conditions.

The second task, the *test*, starts like a preparation: it has a specified procedure that triggers the interaction between the prepared setup and the object under study. However, the test includes a crucial additional step whereby *information*, which was previously unknown, is supplied to an observer, the experimenter. This information constitutes, after a suitable reduction, the *new* physical knowledge sought. Within certain constraints, experimenters are free to *choose* preparations and tests that they wish to perform—this is their prerogative (provided that their research proposals receive the seal of grant approval). However, they are not free to choose the future outcome of a test. They are bound by the information acquired.[1]

Given this characterization of the procedure of experiment as *preparation* and *test* with its resultant flow of *information*, one may focus on the essential feature of this method of inquiry. It is the method of variation. By varying (the *test*) a certain group of elements or a single element within the system under study (the *preparation*), other elements will vary too or perhaps remain unchanged (the binding *information*). This fundamental rule of variation facilitates the severing of the many antecedents from their consequents. To distinguish the real laws—to assign to any cause its effect, or to any effect its cause—"we must be able," as Mill puts it, "to meet with some of the antecedents apart from the rest, and observe what follows from them; or some of the consequents, and observe by what they are preceded" (Mill 1949 [1843], 249, Book 3, Chapter 7, §2).

Thus, however general, the characterization of the procedure of experiment in terms of *preparation* and *test* may be useful for the analysis of experiment. This characterization allows for a clear apprehension of the connection between the actual procedure of the experiment and its implied argument. The preparation stage is principally about presuppositions, whereas the test stage has to do with a mixture of premises and conclusions—the outcome of the experiment. It may be further observed that each of the two stages has two subcategories. The preparation stage has theoretical and practical subcategories, whereas the test stage may be conceived of as the actual measurement (the recording of the information) and its processing.

We have seen that the experimenter aims to secure premises that correspond *correctly* to the states of the system under experimental examination. For that purpose one assumes in the preparation stage a theory that is considered correct; it underpins the experiment. This theory, the background theory and its "daughter" theories (the theories of the instruments and of the setup itself), are therefore taken for granted. They are not tested by the experiment. Then there is the process of realizing the theoretical requirements in practice. Once the theory and its physical realization have been put to work in the preparation stage, information is allowed to flow to the recording instrument (natural or artificial). Finally, a process of reducing this information and interpreting the result takes place. This is the conclusion of the experiment—the outcome.

Four stages may be thus distinguished: (1) laying down the theoretical framework; (2) constructing the apparatus and making it work; (3) taking observations and readings; and (4) processing the recorded data and interpreting them (Hon 1989, 480). Whereas stages (1) and (2) constitute the premises of the argument in the theoretical and concrete sense, the conclusion (4) is inferred from (3): the empirical information obtained from the test.

At each stage of the experiment we may identify different types of hindrances, pitfalls, constraints, and errors. Some of them may be controlled and even eliminated; others persist and indeed set limits on the measurement and ultimately on the experiment. We may think, for instance, of the dichotomy between systematic and random error, of classical constraints such as thermal agitation, and of problems that arise in measuring parameters of quantum mechanical systems. Yet, however perplexing and of whatever considerable philosophical and technical importance, these difficulties are not of concern here (but see Hon 1989). I am rather interested in the larger issue, that is, in methodological limitations and specifically in the process of casting the experiment into an argument. How does one set up the presuppositions? How does one determine the background assumptions? This is the point at which I return to biology and specifically to its microscopic dimension, to the biophysical aspects of the elementary living whole—the cell.

The Nature of Living Systems

What then do we have in cell biology? Here is a recap of some basic facts. The most striking feature of cellularity is the very existence of permeable plasma membranes, which separate all cells from their environment by a barrier to free diffusion. As Morowitz points out, "the necessity of thermodynamically isolating a subsystem is an irreducible condition of life" (1992, 8; see also 56). It appears then that "life in the absence of membranes is impossible" (Luisi 1993, 21). The membrane furnishes the cell with a physical definition; it individuates the cell and indeed provides it with an "identity" insofar as the cell can recognize foreign elements and harmful intruders (Tauber 1994; cf. Oparin 1966, 80–82).

However, an enclosure of matter would not retain its character as living matter were it not for its capacity to be an open thermodynamic system. A static system whose free energy is at a minimum and hence perfectly stable thermodynamically would not enter the gate of the living—it will be a dead lump of matter. A material system in thermodynamic equilibrium whose overall change in free energy is zero ($dF = 0$) is a dead system. As Oparin observes, "open systems . . . [are] maintained . . . because the systems are continually receiving free energy from the surrounding medium in an amount which compensates for its decrease in the systems" (1966, 84; see also 72–73). Therefore a crucial characteristic of a living system is that it is "in commerce with its surround" (Sherrington 1953, 87), where the free energy is changing all the time but at a constant rate (dF/dt = constant).

While the system is furnished with a permeable membrane and its thermodynamic character is secured by homeostatic processes (the reader should note that this project takes a couple of billion years to accomplish), two principal objectives are being achieved: self-preservation and growth as well as reproduction and replication. These two sets of features, metabolism and replication, have been shown by von Neumann in his study of automata in the late 1940s to be logically separable and mutually exclusive; that is, the cell may be understood to have two logically separated components. Computer scientists christened these two central components "hardware" and "software," respectively. The cell can then have either hardware (metabolism) or software (replication). Typically, however, the cell has both components: it grows and it multiplies (Dyson 1988 [1985], 4–9, 60–66, 72–73; cf. Eigen 1995, 9–10).

One can think of a list of attributes that an entity characterized as living should possess: it should have the capacities "to reproduce, to grow, to move, to evolve, to respond to the environment, to remain stable under small environmental change ('homeostasis'), and to convert matter and energy from the environment into the organizational pattern and activities of the organism ('metabolism')" (Lange 1996, 227). Such a list of attributes does not, however, exhaust the nature of a living system. Though these (mostly behavioral) criteria of living matter will be considered here features of "signs of life," as Lange calls them, they are not sufficient for bringing out the nature of the living as distinct from dead lumps of matter. Lange rightly points out that the presentation of such lists is invariably followed by certain provisos,

that some non-living things possess one or more of these attributes, that presumably an entity could possess all of these properties and nevertheless not be alive, and that some living things lack some of these properties. There are canonical examples: a whirlpool or ocean wave assimilates surrounding matter into its form; crystals, clouds, and fires grow whereas many mature organisms do not; iron rusts in response to being surrounded by oxygen; bubbles are stable under various small environmental change; mules . . . cannot reproduce; Penrose's (1959) plywood blocks reproduce; fires, ocean waves, and planets move; and so on. (227)

I maintain that what completes the analysis, what makes the difference between the living and the dead distinct, is the claim that *the living system does not have a state.*

What is a state? A state is "the most informative consistent descriptions of the system" (Posiewnik 1985; quoted by Kampis 1991, 104). A state may be understood as "a characteristic of a dynamical process that determines the future of a process by means of a state variable. A state variable is a static variable interpretable as a static observable of the system" (Kampis 1991, 130). It is commonly assumed that by algorithmic means one may construct the state of the system from its empirical data—the static observables (Kampis 1991, 115). In other words, the state of a system is a set of instantaneous values of all the qualities of the system, external as well as internal, which in the classical case determine the development of the system (Kampis 1991, 104). Though states in the quantum mechanical case do not of course determine the actual evolution of the system, they do yield accurately its probability distributions. I maintain that in contrast to an inanimate entity—classical or quantum mechanical—an animate object does not possess a state. That is, the cell does not have static observables from which a state may be deduced.

Why a Living System Cannot Have a State

Let us get a closer view of the subject of our inquiry. One typically pictures the cell, the elementary living system, as an industrial plant. Paul Weiss analyzes this misleading metaphor well, and it is worth quoting him at some length:

A cell works like a big industry, which manufactures different products at different sites, ships them around to assembly plants, where they are combined into half-finished or finished products, to be eventually, with or without storage in intermediate facilities, either used up in the household of that particular cell or else extruded for export to other cells or as waste disposal. Modern research in molecular and cellular biology has succeeded in assigning to the various structures seen in micrographs specific functional tasks in this intricate, but integrated, industrial operation.

Weiss then proceeds to warn us against the use of this factory metaphor:

While in the [man-made plant], both building and machinery are permanent fixtures, established once and for all, many of the corresponding subunits in the system of the cell are of ephemeral existence in the sense that they are continuously or periodically disassembled and rebuilt, yet, always each according to its kind and standard pattern. In contrast to a machine, the cell interior is heaving and churning all the time; the positions of granules or other details [seen] in the [micrographs], therefore, denote just momentary way stations, and the different shapes of sacs or tubules signify only the degree of their filling at [that] moment. The only thing that remains predictable amidst the erratic stirring of the molecular population of the cytoplasm and its substructures is the overall pattern of dynamics which keeps the component activities in definable bounds of orderly restraints. These bounds again are not to be viewed as mechanically fixed structures, but as [a] "boundary . . . " set by the dynamics of the system as a whole. (Weiss 1973, 39–40; cf. Sherrington 1953, 80–81; Oparin 1966, 14–30, and especially 21–22; Morowitz 1992, 64)

It transpires that the functioning of the cell does not depend upon structure but rather on the rhythm of the reciprocal actions between the elements that make up the reaction pattern of the entire cell. "Structure and function are closely related matters but the knowledge of one by no means solves problems presented by the other" (Dean and Hinshelwood 1966, 422; Hinshelwood 1956, 155; cf. Oparin 1966, 14–16, 21– 22, 110; see also Tauber 1994, 230, 261).

It is typical of the living system that its structure is multifunctional. As Kelso and Haken observe, "the same set of components may self-organize for different functions or different components may self-organize for the same function" (1995, 140). The highly condensed and connected nature of the cell makes it extremely dynamic; in fact, the connections between the parts as well as the configuration can all be remodeled within minutes, subject to signals from the environment, such as the presence of food and light, hormones, growth factors, or mechanical or electrical stimulations. The entire cell acts as a coherent whole in search of stability and adaptability; thus information or disturbance to one part propagates promptly to all other parts (Ho and Popp 1993, 192).

One may argue that to a good approximation the cell may be conceived of as having states, which, however, change swiftly and in rapid succession. Yet that would be misleading, since the cell's dynamic is not just an expression of the net results of some physical laws; rather, this dynamic is regulated as a whole in a directive fashion so as to maintain the cell according to a certain "norm" or "standard"; static observables would not capture these "directives." As Eigen remarks, "all reactions in a living system follow a controlled program operated from an information center. The aim of this reaction program is the self-reproduction of all components of the system, including the duplication of the program itself, or more precisely of its material carrier" (1995, 9). This is indeed the "goal" of the control that guides the behavior of the cell. It therefore can be quite misleading to study experimentally one cell's activity by itself, to single it out and to consider it in isolation. Such an approach has led to many erroneous conclusions (Sinnott 1961, 31–33, 52).

The anatomist habitually exhibits the knowledge of an organism as a kind of display in extensiveness. Once again, this is misleading. The organism itself, as Canguilhem observes, "does not live in the spatial mode by which it is perceived. The life of a living being is, for each of its elements, the immediacy of the co-presence of all" (1991, 253). An organism is not an aggregate; it is first and foremost a triumph of chemical, not mechanical, engineering. This successful engineering is the result of nonequilibrium chemical processes that may be analyzed, as Wolpert remarks, "in terms of fluctuation and instability and, particularly, self-organization of spatial and temporal patterns." Wolpert continues by observing that a characteristic feature of all such nonlinear systems is that "they seem to exclude structure from the initial conditions" (1995, 62). To be sure, parts of the organism may be isolated for analytical purposes, and the different processes it undergoes may be separately studied in a linear fashion, but these processes cannot be understood unless they are referred to the entire, nonlinear dynamic of the system (Sinnott 1961, 21, 40).

It appears, therefore, that the standard distinction between structure and function is not instructive for the understanding of the special behavior of living matter—the distinction simply does not work! Indeed, it is misleading to think of the cell as possessing definite spatial parameters. In other words, applying the structure category to the living system is false; it should not be conceived of solely as a spatial pattern, but rather, as Weiss suggests, a cross section through a process (in Gerard and Stevens 1958, 163, 192; cf. Sinnott 1961, 59–60). Put differently, the strict distinction between structure and function is misleading insofar as living systems are concerned. With the crumbling of this distinction, the concept of state in living systems loses its raison d'être. To be sure, as Lange points out, one can conceive of physical systems that act as a "coherent whole": inanimate systems that exhibit "lifelike" behavior, such as spontaneous pattern formation, pattern change, and the creation and annihilation of forms (1996, 227; cf. Dean and Hinshelwood 1966, 420). However, the assumed strict differentiation between structure and function in such systems reflects faithfully their equilibrium nature and thus gives substance to the concept of state. This is not the case with living systems, which are fundamentally nonequilibrium systems, systems in which new patterns emerge and sustain themselves in a relatively autonomous fashion (Kelso and Haken 1995, 156). What are then the consequences of the claim that living systems do not have states?

On the Difference between the Science of Physics and That of Biology

In his by now classic contribution to biology, *What is Life?*, Schrödinger underlined the shortcomings of contemporary physics when questions of biological systems are at stake. He observed that "living matter, while not eluding the 'laws of physics' as established up to date, is likely to involve 'other laws of physics' hitherto unknown" (1969 [1944], 73). That was in 1944. The physics known at that time included all the foundations of modern linear physics, that is, quantum mechanics. Since then, important developments in understanding many-body, nonlinear, and nonequilibrium systems have become established parts of the physics framework and, as a matter of fact, currently draw much of the attention of the physics community. This new framework for the study of complex phenomena seems to offer a convenient departure point for thinking about biological phenomena. However, this very approach of great sophistication leads us back to the conflict that Schrödinger had already identified: the question "What is life?" may not be answerable within the existing framework, which was conceived in an attempt to explain the inanimate world (Braun and Hon 1996).

The success of physics in explaining the inanimate world may be traced back to its roots in positional astronomy. As Schrödinger noted, there is a direct genealogical link from quantum mechanics in both of its central formulations (the matrix and the waveform); via analytical mechanics of its central theorems, due to Hamilton and Jacobi; over to Newton's general laws of motion and gravitation; and further back to Kepler's celestial physics (1984 [1954], 562–64). The idea of a God-given, pervasive law that links the initial conditions (i.e., the state of the system) with its nature (e.g., its motion) has made mechanics the prototype of exact physical science to be emulated by all the sciences. This is not surprising, since it is a most ingenious solution for connecting the necessary and general element of the law with the contingent and particular aspect of the system. Put differently, Newton's profound and useful idea of dividing the analysis into a dynamical and a static part allows for a coherent and apparently successful connection between the laws of evolution of the system and its state at some point in time. This kind of analysis finds its immediate expression in the infinitesimal calculus in which the solution of the differential equations (the dynamical part) requires constants of integration (the static part), which are nothing else than the initial conditions of the system—its state. Notwithstanding the great—many would say revolutionary— innovations and discoveries since the time of Newton, nothing significant has changed in the comprehensive application of this successful Newtonian methodology.

Where do we stand in biology? How does the physical study of the question "What is life?" fare in this perspective? I suggest that we are still in the pre-Keplerian stage and that we may stay there for good, and indeed for good reason, unless we come up with a new method of inquiry. (For a different view see Gerard and Stevens 1958, 114.)

Two options present themselves. If there were biological laws similar to those of physics, then, I claim, we would not be able to discover them with the current methodology. Since the living system does not possess a state, any attempt to connect a law to the initial conditions of the system would be, in the final analysis, incoherent and therefore doomed to fail. The combination of a law with the system's initial conditions, so successful in the physics of the inanimate, is misleading in biology.

Alternatively, there might be no eternal, pervasive laws to govern the living system; that is, there might be sets of rules which do not presuppose a state and furthermore function in different time scales. To be sure, the living material of the organism obeys, as far as it is known, the laws of physics and chemistry; but, crucially, it may be governed further by sets of flexible rules, rules not laws, whose ultimate guiding principle is natural selection. There again, I claim, we would encounter a limit which the present methodology is unable to remove. Let us examine these possibilities. It will however be instructive to criticize first the common approach.

The Experimental Study of Living Systems: Methodologies and Limits

In his 1992 book, *Beginnings* [note the plural form] *of Cellular Life,* Harold Morowitz writes that "the origin of cells is a natural phenomenon lying within the domain of physics and chemistry. The physical sciences may have to add biogenic laws, but they should in no way be methodologically different from current laws in relation to experimental confirmation." Mark the phrase *in relation to experimental confirmation*. Morowitz continues: "To test . . . [the various] hypotheses, we must first assume that the origin of cellular life is a terrestrial phenomenon obeying natural laws. Only if we face failure after exhaustively pursuing such a set of assumptions should we then turn to other hypotheses that present a whole new set of epistemological problems" (12–13). Enthusiastically, Morowitz declares that "we are now able to do experiments asking the question what is life at a very deep chemical level. Let's move ahead," he urges, "and get on with it" (1993, 187, 190).

Reflect, however, upon what is being urged: we are being pressed to undertake an exhaustive experimental study of living matter with the knowledge that it might eventually transpire that something has been conceptually amiss all along. Might it be that, as far as living matter is concerned, "experimental confirmation" (to use the standard expression) is limited by the very methodology that underlies it?

I claim that in cell biology, and particularly in biophysical experiments on the cell as a whole, it is impossible—in principle—to lay down correctly the presuppositions of the argument that exhibits the experiment. There are insurmountable conceptual problems in setting up the *preparation* stage of the experiment, notably that the system under study has no states to be correlated to the presuppositions of the argument of the experiment. As a result, the experimenter continuously draws inferences from a set of incorrect propositions (cf. Wittgenstein 1966 [1921], §4.023). As we shall see, the problem is neither one of accuracy nor one of precision, and it is not the popular subjective-objective problem of measurement. The problem is methodological.

In the final accord of *The Origin of Species,* Darwin writes in a poetic mode, which captures succinctly the problem with which I am concerned: "There is a grandeur in this view of life, with its several powers, having been originally breathed into a few forms or into one; and that, whilst this planet has gone cycling on according to the fixed law of gravity, from so simple a beginning endless forms most beautiful and most wonderful have been, and are being, evolved" (1976 [1859], 459–60). Mine is not the esthetic concern but rather the striking difference between the fixed law of gravity and the evolving results of natural selection. Biology is indeed different from physics. The basic laws of physics are assumed to hold throughout the universe, spatially as well as temporally. The laws of biology, if they are laws at all, are either broad generalizations that describe rather elaborate chemical mechanisms that natural selection has evolved over billions of years or specific, ad hoc local instructions of a limited time scale. "Evolution is a tinker," says Jacob (quoted by Crick 1988, 5). Indeed, natural selection builds on what went before, so that, as Crick remarks, the primary processes have become encumbered with many subsidiary gadgets (1988, 5). The result is clear: the cell resembles the site of an archaeological excavation (Szent-Györgyi 1972, 6). We are dealing with processes that have history—indeed, a recorded history or, as Crick puts it, "frozen history" (1967, 70; cf. 1988, 101, 109). The contrast between the living and the common objects of science vis-à-vis history is aptly described by Canguilhem: "When we think of the object of a science we think of a stable object identical to itself. In this respect, matter and motion, governed by inertia, fulfill every requirement. But life? Isn't it evolution, variation of forms, invention of behaviors? Isn't its structure historical as well as histological?" (1991, 203; cf. Crick 1988, 137).

A living unit, by definition, has control over a measuring procedure, a procedure that accumulates data, processes them, and records them so that the unit can effect future dynamics. This buildup of an archive, that is, the system's memory, reflects from another perspective the claim that the living unit does not possess a state. Static observables— from which, it may be recalled, a state is being deduced—cannot exhibit by themselves the evolving, accumulated historical data of the system. Being static, the observables cannot be historically informed. Mark it: the history of the living unit is no metaphor; the cell keeps records and it seems that it does better than we do in absorbing fully its historical lesson (Oparin 1966, 98; cf. Gerard and Stevens 1958, 131, 161–64).

At the juncture when life was "breathed into a form," to use Darwin's poetic expression, there occurred a most remarkable phenomenon: the introduction of contingency and stratagem into nature. Nature has since then been experimenting, as it were, on itself. In fact, we witness the biggest experiment ever, except that nature does not fuss about epistemological issues as we do. This experiment churns up much "junk," and a living system that is too finely tuned to carry a large load of it may literally be overwhelmed and not survive the enterprise (Dyson 1988 [1985], 76; cf. Crick 1988, 139). Under no circumstance, at least within the Newtonian outlook, can we attribute to the inanimate world redundancy: "Nature does nothing in vain," proclaimed Newton (1995 [1687], Rules of Reasoning in Philosophy, Rule 1, Book 3). Redundancy, however, is the very mark of history and indeed of natural history; thus, although Occam's razor plays a useful role in the physical sciences, it can be, as Crick puts it, "a very dangerous implement in biology" (1988, 138; Oparin 1966, 98; cf. Rosen in Buckley and Peat 1979, 86). Put differently, physical laws convey simplicity and generality, whereas evolutionary theory requires complexity and individuality (Pattee 1966, 77; cf. Crick 1988, 138). Here then is the mystery: "how, from an *inexorable* physical dynamics, one would ever achieve an *arbitrary* self-coding, self-describing system" (Pattee in Buckley and Peat 1979, 91; my emphasis).

Historically, the watershed in biology took place in this century with the advent of Delbrück's experiments on the bacteriophage. Delbrück realized that the bacteriophage (literally, the bacteria gobbler) is an ideal experimental tool. As Dyson remarks, it is "a biological system stripped of inessential complications and reduced to an almost bare genetic apparatus." In other words, it consists of software and nothing else. The bacteriophage, Dyson continues, "is a purely parasitic creature in which the metabolic function has been lost and only the replicative function survives. It was indeed precisely this concentration of attention upon a rudimentary and highly specialized form of life that enabled Delbrück to do experiments exploring the physical basis of biological replication." Dyson aptly sums up this historical juncture: "the bacteriophage was for biology what the hydrogen atom was for physics" (1988 [1985], 3–4; cf. Dean and Hinshelwood 1966, 421).

Consider a comparison between the two systems. Both the stability of the bacteriophage and that of the hydrogen atom may be explained by resorting to a physical theory—say, quantum mechanics. However, do the constituent parts of the atom play any functional role? Can we discern in the atom a feature that may be characterized as a "norm," a "standard," or indeed a "goal" (Sinnott 1961, 52; cf. Sherrington 1953, 85; Oparin 1966, 10–13)? Has the atom undergone any development of which it retains records? Does it contain a self-describing mechanism? Can we think of the atom as having an ethos, let alone a parasitic one? We agree, I think, that the answers are *No* (see Needham 1968 [1936], 25–26). Over and above these crucial differences, notice further that although the atom has no functional role to play, it does have a structure; the bacteriophage, by contrast, may be said to have both function and structure, but they cannot be pried apart—an indication that the bacteriophage does not possess a state.

To circumvent the difficulty of lack of state—that is, to make the living unit suitable for experiments similar to those carried out on inanimate matter—one must turn the unit as closely as possible into an inanimate system so that to some good approximation it will indeed possess a state. A way of approaching this condition is to attempt to make the living system forget its "records," that is, its memory, so that it will lose its function and become, like an atom, a suitable system for experimentation (Braun and Hon 1996). Making the cell forget its memory is a study of limit. But note the difficulty of executing such an experiment: although the memory is naturally dependent on the variables of the system, it must also have *independent* variables that control to some extent the dynamic of the system (Pattee 1966, 78). Where exactly should we look for these independent variables?

Another study of limit, to regain the ground of unproblematic experimentation, is the attempt to strip the cell gradually of all its inessential molecular components—to dump its junk, as it were. One would hope in this way to find the irreducible minimum degree of complication of a homeostatic apparatus (Dyson 1988 [1985], 64; see also Morowitz 1992, 59). As in the *Tractatus,* we make our way toward a limit (§4.114); but, unlike the *Tractatus,* we receive assurances from people like Crick that "there is nothing unusual at this borderline between the living and the nonliving" (1967, 63).

However, technical difficulties apart, it seems that the conceptual problems may not be surmountable. As I noted, my concern is methodological: since one does not know which part of the homeostatic apparatus is inessential, Dyson (1988 [1985], 64) suggests that one should proceed in a trial and error fashion. If this were the case, then there would be no end to either the error or the trial. In the first place, the living system practices a variety of dynamics whose main features are plasticity and adaptability, not to mention the use of special tricks (Mittenthal et al. 1993, 68). To be sure, as Canguilhem observes, "among all the reactions of which an organism is capable under experimental conditions, only certain ones are used as those preferred." The preference is for those reactions that include the most order and stability, the least hesitation, disorder, and catastrophe (1991, 184–85). However, the cell is resourceful; it may come up with a specific, new, and unexpected dynamic designed to thwart the onslaught of the experimenter (Oparin 1966, 177; cf. Sinnott 1961, 33; Dean and Hinshelwood 1966, 33; Mittenthal et al. 1993, 68). Second, it is typical of the biophysical approach to disturb the living system, observe its resistance, and then the transition to some other form that is no longer living (see, e.g., Gerard and Stevens 1958, 184). Pursuing this analysis, one must be careful, as Szent-Györgyi cautions, not to kid oneself into believing that one understands when one does not:

Analyzing living systems we often have to pull them to pieces, decompose complex biological happenings into single reactions. The smaller and simpler the system we study, the more it will satisfy the rules of physics and chemistry, the more we will "understand" it, but also the less "alive" it will be. So when we have broken down living systems to molecules and analyzed their behavior we may kid ourselves into believing that we know what life is, forgetting that molecules have no life at all. (1972, 5)

To make this standard practice coherent, Dyson suggests further that one should proceed not only from the top down but also from the bottom up. He writes that

we need to experiment with synthetic populations of molecules confined in droplets in the style of Oparin, adding various combinations of catalysts and metabolites until a lasting homeostatic equilibrium is achieved. If we are lucky, we may find that the experiments from the top and those from the bottom show some degree of convergence. In so far as they converge, they will indicate a possible pathway which life might have followed. (1988 [1985], 64–65; cf. Gerard and Stevens 1958, 132–33; see also "The Cell as a Solid State System," in Ho and Popp 1993, 191–92)

Oparin, however, thought differently. Commenting on the attempt to construct artificially a semipermeable membrane, he concluded that the results were "far more valuable towards establishing the general physico-chemical laws of diffusion of dissolved substances formulated by van't Hoff than towards solving biological problems" (1966, 164). So much for the bottom-up approach.

Conclusion

There are, as I have suggested, two possibilities. First, let us suppose with Schrödinger that the laws of biology are like those of physics. If this is the case, then my claim is that a crucial and one of the most central distinctions in physics, namely the separation between the state of the system and its nature, does not work in biophysics. The living system, I claim, does not have a state, or rather, a state cannot be pried apart from the nature of the system.

Consider by way of comparison the preparation stage in quantum mechanics. The physical system is made to pass, for instance, through a polarizer of, say, the Stern-Gerlach type, and is given thereby a certain state before the actual measurement is carried out. However problematic, preparations in quantum mechanical experiments are of a different kind altogether than those in biophysical experiments. My point is that in the latter case the experimenter invariably builds on an incorrect set of background assumptions, which originates in the false belief that the system has a state. As we have seen, the crumbling of the distinction between function and structure in living systems on the one hand, and the buildup of a historical record that affects the future dynamic of the system on the other, make the fundamental assumption of state highly dubious. It is this characteristic of the living unit that limits the ability to get the premises of biophysical experiments right. It is as if the cell has a Hamiltonian (i.e., a specific nature) but no integration constants (i.e., no initial conditions) to solve the ensuing differential equations.

It is noteworthy that Schrödinger opens the final chapter of his influential book with the conclusion that "from all we have learnt about the structure of living matter, we must be prepared to find it working in a manner that cannot be reduced to the ordinary laws of physics" (1969 [1944], 81). It seems then that by attempting to reduce the workings of the living system to ordinary laws of physics we may be barking up the wrong tree. The alternative—that is, that the laws of biology are not really laws but sets of rules—seems indeed more promising; but there again the experimenter encounters logical and methodological difficulties. This possibility is, however, more promising. Laws should not have control over evaluative procedures: laws, unlike rules, cannot be broken, cannot be changed; in fact, laws undermine the very notion of assessment. Since the principal practice of the living system is evaluation and assessment, laws will not be suitable for underwriting its activity (see Pattee in Buckley and Peat 1979, 120–21).

I therefore suggest the following analysis. It appears that the living system does not obey the standard mechanical law of "action and reaction." A living system, as I have noted, acts on the basis of a historical record that constitutes a database (see, e.g., Morowitz 1992, 52–53). The system assesses the situation by doing a comparative analysis of the circumstances on the basis of the historical records it retains. It reaches a decision not by algorithm but by analogy, and it chooses a course of action that should optimize its chances of survival; this is the "norm," the "standard," or indeed the "goal" that the living unit seeks. After all, it's a matter of life or death. "Life is . . . a normative activity. . . . Life is not a monotonous deduction, a rectilinear movement, it ignores geometrical rigidity, it is discussion or explanation (what Goldstein calls *Auseinandersetzung*) with an environment where there are leaks, holes, escapes and unexpected resistance" (Canguilhem 1991, 126, 198). When the system is being interfered with—indeed at any moment of its existence as a living system—what we witness is not "action and reaction" but rather "action and proportionate reaction." In other words, the system makes an evaluative judgment. This process is not algorithmic; the proportionate reactions must be decided upon on the basis of a vast (albeit finite) database and in a finite (preferably short) time. To effect this process successfully, the living system must not follow an algorithm but pursue an analogical mode of reasoning. That process is, however, indeterminate. The system may reach different decisions under similar circumstances, and indeed it may err with respect to the optimal course of action it should take: cells do suffer, as it were, unnatural death. The error, this failed evaluative judgment, seems to indicate the existence of a function that may be described as a prototype of thinking.

The indeterminacy that marks the proportionate reaction of the living system limits experimenters' ability to set up correctly the background assumptions of the preparation stage. Experimenters seek laws of nature, but how does one experiment on analogies, on comparative patterns, whose guiding principles, apart from the general dynamical pattern, shift continuously? As in the previous situation, the experimenter cannot cast correctly the presuppositions of the preparation stage and therefore continuously draws inferences from false propositions.

Nature, as Darwin would have us believe, takes pains to reveal her schemes. We have had some success with its inanimate domain; but as to the animate domain, I have my doubts. We may have reached a methodological limit. If the experiment rests on a false set of background assumptions, then silence is a better investment, and the time has come to take a break and go back to first principles (see, e.g., the work of the Biology Council in Gerard and Stevens 1958). Reconsidering the crucial separation between the initial conditions of a defined system and the dynamical laws that describe its development constitutes such an attempt.

As I noted, the demand to draw a separation between the system and its boundary conditions—that is, the state of the system—conforms to a mathematical property of a particular type of equation, namely differential equations, which has been successfully applied in physics since the time of Newton. Is this separation here to stay? Are we to be constrained by the demand of mathematics? Could we conceive of a new methodology couched within a suitable mathematics that does away with the presupposition of the state? "In my opinion," writes Roger Penrose,

when we come ultimately to comprehend the laws, or principles, that *actually* govern the behavior of our universe—rather than the marvelous approximations that we have come to understand—we shall find that this distinction between dynamical equations and boundary conditions will dissolve away. Instead, there will be just some marvelous consistent comprehensive scheme. . . . Many . . . might not agree with it. But [it is] such [a viewpoint] . . . I have vaguely in mind when trying to explore the implications of some unknown theory. (1989, 352)

I conclude with a cautionary tale. In the film *Some Like It Hot,* when "Josephine" (Tony Curtis as saxophonist Joe) and "Geraldine" (Jack Lemmon as bass player Jerry) first see the voluptuous, hip-swinging, twenty-four-year-old singer Sugar Kane Kowalczyck (Marilyn Monroe) moving down the train platform, where she is squirted by a jet of steam, Geraldine/Lemmon remarks with great astonishment: "Look at that! Look how she moves. That's just like Jell-O on springs. She must have some sort of built-in motors. I tell you, it's a whole different sex!"

NOTES

This work was carried out at the Center for Philosophy of Science, University of Pittsburgh. I wish to thank the center and its director, James Lennox, as well as its former director, Gerald Massey, for their warm and generous hospitality. I am indebted to Bernard Goldstein for incisive criticism and instructive comments. I thank Martin Carrier for his critical comments and Nicholas Rescher and Gereon Wolters for their invitation to attend the fourth meeting of the Pittsburgh-Konstanz Colloquium in the Philosophy of Science.

1. This is an ideal analysis. In practice, problems of reduction and questions of interpretation make the issue much more complex. It should be further noted that the present study excludes the discussion of fraud in science. This discussion is irrelevant to the epistemology of experiment.

REFERENCES

Anscombe, G. E. M. 1967. *An Introduction to Wittgenstein's Tractatus.* London: Hutchinson University Library.
Bacon, F. 1859 [1620]. *Novum Organum.* A. Johnson, transl. London: Bell and Daldy Fleet St.
Braun, E., and G. Hon. 1996. "Are There Limits to Experimentation? A Philosophical Inquiry into the Method of Experimentation: The Case of Experiments in Biophysics." Research Proposal for the Technion-Haifa University Fund. Haifa.
Buchwald, J. Z., ed. 1995. *Scientific Practice: Theories and Stories of Doing Physics.* Chicago: University of Chicago Press.
Buckley, P., and F. D. Peat, eds. 1979. *A Question of Physics: Conversations in Physics and Biology.* Toronto: University of Toronto Press.
Canguilhem, G. 1991. *The Normal and the Pathological.* New York: Zones.
Crick, F. 1967. *Of Molecules and Men.* Seattle: University of Washington Press.
———. 1988. *What Mad Pursuit: A Personal View of Scientific Discovery.* New York: Basic Books.
Darwin, C. 1976 [1859]. *The Origin of Species.* Middlesex: Penguin.
Dean, A. C. R., and C. Hinshelwood. 1966. *Growth, Function and Regulation in Bacterial Cells.* Oxford: Clarendon Press.
Dyson, F. 1988 [1985]. *Origins of Life.* Cambridge: Cambridge University Press.
Eigen, M. 1995. "What Will Endure of 20th Century Biology?" In Murphy and O'Neill 1995, 5–23.
Galison, P. 1997. *Image and Logic.* Chicago: University of Chicago Press.
Gerard, R. W., and R. B. Stevens, eds. 1958. *Concepts of Biology.* Publication 560. Washington, D.C.: National Academy of Science, National Research Council.
Heidelberger, H., and F. Steinle, eds. 1998. *Experimental Essays—Versuche zum Experiment.* Baden-Baden: Nomos.
Hinshelwood, C. 1956. "Address of the President." *Proceedings of the Royal Society of London, Series B* 146: 155–65.
Ho, M.-W., and F.-A. Popp. 1993. "Biological Organization, Coherence, and Light Emission from Living Organisms." In Stein and Varela 1993, 183–213.
Hon, G. 1989. "Towards a Typology of Experimental Error: An Epistemological View." *Studies in History and Philosophy of Science* 20: 469–504.
———. 1998. "'If This Be Error': Probing Experiment with Error." In Heidelberger and Steinle 1998, 227–48.

Kampis, G. 1991. *Self-Modifying Systems in Biology and Cognitive Science*. Oxford: Pergamon Press.
Kelso, J. A. S., and H. Haken. 1995. "New Laws to Be Expected in the Organism: Synergetics of Brain and Behaviour." In Murphy and O'Neill 1995, 137–60.
Lange, M. 1996. "Life, 'Artificial Life,' and Scientific Explanation." *Philosophy of Science* 63: 225–44.
Luisi, P. L. 1993. "Defining the Transition to Life: Self-Replicating Bounded Structures and Chemical Autopoiesis." In Stein and Varela 1993, 17–39.
Mill, J. S. 1949 [1843]. *System of Logic*. London: Longmans, Green.
Mittenthal, J. E., B. Clarke, and M. Levinthal. 1993. "Designing Bacteria." In Stein and Varela 1993, 65–103.
Morowitz, H. J. 1992. *Beginnings of Cellular Life*. New Haven, Conn.: Yale University Press.
———. 1993. *Entropy and the Magic Flute*. New York: Oxford University Press.
Murphy, M. P., and L. A. J. O'Neill, eds. 1995. *What Is Life? The Next Fifty Years: Speculations on the Future of Biology*. Cambridge: Cambridge University Press.
Needham, J. 1968 [1936]. *Order and Life*. Cambridge, Mass.: MIT Press.
Newton, I. 1995 [1687]. *The Principia*. Andrew Motte, transl. New York: Prometheus.
Oparin, A. I. 1966. *Life: Its Nature, Origin and Development*. New York: Academic Press.
Pattee, H. H. 1966. "Physical Theories, Automata, and the Origin of Life." In H. H. Pattee, E. A. Edelsack, L. Fein, and A. B. Callahan, eds., *Natural Automata and Useful Simulations*. London: Macmillan, 73–105.
Penrose, L. S. 1959. "Self-Reproducing Machines." *Scientific American* 200(6): 105–13.
Penrose, R. 1989. *The Emperor's New Mind*. Oxford: Penguin.
Peres, A. 1990. "Axiomatic Quantum Phenomenology." In P. Lahti and P. Mittelstaedt, eds., *Symposium on the Foundations of Modern Physics, 1990*. Singapore: World Scientific, 317–31.
Posiewnik, A. 1985. "On Some Definition of Physical State." *International Journal of Theoretical Physics* 24: 135–40.
Radder, H. 1995. "Experimenting in the Natural Sciences: A Philosophical Approach." In Buchwald 1995, 56–86.
Schrödinger, E. 1969 [1944]. *What Is Life? Mind and Matter*. Cambridge: Cambridge University Press.
———. 1984 [1954]. "The Philosophy of Experiment." In *Collected Papers*, vol. 4: *General Scientific and Popular Papers*. Vienna: Austrian Academy of Sciences, 558–68.
Sherrington, C. 1953. *Man on His Nature*, 2nd ed. New York: Doubleday Anchor.
Sinnott, E. W. 1961. *Cell and Psyche: The Biology of Purpose*. New York: Harper & Row.
Stein, W., and F. J. Varela, eds. 1993. *Thinking about Biology*. Lecture notes, vol. 3, Santa Fe Institute, Studies in the Sciences of Complexity. Reading, Mass.: Addison-Wesley.

Stenius, E. 1960. *Wittgenstein's* Tractatus: *A Critical Exposition of Its Main Lines of Thought.* Ithaca, N.Y.: Cornell University Press.

Szent-Györgyi, A. 1972. *The Living State, with Observations on Cancer.* New York: Academic Press.

Tauber, A. I. 1994. *The Immune Self: Theory or Metaphor?* Cambridge: Cambridge University Press.

Weiss, A. P. 1973. *The Science of Life: The Living System—A System of Living.* New York: Futura.

Wittgenstein, L. 1966 [1921]. *Tractatus Logico-Philosophicus.* D. F. Pears and B. F. McGuinness, transls. London: Routledge & Kegan Paul.

———. 1984 [1921]. *Tractatus Logico-Philosophicus.* Werkausgabe Band I. Frankfurt: Suhrkamp.

Wolpert, L. 1995. "Development: Is the Egg Computable or Could We Generate an Angel or a Dinosaur?" In Murphy and O'Neill 1995, 57–66.

18

On Some Alleged Differences between Physics and Biology

Andreas Hüttemann
Department of Philosophy, University of Bielefeld

In his stimulating and provocative chapter, Giora Hon argues that the range of application of the experimental method is limited. Even though it is a successful method with regard to inanimate systems, it cannot be extended to living systems. This methodological claim is based on an ontological claim. Living systems as opposed to inanimate systems are said not to possess a state. However, the experimental method presupposes that the systems under investigation possess such a state.

In the first part of my commentary I focus on what I take to be the main line of argument for this thesis. In the second part I examine some of Hon's suggestions that seem to have some motivational force with respect to the main argument.

The Main Line of Argument

Hon outlines a theory of experimentation that is relevant for our purposes only insofar as he assumes that for the experimental method to work the system under investigation must possess a definite state and that its time evolution is determined by a law. If the system did not possess a state or its evolution were not determined by a law, an experiment would not be able to give us information about the system. In the case of living systems, it is argued, this presupposition is not fulfilled. Hon's point thus is that there is an ontological difference between physical systems on the one hand and living systems on the other. It is this ontological difference that is responsible for the limitation of the experimental method.

So what is characteristic of physical systems and physics? "The idea of a God-given, pervasive law that links the initial conditions (i.e., the state of the system) with its nature (e.g., its motion) has made mechanics the prototype of exact physical science to be emulated by all the sciences. . . . Newton's profound and useful idea of dividing the analysis into a dynamical and a static part allows for a coherent and apparently successful connection between the laws of evolution of the system and its state at some point in time." Physical systems, according to Hon, possess a definite state and their evolution is determined through laws. Living systems are categorically different. There are two alternative routes that may warrant this claim. Either there are no biological laws or there are no states that living systems possess. Hon focuses almost entirely on the latter claim, so we will follow him in this.

Let us start with his characterization of a living system and of states. Hon mentions that one important characteristic of such a system is its being an open thermodynamic system. Furthermore living systems are able to reproduce, to grow, to move, to evolve, to respond to the environment, and so on. However, with respect to all of these features it must be admitted that "they are not sufficient for bringing out the nature of the living as distinct from dead lumps of matter." Hon's conclusion is this: "I maintain that what completes the analysis, what makes the difference between the living and the dead distinct, is the claim that *the living system does not have a state.*" At this point it is not entirely clear whether this remark is meant as a definition of living systems or as a hypothesis to be proved. Given the subsequent arguments, the latter is more likely. However, in that case the characterization of living systems is not yet complete at this stage of the argument.

Hon defines a state of a system as "a set of instantaneous values of all the qualities of the system, external as well as internal, which in the classical case determine the development of the system." He does not give an example, but presumably the state of a point particle is the set of impulses and coordinates that describe its position at a certain time in phase space. At one point Hon equates the initial conditions and the state of a system. Its state (together with a law of evolution) then determines its future development. What is added is that a state variable "is a *static* variable interpretable as a *static* observable" (quoting Kampis 1991, 130; emphasis added). No explanation is given as to what "static" means in this context. We will return to this point.

How can one argue that living systems do not possess a state? How can one argue for a claim of the form: "System of kind k possess feature f"? Roughly there seem to be three possibilities:

1. Systems of kind k are defined so as to possess feature f.
2. It can be shown empirically that systems of kind k possess feature f.
3. Systems of kind k possess certain features f^* that imply that they possess feature f as well.

Strategy (1) would rely merely on a tautology. The fact that Hon gives an elaborate argument for his no-state thesis indicates that this is not his strategy. The above-mentioned characterization of living systems is not meant as a definition. Neither can Hon pursue strategy (2), because that would presumably presuppose the experimental method, the very method whose applicability to living systems he denies. In fact, Hon pursues strategy (3): those systems we intuitively classify as living systems possess features or properties that imply that the systems do not possess a state.

There are two problems with this argument, each of which on its own is sufficient to undermine it:

1. All the features he mentions pertain to physical systems as well. However, physical systems possess states—even according to Hon.
2. It is not obvious why the possession of the properties in question implies that the systems do not possess a state.[1]

Let us now examine the arguments in more detail. First we have the attempt to argue that a cell does not possess a state because "the highly condensed and connected nature of the cell makes it extremely dynamic," "the entire cell acts as a coherent whole," and "the configuration can . . . be remodeled within minutes, subject to signals from the environment." "Information or disturbance to one part propagates promptly to all other parts." Two things must be said. First, it seems that there are examples of physical systems, such as the solar system, that act as coherent wholes as well. If the solar system or other physical systems that can be characterized through the foregoing descriptions are supposed to possess states, there is no reason to take this as an argument for an ontological difference between living systems and inanimate systems. Second, even if these features pertained exclusively to living systems, it would not be clear why they prevent the system from possessing a state. Maybe cells just have states that develop rapidly in time.

Hon takes this objection into account and provides what seems to me an entirely new argument: cells do not possess rapidly changing states because they are parts of a whole that are regulated in a directive fashion. "Static observables would not capture these 'directives.'" Why not? Is this supposed to mean that because the observables are *static,* they cannot, so to speak, capture the rapidly changing state? The latter claim only works if "static" is opposed to "changing." However, no reason is given why state variables should be taken to be static in this sense.[2]

Or is it supposed to mean that the cell, if it had a state, would not be able to be regulated or directed? Hon quotes Eigen at this point: "all reactions in a living system follow a controlled program operated from an information center" (1995, 9). Even if it is granted that this is a feature physical systems do not share, it is by no means clear why the living system in question cannot have a state. It would be rather more astonishing if a cell could be regulated despite the fact that it has no state.

There is a further argument in terms of structure and function. The structure of a living system seems to be closely correlated with the system's state. Hon first claims that "It is typical of the living system that its structure is multifunctional." He then takes the fact that living systems apparently obey a nonlinear dynamic as an argument for their not possessing spatial definite parameters. Therefore, "the strict distinction between structure and function is misleading insofar as living systems are concerned. With the crumbling of this distinction, the concept of state in living systems loses its raison d'être." The problem with this argument is the transition from multifunctionality of structure to no structure at all. It is by no means clear how the fact that a system obeys a nonlinear dynamic can be adduced as an argument for it not having a state.[3]

At the end of the section there is yet a further attempt to argue for the no-state thesis. Admitting, once again, that there may be physical systems that act as "coherent wholes," Hon points out that "the assumed strict differentiation between structure and function in such systems reflects faithfully their equilibrium nature and thus gives substance to the concept of state. This is not the case with living systems, which are fundamentally nonequilibrium systems, systems in which new patterns emerge and sustain themselves in a relatively autonomous fashion." Once again, there are physical nonequilibrium systems. Haken, whom he cites in support, has developed his nonequilibrium theories with respect to physical systems. Furthermore, why should nonequilibrium systems possess no state? Kelso and Haken (1995, 141), whom Hon mentions, give examples of nonequilibrium systems, whose behavior can be mathematically modeled and predicted successfully.

Thus none of the arguments presented for the thesis that living systems do not possess a state has been really convincing. What seems to me to be the only scenario in which one would have to admit that the experimental method cannot be applied to living systems is the case of experimental results being unreliable despite all kinds of attempts to stabilize them. However, the fact that living systems can even be used to manufacture industrial products does not make this scenario very likely.

Laws and Contingencies

Besides the explicit arguments just discussed, there are some claims to which Hon at times appeals that seem to warrant the conviction that living systems are significantly different from inanimate systems.

Thus he mentions, for example, the striking difference between the law of gravitation and those rules or laws that play a role in biology: "Biology is indeed different from physics. The basic laws of physics are assumed to hold throughout the universe, spatially as well as temporally. The laws of biology, if they are laws at all, are either broad generalizations that describe rather elaborate chemical mechanisms that natural selection has evolved over billions of years or specific, ad hoc local instructions of a limited time scale." Contingencies and time-scale limits have their place in physics, too. Physical laws depend on contingencies in various ways. In the first place, the law-governed behavior of a physical system at time t depends on the initial conditions of the system (see Carrier 1995, 86). In the second place, laws in physics come along with ceteris paribus clauses; that is, a law describes a system's behavior provided there are no interfering factors. Even though Newton's law of gravitation may be all-pervasive, it is not able on its own to describe how an electron moves if electromagnetic forces are present. In cases such as these the laws must be adapted to the particular situation by taking into account other causal factors and rules that tell us how they interfere with one another. So contingencies can be found both in physics and in biology.

What about the time-scale argument? Biological systems have a history. The generalizations with respect to them allegedly become valid only when the systems in question have evolved. However, the same is true in physics. Most of the systems physicists deal with did not exist thirty seconds after the Big Bang took place. Crystals, for example, took a while to develop. Nowadays physicists experiment most of the time with systems that have been *manufactured*. Certain kinds of amorphous solids and some of the elements did not exist before they were synthesized. So we have the same kind of time limitation as in biology. Furthermore—and this is a third kind of contingency that has its place in physics—if evolution had taken another course (e.g., if mankind or similarly intelligent beings had not evolved), hassium and mendelevium would never have been synthesized. The same is true for other manufactured systems. Whether or not a physical law that concerns these kinds of systems is instantiated thus depends on the outcomes of evolution.

What was presupposed in Hon's argumentation is a conception of laws that is too impressed by the pervasiveness of Newton's law of gravitation. This is a very special law because it applies to all entities that possess mass. Other laws in physics, such as the ideal gas law, are much less pervasive. In order for the law to be instantiated, certain conditions must be fulfilled: given that there exists an ideal gas and given that nothing interferes and given that its pressure is P_1 and its volume is V_1, then every such system will have the temperature $P_1 V_1/R$.

Further Remarks

At various points in the chapter Hon points to problems that might arise if complex living systems are analyzed and decomposed, suggesting that whereas this might be an unproblematic procedure in mechanics or physics, it is not so in biology.

Hon is right in pointing out that one must be careful in extrapolat-
ing from a system's behavior in isolation to its behavior as part of a
complex system. However, this is not an exclusively biological problem. As C. D. Broad has already pointed out, "It is clear that in no case
could the behavior of a whole composed of certain constituents be
predicted merely from a knowledge of the properties of these constituents taken separately" (1925, 63). There is no a priori way of combining, say, two quantum mechanical systems. Thus, in quantum mechanics there are different rules of combination for identical and
nonidentical subsystems of a complex system. So in order to explain
the behavior of a complex system on the basis of its constituents one
must know not only how the constituents would behave in isolation,
but also the rules of combination. In this respect there is no difference
between physical and biological systems.

A further difference between living and inanimate systems that Hon
alleges is that only the former are able to react on the basis of memory:

A living unit, by definition, has control over a measuring procedure, a procedure that accumulates data, processes them, and records them so that the
unit can effect future dynamics. This buildup of an archive, that is, the system's memory, reflects from another perspective the claim that the living unit
does not possess a state. Static observables—from which, it may be recalled, a
state is being deduced—cannot exhibit by themselves the evolving, accumulated historical data of the system. Being static, the observables cannot be
historically informed.

Every complex physical system that is not in thermodynamic equilibrium has a memory (a history). The future behavior of these systems
critically depends on what has happened in the past. Reacting on the
basis of history or memory is thus not a distinctive feature of biological
systems.

Finally, I remark on the question of whether experimentation presupposes that the system under investigation is in a definite state. The
argument is that in order for an experiment to be informative it is
necessary that the experimenter be able to give a state description of
the system under investigation. An experiment, Hon maintains, "implies an argument the premises of which are assumed to correspond to
the states of the physical systems involved, for example, the initial
conditions of some material systems and their evolution in time."

How much do we have to know about a physical system if we want
to weigh it? What is out of the question is that we have to give a

complete state description. If we want to measure the specific heat of an amorphous solid all we need to know is that the system in question is an amorphous solid. We do not need to know anything about its microphysical constitution (besides that it is not of a crystalline structure), let alone the initial conditions of all the particles that constitute this system. Experiments are very often performed without such knowledge and precisely in order to *gain* it. We can very well measure the electrical conductivity of high-temperature superconductors without knowing the complete state of the system. Experiments on macroscopic systems would be impossible if it were necessary to determine more than 10^{23} initial conditions.

According to Hon's main line of argument the experimental method as applied in physics cannot be applied to living systems. The experimental method presupposes that the systems under investigation possess a state. Biological systems, however, fail in this respect. The arguments Hon provides for this fundamental difference between the physical and the biological rely too heavily on a conception of physics that is adequate for collision mechanics. In fact, physics deals with systems of vastly differing type and complexity. This seems to me the main reason why it is hard to see how a principled distinction between the physical and the biological could be argued for.

NOTES

1. One might want to argue that there is a further problem with Hon's strategy: It seems that he has to presuppose that the experimental method does work for his strategy (3), in order to establish that the systems in question possess those features that in turn imply that the system does not possess a state. Against this, however, Hon might argue that the results of the experiments are eventually self-defeating. They imply that the presuppositions of the experimental method are not fulfilled.

2. Hon at one point reformulates his claim that animate objects do not possess a state as "the cell does not have static observables from which a state may be deduced."

3. In this part of his argument Hon seems to misunderstand one of the authors he cites in support of his view. Hon remarks, "Wolpert continues by observing that a characteristic feature of all such nonlinear systems is that 'they seem to exclude structure from the initial conditions.'" According to Wolpert this is a reductio ad absurdum. Dynamical models, according to him, should not be used to model cells. This is most evident if we look at the sentence that follows immediately the one quoted: "Yet both cells and embryos are highly structured" (Wolpert 1995, 62). This is not exactly a claim that fits into Hon's main line of argument.

REFERENCES

Broad, C. D. 1925. *Mind and Its Place in Nature*. London: Kegan Paul.

Carrier, M. 1995. "Evolutionary Change and Lawlikeness: Beatty on Biological Generalizations." In G. Wolters and J. Lennox, eds., *Concepts, Theories, and Rationality in Biological Sciences*. Pittsburgh: University of Pittsburgh Press, 83–97.

Eigen, M. 1995. "What Will Endure of 20th Century Biology?" In Murphy and O'Neill 1995, 5–23.

Kelso, J. A. S., and H. Haken. 1995. "New Laws to Be Expected in the Organism: Synergetics of Brain and Behaviour." In Murphy and O'Neill 1995, 137–60.

Kampis, G. 1991. *Self-Modifying Systems in Biology and Cognitive Science*. Oxford: Pergamon Press.

Murphy, M. P., and L. A. J. O'Neill, eds. 1995. *What Is Life? The Next Fifty Years: Speculations on the Future of Biology*. Cambridge: Cambridge University Press.

Wolpert, L. 1995. "Development: Is the Egg Computable or Could We Generate an Angel or a Dinosaur?" In Murphy and O'Neill 1995, 57–66.

19

Models of Error and the Limits of Experimental Testing

Deborah G. Mayo
Department of Philosophy, Virginia Polytechnic Institute and State University

Experimental practice, if viewed as more than a source for experimental narratives or illustration, could produce a decisive transformation in the image of scientific inference by which we are now possessed.[1] Anyone who looks upon the nitty-gritty details of most experimental testing in science soon discovers that our neat and tidy models of experimental inference have very limited applicability. The data are inexact, noisy, and incomplete; extraneous factors are uncontrolled or physically uncontrollable; it may be impossible to replicate and nearly so to manipulate; and there may be huge knowledge gaps between the scope of theories and any experiment we can actually perform. Theories, even theories plus auxiliaries, fail to tell us how to test them and are silent about what to expect in the experiment before us. Yet despite these limitations, we do often manage to obtain reliable knowledge from data, and a question of interest to philosophers of science is "How?" How do we learn about the world in the face of limited information, uncertainty, and error?

An Adequate Account of Learning in the Face of Limits and Error

Is it possible to have a general account of scientific inference or testing that shows how we learn from experiment despite these limitations? One way that philosophers have attempted to answer this question affirmatively is to erect accounts of scientific inference or testing in which the limitations and uncertainties would be accommodated by appealing to probabilistic or statistical ideas. Leading attempts take the form of rules or logics relating evidence (or evidence statements) and hypotheses by measures of confirmation, support, or probability. Granting that hypothetico-deductive models are too limited to account for scientific inference, many propose that science can have a logic based on probability theory. We can call such accounts *logics of evidential relationship* or *E-R logics*. The leading example of such an E-R logic on the contemporary philosophical scene is the subjective Bayesian approach. Beginning with an accepted statement of evidence *e*, scientific agents are to assign prior probabilities, understood as measuring subjective degrees of belief, to an exhaustive set of hypotheses, which are then updated according to Bayes's formula from the probability calculus. What we have learned about a hypothesis *H* from evidence *e* is to be a function of the difference between the agent's posterior degree of belief in *H* given *e* and the prior degree of belief in *H*. Take Howson and Urbach: "The Bayesian theory of support is a theory of how the *acceptance as true of some evidential statement* affects your belief in some hypothesis. How you came to accept the truth of the evidence, and whether you are correct in accepting it as true, are matters which, from the point of view of the theory, are simply irrelevant" (Howson and Urbach 1989, 272; emphasis added). This is the Bayesian's way of doing something akin to deductive logic while taking into account the limitations of data in scientific inquiry.

The increased interest in studying actual experimental episodes in science, however, has led many philosophers to regard such logics of evidential relationship as failing to do justice to the actual limits and errors of inquiry. Scrutinizing evidence, in practice, is not a matter of logical or probabilistic relationships between statements, but turns on *empirical* information about how the data were generated and about the overall experimental testing context. It depends not on appraising full-blown theories or hypotheses against rivals but rather on *local experimental tests* to estimate backgrounds and to distinguish real effect from artifact and signal from noise. Philosophers, by and large, have either despaired of coming up with a systematic account of evidence and testing or else pursued minor tinkering with the existing logics of evidential relationship. The former group has concluded that the complexities and context-dependencies of actual experimental practice seem recalcitrant to the kind of uniform treatment dreamt of by philosophers; the latter seems to have accepted the false dilemma of "Bayes or Bust."[2]

I urge a move away from both these paths and propose a third way. The "data" from experimental practice should serve, not just as anomalies for philosophical models of evidential support, but as evidence pointing to a substantially different kind of account of experimental testing. What the evidence from practice shows is that where data are inexact, noisy, and incomplete, there is often disagreement and controversy as to whether they provide evidence for (or against) a claim or hypothesis of interest. Thus an adequate account must not begin with given statements of evidence but should provide methods for determining if we even have evidence for a hypothesis to begin with. An adequate account should also be able to motivate the ways in which scientists actually struggle with and resolve disagreements about evidence in practice: and in these struggles it is clear that scientists are in need, not of a way to quantify their beliefs, but of a way to check if they are being misled by beliefs and biases in interpreting their own data and that of other researchers. Yet to perform this check requires something else that an E-R logic will not give them—it requires some way to assess reliability. In other words, even when the E-R logic has done its work—say high support for *H* is found—we still need to know something more: how frequently would your measure of support give this much credit to hypothesis *H* even it would be a mistake (to regard the data as good evidence for *H*)? If it would often be a mistake, then the test has not passed a reliable test (the test lacks severity). The probability that a procedure will be wrong over a series of applications is called an *error frequency* or *error probability*. A philosophy of scientific inference based on error frequencies may be called an *error-statistical account.*

Nevertheless, simply pointing to the widespread use of error-statistical methods in scientific practice (e.g., Fisherian and Neyman-Pearson tests) is not yet to motivate my claim that they contain essential features for an adequate philosophy of evidence. Indeed, there has been a good deal of controversy as to the relevance of a test's long-run frequencies of error in interpreting evidence—leading many philosophers to prefer Bayesian and other E-R logics (see, e.g., Mayo 1996 and Mayo and Kruse, forthcoming). My goal in this chapter is both to elucidate key aspects of the error-statistical account that I favor and at the same time to offer a rationale for making error probabilities central to the interpretation of data. Focusing on just one key type of statistical tool (based on "null hypothesis" testing), I will show how error-statistical reasoning effectively deals with limited information in very different contexts.

Some Examples

Consider these examples:

- Is the observed correlation between a given (ApoE4) gene and Alzheimer's disease (AD) evidence for a genetic theory of AD (as against the leading neurological theories)?
- Is the failure of a spectrometer to detect muons in a proportion of events evidence of the existence of neutral currents?
- Are observed numbers of extinctions in given stages of the fossil record evidence of mass extinction events (as postulated by a given theory in paleontology)?

In each case the evidence accords with or "fits" a given hypothesis against a rival. One might express this by means of one of the E-R measures of fit, say, by asserting that a hypothesis H makes e more likely than do (extant) rivals. Although the E-R logic has done its job, the work actually required to answer the question "Is e evidence for H?" is not thereby ended but has only just begun. Something beyond the measures of "fit" offered by E-R logics is required, or so I claim, in order to tackle this question. Two pieces of data that would equally well support a given hypothesis, according to logical measures of evidential relationship, may in practice be regarded as differing greatly in their evidential value because of differences in *how reliably* each was produced. More specifically, scientists, it seems, seek to scrutinize if the overall experiment from which the data arose was a reliable probe of the ways we could err in taking e as evidence for (or against) hypothesis H—that is, the ways we can *mis*take e as evidence for H. Scientists seem willing to forgo grand and unified schemes for relating their beliefs in exchange for a hodgepodge of methods that offer some protection against being misled by their beliefs.

Modeling Mistakes

This does not mean we have to give up saying anything systematic and general, as many philosophers nowadays fear. The hodgepodge of methods gives way to rather neat statistical strategies, and a handful of similar models may be used to probe a cluster of mistakes across a wide variety of domains. Although limited information may lead to uncertainty as to whether we have any kind of evidence for a hypothesis, we may know a good deal about *how* the type of evidence can be mistaken as evidence for *H*. On this knowledge, we base models of mistaken construals of evidence.

The mistakes arise because although the data may accord with a hypothesis, we cannot be sure they are not actually the result of "background" or "noise," of artifacts for which we have not controlled, or of faulty experimental and theoretical assumptions. Rather than be stymied by our limited control, we may instead learn enough about background factors to "subtract them out" in comparing hypotheses to data or to estimate the likely upper bound of their influence. In coping with potentially faulty assumptions whose influences we cannot estimate, we may deliberately run tests with varied assumptions and check for converging results.

So what we need are models and methods for estimating what it would be like were it a mistake to regard *e* as evidence for *H,* and strategies for discerning whether the actual situation is one of these mistaken ones. The history of mistakes made in a type of inquiry gives rise to a list of mistakes that researchers either work to avoid (before-trial planning) or check if committed (after-trial checking). For example, when inferring the cause of an observed correlation, such a repertoire of errors might include a set of questions: Is the correlation spurious? Is it due to an extraneous factor? Are we confusing cause and effect? Corresponding to such a repertoire of errors is a "reservoir of models."[3] I call them *models of error.*[4]

Null Hypotheses

A standard source of models of error may be found in what are commonly called *null hypotheses* or null models in statistics.[5] These null, or error, hypotheses let us model, in effect, the situation of being fooled into thinking *e* is evidence for *H.* Hence we have a warrant for taking *e* as evidence for *H* only to the extent that we can reject the null hypothesis. The term *null* comes from the fact that the situation of being fooled may often be expressed as *there is really "no effect,"* whereas an alternative hypothesis *H* asserts that some real effect or phenomenon exists. (We can more generally call them error hypotheses, but I will usually keep to the familiar term here.)

Take the classic null hypothesis that says a given pattern (in *e*) is or may be regarded as nongenuine or merely "due to chance." By providing us with a contrast against which to compare the observed pattern, the null model[6] lets us see if the data can *easily be accounted for* by mere chance. These null or error models provide ways to show just how easily (i.e., frequently) results that may appear to show the *absence* of the error can be produced even when the error is present. Tests can be designed so that with high probability they would yield a result deemed "reasonably typical of a process in which the error is committed," if in fact it is committed—but not otherwise. Thus if a result is one that the test deems practically incapable of arising under the assumption of error, it is a *good indication* that the error is absent. Many different kinds of problems in learning from limited data may be tackled by comparisons with a null or error model, and a considerable amount of effort goes into constructing useful null models. Models of error are not limited to statistical models. However, because the nonstatistical variants may often be seen as attempts to approximate statistical models, understanding the statistical case is a useful springboard for extracting a general kind of experimental argument.

Arguing from Error

The argument I have in mind follows an informal pattern of reasoning that I call an *argument from error* or *learning from error*. The overarching structure of the argument is guided by the following thesis, which I will state in two equivalent ways:

(a) *It is learned that an error is absent when (and only to the extent that) a procedure of inquiry signals the absence of the error, despite the procedure having a very low probability of doing this if in fact the error is present.*

That a test *signals the absence of the error* means that the test produces a result that is classified by the test as "no error." The job of a statistical test is to indicate which of the possible test results to classify as indicating (or signaling) the absence and which results to classify as indicating the presence of the error of interest, and to do so in such a way as to substantiate the above kind of argument from error. Since the set of results classified as indicating "no error" is the complement of the set indicating the presence of the error, the above argument from error can be written equivalently as follows:

(b) *It is learned that an error is absent when (and only to the extent that) a procedure of inquiry signals the absence of the error, despite the procedure having a very high probability of signaling the presence of the error, if the error is present.*

Although form (a) of the argument from error may seem easier to parse, the equivalent form (b) brings out why such a procedure of inquiry may be called a reliable or highly *severe* error probe. The value of this "high probability" would measure the severity with which the test result indicates the error is absent. According to this thesis, we can argue that an error is absent if it fails to be detected by a highly severe error probe. A corresponding argument may be given to learn that an error is present,[7] and indeed, one of the best-known "null experiments" in science, that of Michelson and Morley, involved identifying the presence of an error by failing to reject a null hypothesis.[8]

The pattern of arguing from error underlies many different experimental arguments. The kind of example on which I focus can be described in terms of a test of a hypothesis H, where hypothesis H asserts that a given error is *absent*—or is less than a given amount. Correspondingly, a null or error hypothesis, often written as H_0, may be used to express the *presence* of the error (i.e., not-H). Experimental results are good evidence for H to the extent that H passes *a severe test* with these results—that is, a test with a high probability of failing H (and detecting the presence of the error) just in case H is false (and the error is present, i.e., H_0 is true).

To elaborate by means of a familiar example, suppose we are testing whether a coin is biased for "heads" on the basis of observing e, the proportion of heads in one hundred tosses (appropriately made). The null hypothesis H_0 is:

H_0 *It is an error to take e as due to genuine bias (i.e., any observed discrepancy from 50 percent heads is merely "due to chance").*

while H asserts:

H *It is* not *an error to take e as due to genuine bias.*

If the test classifies results between 30 and 70 percent heads as still consistent with the presence of the error, that is, as still not so far from 0.5 to indicate bias, then the test will correctly signal the presence of the "error" more than 99.9 percent of the time. It is a severe error probe (where the error consists of taking the coin as biased for heads when it is actually fair). Thus, an outcome e such as 71 percent heads would be classified as signaling H, and H would thereby be said to pass a severe test with e.

A question that immediately arises is whether such severe tests may be constructed for more general hypotheses in science. In particular, an important challenge in setting out this notion of severity is to show how it avoids the obstacles that vitiate other, broadly analogous, notions of severity (e.g., that of Popper). The most serious obstacle is this: how can one satisfy the severity requirement—that there be a high probability of detecting H's errors if H is false (i.e., H_0 is true)—when there are always alternatives to H that have not even been thought of? What enables my account of severity to answer this objection is that it is a *piecemeal* account of testing. Where we cannot test everything at once, we may be able to test piecemeal—and it is precisely the role of null models, together with a methodology of testing, to enable such piecemeal testing to proceed.

Models of Inquiry

For each experimental inquiry we can delineate three types of models (figure 19.1): *models of primary scientific hypotheses, models of data,* and *models of experiment* that link the others by means of test procedures.

A substantive scientific inquiry is to be broken down into one or more local primary hypotheses, and the experimental models serve as the key linkage models connecting the primary model to the data models. To determine if data provide strong evidence for a hypothesis requires considering how reliably a test is able to detect a given type of error, that is, the test's *error probabilities* or error characteristics. The need for one or more null models or hypotheses arises to give us some estimate of these error probabilities, that is, to assess how good the test is at revealing if the situation is one of the erroneous ones. The null

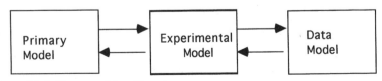

Figure 19.1 Three models of inquiry.

model is defined within the experimental model for the test, which is very often statistical.

The thrust of experimental design is deliberately to create contexts that enable questions to be asked one at a time. Within an experimental testing model, a null hypothesis asserts that "H is false," or the situation is as if H is false. "H is false" (H_0 is true) refers to a specific error that the hypothesis H is *denying*. If H states that a given effect is systematic—of the sort brought about more often than by chance—then not-H states that it is due to chance; if H states that a parameter is greater than some value c, not-H states that it is less than c; if H states that factor F is responsible for at least p percent of an effect, not-H states that it is responsible for less than p percent, and so on. *In each case, the not-H may be expressed as a null hypothesis* H_0 that asserts it would be an error to take the data as evidence of H. More than that, H_0 may be used to model *what it would be like* if it were an error to take the data as evidence of H. How specific the error hypothesis is depends upon what is required to ensure a good chance of learning something of interest.

Looking at the problem in terms of the logical or probabilistic relationships between given evidence and hypotheses overlooks all of the active intervention that provides the basis for arguing that if a specific error is committed, it is almost certain to show up in one of the results of a given probe or series of tests—each with deliberately varied assumptions. By such active intervention one can substantiate the claim that we should expect (with high probability) hypothesis H to fail a given test, if H is false. Equivalently, there is a high probability of accepting the error explanation asserted by the null hypothesis, if (and only if) in fact the error is committed. This "active intervention" need not require any literal manipulation—far from it. The key service performed by the null hypothesis is to let us model what would be expected assuming the error when this necessitates a hypothetical or simulated set of possibilities. We will see this in our examples. So long as the effects of different factors can be sufficiently *distinguished* or *subtracted out,* then the inferences are not threatened by our limited ability to control for them.

Standard (Error) Statistics

The key value of standard error-statistical methods, as I see them, is that they provide very effective tools for modeling what would be expected (statistically) given that we are erring about some aspect of the underlying cause or process of interest. They do so, first, by providing standard or *canonical* hypotheses to model the presence of an error (e.g., by means of standard null hypotheses) and, second, by providing tests that make it very hard, or extremely improbable, for us to rule out the error incorrectly. In referring to error statistics I include the familiar techniques of statistical analysis we read about every day in polls and studies, such as *statistical significance tests* and *confidence interval estimates,* although I adapt them in ways that go beyond what is strictly found in statistics texts.[9] (Here I limit myself to discussing the former.) The name *error statistics* or *error probability statistics* comes from the fact that what fundamentally distinguishes this approach from others is that to determine what inferences are warranted by evidence requires considering the error probabilities of the overall testing procedure. (For an in-depth discussion of how this distinguishes error statistics from the Bayesian approach, see Mayo and Kruse, forthcoming.) In other words, it requires, *in addition* to a measure of how well *e* fits *H,* a measure of the probability of achieving so good a fit, even if *H* is false.

Anyone who opens an issue of *Nature* or *Science* sees significance tests and interval estimates used to scrutinize evidence in widely different fields. Yes, there are controversies surrounding the interpretation and justification of null hypothesis significance tests, but it seems to me that it is precisely the job of the philosopher of science to help resolve, or at least help clarify, these controversies. Certainly it has not helped matters for philosophers of science to dismiss the widespread use of standard statistical tests as the result of confusion and/or brainwashing (as Howson 1997 and some other Bayesian critics have alleged[10]). Putting aside the textbook rationale for significance tests (in terms of low long-run error rates) and their well-worn criticisms, I propose that philosophers undertake a serious reexamination of these tests and consider whether—and if so how—they may be serving to provide, if not a uniform evidential-relation logic, a valuable set of tools for coping with limitations, uncertainties, and errors in learning from data.

To illustrate the kind of examination I have in mind, let us consider aspects of the three examples delineated earlier.

Alzheimer's Disease

Suppose we have recorded that a high proportion of people suffering from a disease, say AD, are found to have some factor, such as the presence of a special gene (E4). This is our data *e*. Is this evidence for a genetic theory of Alzheimer's disease?

The Bayesian rule asks us to compute P(*e*, given the genetic theory is false), but our information is much too limited for us to proceed with such a computation. Even restricting "the genetic theory is false" to the existing alternative theory (at the time of this evidential appraisal), the computation requires us to know the probability of *e* given the nongenetic, neurological theory of AD. But that theory says nothing about genes, so how can we even begin? The error statistician, by contrast, proceeds by considering, not alternative Alzheimer's theories, but alternatives to the claim that *e* is good evidence for the *given* (genetic) Alzheimer's theory.

In other words, the error statistician begins by recognizing that although a genetic hypothesis may accord with the observed proportion *e*, in fact *e* might be very poor evidence for *H* because the inquiry that brought forth *e* has not even begun to rule out several ways that *H* can be in error. These errors can be stated and probed quite apart from a specific alternative theory of AD.[11] The very limitation of the available evidence at this stage alerts the error statistician to the absence of a comparative group: that is, the observed proportion *e* gives no comparison of the proportion (with this genetic factor) among those who are *not* afflicted with AD. As one set of researchers put it: "Since the ApoE4 allele is . . . common . . . in the general population . . . one would expect that a substantial proportion of cases of the coincidence of AD and inheritance of the ApoE4 allele is due to chance and not inherited predisposition to the disease" (Tanzi et al. 1996).

They could, however, appeal to a model of "chance": a model of what it would be like to err in taking the observed proportion as evidence of a genuine connection (between the allele and AD). More specifically, they could represent the proportion *that would be expected* to have this gene if they were observing non-AD sufferers (in the given experiment). But they could not strip away their disease and see if they still have the gene. What they could do (and did) was to obtain an appropriate comparison group all of whom lack AD—a *control group*—and observe the proportion of the suspect gene among them. All of this is very familiar, and I begin with such a familiar point to facilitate our understanding of some less familiar ideas.

We can employ a standard "null" hypothesis H_0 to assert: there is no genuine correlation between the disease and the presence of the special gene (E4). Equivalently, H_0 *says it is an error to suppose a genuine correlation is responsible for any observed correlation.* It is not enough to report a difference between the two proportions. The null model must enable us to say, approximately, how often there would occur a difference in proportions (with the gene) as large as the difference observed, if in fact H_0 were the true state of affairs (i.e., if taking the result as evidence of a real correlation would be a *mis*take). That is, the null model must give us an *error probability;* and in particular, it must give us what is called the *statistical significance level* of the observed difference in proportions.

However, in order for models of error to give us error probabilities, there typically must be a deliberate *introduction* of statistical considerations into the data generation (e.g., by randomization, by approximations using matched controls) or by means of the data modeling (e.g., as with simulations). Only then can we carry out the comparison with the null model, by calculating the statistical significance of the data.

The Significance Question

Picturing our scheme of models, we can locate the hypothesis H, about the correlation, in an experimental model of an inquiry. To probe H, in our example, the researchers developed a design that would let them use the standard statistical null hypothesis H_0, which approximates to the familiar normal distribution of error.[12] This allowed the researchers to direct a question to the given observed difference, which we may abbreviate as d. The question they posed of d may be called a *significance question.* It is this:

How frequently would we obtain a result that accords with H *as well as* d *does (or better) if in fact* H *is false and* H_0 *is true (i.e., the error is committed)?*

What is being asked, in statistical terminology, is: what is the *statistical significance level* of the observed difference d? The null model is what let the researchers answer this question. The answer (i.e., the statistical significance level) is of interest for the simple reason that they want to be assured—before they take the observed fit (between d and hypothesis H) as evidence of a real effect—that d would *not* be expected to occur fairly frequently (in samples of this size) if null hypothesis H_0 were true. So if the significance level says it *is* fairly frequent—that is, if the significance level is not very small—the researchers do not take the data as ruling out the null or error explanation. But in this particular case study, the significance levels observed were very small, that is, they were *statistically significant.*

To illustrate, suppose a statistically significant correlation is observed between those with AD and the suspect gene: perhaps the *significance level* (or p-*value*) is .01. This report might be taken to reject the *null hypothesis* H_0 and imply there is a genuine correlation between the two. Notice that in rejecting H_0 we are rejecting or denying the error asserted in H_0. But the significance level of .01, with which we reject this error, is *not* an assignment of probability to the null hypothesis—we cannot say H_0 has probability .01. Rather, .01 is the probability that such a test *procedure* would reject the null hypothesis erroneously, thus the term *error probability*. It asserts, in particular, that *were* it an error to infer the genuine correlation (i.e., were H_0 true), so large an observed correlation (as we have in *d*) would occur only 1 percent of the time.

Hence, following the argument from error, the low significance level, .01, lets us pass the hypothesis *H,* that the correlation is genuine. Evidence from any one test might at most be taken as evidence that the correlation is genuine. But after several such failed null hypotheses, hypothesis *H* passes a *severe test* because, were *H* false and each null hypothesis H_0 true, we would very probably (probability >.99) have obtained results that accord *less well* with *H* than the ones we got. Note that it is the entire *procedure* of the various subexperiments that may properly be said to have the probative power—the high probability of detecting a spurious correlation by *not* yielding such consistently statistically significant results. In this way, the significance test informed the researcher (Alan Roses at Duke University) whether he would be wasting his time trying to find the cause of something that might be regarded as accidental.

I admit that my discussion here remains sketchy and at a rudimentary level, but this should suffice for my present goal, which is to identify key features of how I think error-statistical reasoning enters to cope with limitations and errors in a cluster of cases. These inferences are, by and large, directed to models of experiment intermediate between the data and substantive scientific hypotheses or theories. Nevertheless, inferences about such experimental models are often taken as input into models of primary hypotheses of interest, thereby providing *indirect* evidence for hypotheses in the primary scientific model. One can of course attach many additional models to the scheme in this same fashion.

Information from one experimental model is often used to supply a piece of information that can be used as a clue for further investigations. For instance, having severely passed the correlation hypothesis between the gene E4 and AD, researchers had clues about a possible genetic explanation, but they had not yet reliably demonstrated a phenomenon that rival causal hypotheses had to confront. Not only did the genetic hypothesis of AD go against the neurological hypotheses generally accepted at the time (which regarded AD as caused by a buildup of plaque in the brain), the researchers had simply not yet ruled out numerous ways in which they would have been *mistaken* to infer a causal connection from such a correlation—however real that correlation was found to be. Specifying which errors are ruled out at a given stage also reveals which errors have *not* yet been considered; and that suggests what would be needed to go further in investigating the phenomenon of interest. In this case, it is evident that one must next go on to confront causal errors.

The first causal error they detected was that it was not ApoE4 itself that caused AD, but the *lack* of better versions of this gene (E3 or E2) that was responsible for the early death of nerve cells, which *then* produced the amyloid tangles. The tangles, it seemed, were an effect and not a cause of AD. This indicated that the rival amyloid plaque hypotheses might have the causal story, and even the causal order, wrong.[13]

Using Data to Probe Hypotheses that Say Nothing about Them

As I remarked earlier, the neurological (plaque) theories said nothing whatsoever about genes (or the associated protein ApoE) involved in the causal process. They neither predicted nor counterpredicted the genetic evidence. But this evidence turned out to be probative for understanding what happens in the brains of Alzheimer's patients, and thus for identifying (unexpected) anomalies in existing neurological hypotheses. (It was said that the link had never even been hinted at by any of the world's foremost Alzheimer's scientists.)

This recognition is important for seeing how the current account circumvents a familiar criticism of accounts of testing—one that Larry Laudan (forthcoming) has reemphasized. Theories of testing, Laudan alleges, cannot account for the fact that in practice theories are appraised by means of phenomena that they do not explicitly address. Such an appraisal, Laudan thinks, is outside the scope of testing, and he concludes that it requires some nonepistemic analysis, thereby showing the in-principle limits of epistemology.

I allow that Laudan's criticism hits its mark with respect to Bayesian theories, which are his principal target, because, as noted earlier, it is not clear how one can assign the needed probability of the data given a rival hypothesis that says nothing about the data domain observed. Nevertheless, we have just illustrated how error-statistical testing avoids his charge. One way to express the current state of knowledge regarding a phenomenon or domain, I propose, is in terms of the errors that have and have not been ruled out by severe tests. Although certain *aspects* of the amyloid plaque theories had passed severe tests (for example, there is evidence of *some* genuine connection between these tangles and AD), none of the hypotheses about how amyloid causally contributes to AD had yet passed severe tests. This is why Roses was led to question the generally held causal story in the first place. So by passing even some aspects of his causal account severely (even lacking anything like a complete theory of AD), Roses was able to obtain evidence that the amyloid account gets the causal story wrong; and he was able to do this despite the fact that plaque theories said nothing about genes.

Again, I admit to being skimpy with the details of a case with an ever-changing evidential base, but my aim is just to show the power of the methods and logic of this account to circumvent obstacles due to common information gaps.[14] It seems to me that a strikingly similar pattern of arguing from error occurs in widely diverse fields, often using the same models of error. Let us now consider an example from twentieth-century physics.

Distinguishing Effects from Artifacts: Galison and Neutral Currents

Although by the end of the 1960s, Peter Galison (1987)[15] tells us, the "collective wisdom" was that there were no neutral currents (164, 174), soon after (from 1971 to 1974) "photographs . . . that at first appeared to be mere curiosities came to be seen as powerful evidence for" their existence (135). How did experimentalists themselves come to regard this data as evidence that neutral currents existed? "What persuaded them that they were looking at a real effect and not at an artifact of the machine or the environment?" (136).

To give the bare bones of the analysis, experimental outcomes here are modeled using a (dichotomous) statistic somewhat like that in the AD case. There each subject either had the E4 gene or did not. Here each event is described as muonless or muonful. In the AD case, the recorded result was the difference in proportions (with the suspect gene); here the recorded result is the ratio R of the number of muonless and muonful events. (Neutral currents are described as those neutrino events without muons.) The main point is that the more muonless events recorded, the more the result favors the existence of neutral currents. The worry is that recorded muonless events are due not to neutral currents but to inadequacies of the detection apparatus.

Experiments were conducted by a consortium of researchers from Harvard, Wisconsin, Pennsylvania, and Fermilab, the HWPF group. They recorded 54 muonless events and 56 muonful events, giving a ratio of 54/56. Rubbia, a researcher from Harvard, emphasized that "'the important question . . . is whether neutral currents exist or not. . . . The evidence we have is a 6-standard-deviation effect'" (220). The "important question" revolved around the question of the statistical significance of the effect.

The Significance Question

The *significance question* in this case is: What is the probability that the HWPF group would get as many as (or more than) 54 muonless events, given there are no neutral currents? We could answer this if we had a null model to tell us how often, in a series of experiments such as the one performed by the HWPF group, we would expect the occurrence of as many muonless events as were observed, given that there are no neutral currents. This is precisely what the researchers sought to determine.

It may be objected that there is only this *one* experimental result, not a series of experiments. True, the frequency of outcomes in a series of experiments supplied by the null model is a kind of hypothetical construct—just as in the case of AD. I am trying to bring out why it is perceived as so useful to introduce this hypothetical construct into the data analysis.

The answer, as I see it, is that such a null model provides an effective way to model the distribution of outcomes assuming it was a *mistake* to regard the data as evidence of the existence of neutral currents. It thereby provides an effective way to distinguish real effect from artifacts (correct from incorrect interpretations of the given result). Were the experiment so well controlled that the only reason for failing to detect a muon is that the event is a genuinely muonless one, then artifacts would not be a problem and this statistical construct would not be needed. But from the start a good deal of attention focused on the backgrounds that might fake neutral currents. A major problem was escaping muons. "From the beginning of the HWPF neutral current search, the principal worry was that a muon could escape detection in the muon spectrometer by exiting at a wide angle. The event would therefore look like a neutral-current event in which no muon was ever produced" (217).

The problem, then, is to rule out a certain error: construing as a genuine muonless event one in which the muon simply never made it to the spectrometer and thus went undetected. If we let hypothesis H be

H   *Neutral currents are responsible for (at least some of) the results.*

then, *within this piece of data analysis,* the falsity of H is the artifact explanation:

H *is false (the artifact explanation): Recorded muonless events are due, not to neutral currents, but to wide-angle muons escaping detection.*

The null hypothesis asserts, in effect, that the observed ratio arose from a universe where H is false. Using the familiar abbreviation, H_0, we have

H_0 *It would be an error to regard the observed ratio as evidence of neutral currents.*

Our *significance question* becomes: What is the probability of a ratio (of muonless to muonful events) as great as 54/56, given that H is false (that is, given H_0)? The answer is the *significance level* of the result.

But how do you get the significance probability or even approximate it? It requires estimating what would be usual or typical (statistically speaking) due to the background alone. The HWPF group, for example, created a computer simulation called a Monte Carlo program in order to model statistically how muons could escape detection by the spectrometer by exiting at a wide angle. "By comparing the number of muons expected not to reach the muon spectrometer with the number of measured muonless events, they could determine if there was a statistically significant excess of neutral candidates" (Galison, 217).

Note that probability arises in this part of the analysis not because the hypothesis about neutral currents is a statistical one, much less because it quantifies credibility in H or in H_0. Probabilistic considerations are deliberately *introduced* into the data analysis because they offer a way to model the expected effect of the artifact (escaping muons). In the AD example, we saw how statistical considerations were introduced by specially selected control groups. Here they are introduced by considering a statistical simulation of the background by a computer. Statistical considerations—we might call them "manipulations on paper" (or on computer)—afford a way to subtract out background factors that cannot literally be controlled for.

The Data

The data used in the HWPF paper are as follows (Galison, 220):

Visible muon events	56
No visible muon events	54
Calculated muonless events	24
Excess	30
Statistically significant deviation	5.1

The first two entries just record the HWPF result. The third entry refers to the number calculated or expected to occur because of escaping muons, as derived from the Monte Carlo simulation.

The simulation can be seen as allowing us to construct a null model: a model of the relevant features of what it would be like (statistically) if the researchers were actually experimenting on a process in which the artifact explanation H_0 is true. It tells us, in particular, that given outcomes (observed ratios) would occur with certain probabilities. (Most experiments would yield ratios close to the average (24/56); the vast majority would be within two standard deviations of it.) The null hypothesis, in effect, asserts that the observed result—54 out of 56—is *not* out of the ordinary even if *all* events are due to escaping muons. The difference between the ratio observed and the ratio expected (due to the artifact) is 54/56 − 24/56 = 0.536. The null model tells us how improbable such a difference is even if the experiment *were* being performed on a process in which the artifact explanation is true (i.e., in which recorded muonless events were all due to escaping muons). In this way it provides the statistical significance level that was sought.

Putting an observed difference between recorded and expected ratios in standard deviation units allows one to use a chart to read off the corresponding error probability. Any difference exceeding two or more standard deviation units is one that is improbably large (occurring less than 2 percent of the time). Approximating the standard deviation of the observed ratio showed the observed difference to be 5.1 standard deviations! This is so improbable as to be off the charts; so, clearly, by significance-test reasoning, the observed difference indicates that it is practically impossible for so many muonless events to have been recorded, were they *all* due to the artifact of wide-angle muons. The procedure is a highly reliable artifact probe.

Only after varying the analysis in many ways was the hypothesis of a real effect regarded as having passed a severe test. For example, Galison explains, the researchers deliberately used three distinct methods to calculate the ratio R, and "since each of the three methods used the Monte Carlo program in very different ways, the *stability* of the data suggested that there was no gross error in the subtraction method" (219). In high-energy physics, as on the laboratory bench, Galison tells us, "the underlying assumption is the same: under sufficient variation any artifact ought to reveal itself by causing a discrepancy between the different 'subexperiments'" (219). And even this multivaried analysis was just one small part of a series of experimental arguments for the existence of neutral currents that took years to build up. My point is that each involved this kind of statistical reasoning to distinguish real effects or signals from artifacts, to estimate the maximum effect of different backgrounds, and to rule out key errors *piecemeal*. They were put together to form the experimental arguments that showed that the experiment could end (to allude to the title of Galison's book).

Learning despite Limits in Paleobiology

In the previous two examples the statistical models needed to answer the "significance question" in analyzing data were arrived at either by techniques of data generation or by special computer simulations. The need for, and the ability to apply, each technique was a function of the particular limits in the available evidence with which researchers were confronted. In the current example the limitations are even more severe. Here knowledge of the statistical distribution of outcomes (given that the error of interest is present)—the knowledge that the statistical null model is to provide—cannot be obtained in either of the ways illustrated earlier. Answering the significance question may still be possible, however, thanks to some newer (computer-driven) analytical techniques. With these newer techniques, there is an attempt actually to construct the statistical distribution of possible results, given that the error is present (i.e., given that one is sampling from a population in which the error is committed). Doing this often takes "brute force," and indeed a method of growing popularity is sometimes called the "brute force" method.

Paleontology offers good examples to illustrate. Here we have a science limited by the inability both to replicate and to manipulate variables in any literal way: "We have only one [fossil record for a given period]; we cannot replay [it] twenty times to see how much variation *could* have been produced by the system. . . . However, these difficulties do not lead us to abandon the scientific approach to the fossil record." What prevents them from abandoning the scientific approach, they explain, is their ability to use to a "toolkit" of statistical techniques in order "to evaluate the statistical significance of historical patterns without replicate samples or repeated experiments" (Signor and Gilinsky 1991, 2).

Resampling Statistics

One technique that is increasingly used is *resampling statistics* (Efron 1979).[16] Thanks to the use of computers, this new method enables the generation, on the basis of a given finite sample, of a population of data sets that *could have resulted* in repeated sampling from the given set. Rather than employing a parametric statistical model (e.g., a normal distribution), the idea is to use the data themselves to generate the hypothetical (statistical) distribution needed for the null model, and in this way obtain the corresponding assessment of statistical significance.

Resampling is one of the more imaginative ways by which researchers appeal to error-statistical ideas to *circumvent* limitations of knowledge: if we cannot (literally) manipulate or replicate, we may still be able to simulate, and thereby learn *what it would be like* to do those things. This same theme underlies all of the error-statistical techniques and arguments that I am trying to articulate by means of examples.

Let us consider a specific application of resampling. To obtain evidence of a mass extinction—understood as an unusually steep increase in extinction rates during a given period—scientists appeal to a simulation model of the expected or "background" extinction rates. The simulation is based on repeatedly resampling values from part of the fossil record. This simulation serves as the null (or error) hypothesis to which data can be compared.

In order to illustrate their method, the researchers describe a very simple, artificial case in which three taxonomic orders are observed and the number of familial extinctions recorded at different stratigraphic stages (figure 19.2). In this illustrative example, the highest number of extinctions was found to occur at stage 20 (i.e., 13 extinctions), and the researchers need to ask: is this evidence of a mass extinction stage? "To decide whether stage 20 should be characterized as a mass extinction stage, we need to ask whether this number of familial extinctions, 13, is [statistically] significantly larger than the number . . . that would normally be expected to occur during stage 20, given the extinction histories of the three orders that were present" (Hubbard and Gilinsky 1992, 153).

To answer this question, they consider the null hypothesis that stage 20 was actually *not* unusual, and that the 13 extinctions occurred during that stage "by chance" (Hubbard and Gilinsky 1992, 153). That is to say, the null hypothesis, H_0, asserts:

H_0 *It would be an error to take the 13 extinctions as good evidence of a mass extinction at stage 20.*

Hypothesis H_0 asserts, in effect, that stage 20 could just as well have experienced the *other* extinction values observed during the history of each order. But now we need a probability assignment for each of the possible extinction values (i.e., each set of three numbers—one for each order), and we are assuming, recall, that we cannot obtain this by relying on one of the known statistical distributions.

To this end, the computer will *generate* the distribution for us. It chooses an extinction number randomly and independently from each of the three orders, and adds them to get the total number of extinctions that could have occurred during stage 20 (or during any other stage in which all three orders exist). This number becomes the first "bootstrapped" number for generating the distribution needed (see note 16). The computer then chooses a random extinction value again from each of the three orders, sums them, and gets the second bootstrapped value. The researchers performed 10,000 such bootstraps. We can graph (see figure 19.2) the frequencies for each sum, yielding the distribution that will serve as the null or error model. (It is a

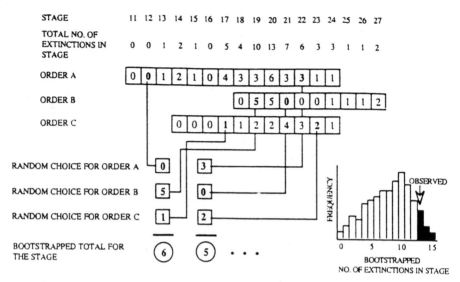

Figure 19.2 Schematic depiction of the bootstrapping method as applied to the problem of mass extinctions. The "observed" total numbers of familial extinctions for hypothetical stages 11–27 are shown in the second row of numbers. Below this row are three rows of numbers depicting the "observed" numbers of familial extinctions for hypothetical orders A, B, and C, each of which contributed a portion of the total number of extinctions. Under the bootstrapping method, one extinction number is chosen randomly and with replacement from each of the orders, and these numbers are summed to create bootstrapped extinction numbers for the stage. The creation of two such numbers is shown. By repeating the process many times (they repeat the process 10,000 times), a distribution of bootstrapped extinction numbers (shown on the right side of the figure) is built up. This can be thought of as the distribution of possible extinction numbers for the stage, given the known extinction histories of the orders in existence during the stage. The observed number of extinctions for the stage under scrutiny (in this case 13 extinctions occurred during stage 20; see the text) is then compared to the distribution of bootstrapped numbers to decide whether the observed number of extinctions could readily occur by chance. The observed number of familial extinctions in stage 20, which was 13, is shown by an arrow on the graph of the distribution, and the substantial amount of area in the tail of the distribution bounded by the observed value indicates that the observed number of extinctions could indeed occur readily by chance. Therefore, no mass extinction is indicated. (From Signor and Gilinsky 1991. Reprinted by permission.)

histogram that approximates the distribution.) We can then compare the observed outcome, 13, to the distribution of possible outcomes (numbers of extinctions): the distribution that was obtained by re-sampling. We can then pose the significance question.

The Significance Question

The significance question in this case asks: is the observed extinction value (13) statistically unusual assuming the extinction values are distributed randomly in time? By finding the frequency with which bootstrapped extinction values equal or exceed 13, the computer answers this question; that is, it tells us the significance level of the observed outcome, 13.

Only if the observed frequency (13) occurred fewer than 100 times in the 10,000 simulations did the researchers consider it statistically significant. In this illustrated case, however, the significance level for 13 was about .1. Hence the researchers do *not* reject the null hypothesis. Rather, they regard the data as "demonstrating that the observed number of extinctions could occur sufficiently often by chance that stage 20 should not be regarded as a stage of mass extinction" (Hubbard and Gilinsky 1992, 155).

The researchers also applied this new analytic method to reanalyze some older data interpretations. They found that two of the five stages that previous researchers had considered indicative of mass extinctions were, according to their reanalysis, easy to explain by the hypothesis of chance error. Whether these past studies really had unearthed evidence of mass extinctions had been controversial, and the controversy was dealt with by means of this innovatively derived null model. As with our earlier examples, uncertainties in the null models led the researchers to perform the analysis using deliberately varied assumptions. In order for a stage to be regarded as showing evidence of a mass extinction, they required it to be found statistically significant using four different null models with 10,000 resamplings each. (They classified three stages as indicating mass extinctions.)

Granted, in this example and the others there is plenty of (subject-specific) background information that is called upon to implement error-statistical tests, but my point is to show that the reasoning in each case follows the standard pattern of arguing from error: we want to assess whether a set of observations is statistically distinguishable from what would be expected were it generated from a null process (here, from the "background" extinction numbers). If the data could be seen as readily (i.e., frequently) generated by this null process, then it would be an error to construe the data as evidence of the hypothesis of interest, H (e.g., mass extinction). If the specific outcome is expected to occur fairly frequently under this null or error process, then it does *not* constitute evidence for H.

R. A. Fisher had said, "In relation to the test of significance, we may say that a phenomenon is experimentally demonstrable when we know how to conduct an experiment which will rarely fail to give us a statistically significant result" (Fisher 1947, 14). Echoing Fisher, we might say: we have evidence of the phenomenon described in H when we know how to conduct an experiment that will very rarely fail to reject the null hypothesis—that is, very rarely fail to reject the claim that we are wrong (to infer we have evidence for H). Although this seems like a lot of negatives—and I suppose it is—a little practice shows that the most mundane day-to-day examples of learning from error are actually instantiations of this reasoning.[17]

Two Important Points

Our examples illustrate two points that relate directly to central disputes in the philosophy of statistics. First, it is clear in these examples that the standard provided by a low significance level is of use not simply because we want to avoid too often interpreting data erroneously in the long run of experience. Rather, it is of use because of its role in informing us of the process that produced the data in front of us. We use it as a standard to weed out erroneous interpretations of data. If I am correct that the rationale for controlling a test's error probabilities is to have a standard for reliably interpreting evidence, a long-standing set of criticisms and confusions surrounding null hypothesis significance testing will have been resolved.

Second, note that in each case arguing from error turns on considering outcomes *other than* the ones actually observed. To use the observed data to learn about their underlying cause we need to think beyond what happened to occur in this case to consider *what else* might have occurred. By contrast, logics of evidential relationship (e.g., Bayesianism) discount this consideration of outcomes other than the one observed for the purpose of reasoning from the data:[18] "The question of how often a given situation would arise is utterly irrelevant to the question how we should reason when it *does* arise" (Jaynes 1976, 247). But for those who use error statistics, reasoning from the result that did arise *is* crucially dependent upon how often it would arise. Lacking such information prevents us from ascertaining which inferences can be reliably drawn from limited data. It follows that if the goal is reliability, error-statistical methods appear in a much more favorable light than those approaches that do not take account of a procedure's error probabilities.

Conclusion

Where we cannot test directly a theory *T*—which contains various hypotheses and may err in different ways—it is often possible to test severely, instead, one or more hypotheses of error: hypotheses that indirectly model "what it would be like" were it a mistake to construe data *e* as evidence for a specific hypothesis *H*. Whether it is by pointing to a statistical calculation, a pictorial display, or a computer simulation, the "what would it be like" question, as I see it, is answered by means of an experimental model that approximates the relative frequency with which certain results would occur in an actual or hypothetical series of experiments. This experimental construct serves as the null or error model. The actual data can then be compared with the data that would be expected assuming the error. By controlling, at some small value, the probability of rejecting the hypothesized error erroneously, one can achieve the goal of making it very difficult to take data *e* as good evidence for *H*, when in fact that would be unwarranted. One can thereby learn about theory *T,* keeping track of the errors that have and have not yet been severely ruled out. Although limited evidence may prevent us from reliably discriminating between a primary scientific hypothesis and its substantive rivals, we may be able to discriminate between correct and erroneous interpretations of *this* data with respect to *this* hypothesis.

These standard or canonical models of error may be located somewhere between the substantive scientific hypotheses and the particularities of the experimental items and processes; they are part of the experimental models that link actual data to substantive primary claims. It is here that the philosopher of experiment might stand to develop a systematic account of evidence. Assessing whether a model is good for the purposes of a null or error model may be quite distinct from the criteria used in testing the adequacy of some substantive model. By developing an account in which inductive inference in science is viewed as reliably distinguishing and ruling out errors, we may begin to resolve a cluster of problems of how to learn from limited evidence that confront philosophers as well as practitioners.

NOTES

1. This is, of course, an adaptation of the first sentence of Kuhn (1962).

2. *Bayes or Bust?* is the title of John Earman's 1992 book.

3. The idea that statistics provides a "reservoir of models" comes from Erich Lehmann's (1990) discussion of R. A. Fisher and Jerzy Neyman.

4. A fuller discussion of these models and the associated "error statistical" philosophy may be found in Mayo (1996).

5. Null models also occur outside statistics. But their uses in statistical testing provide apt analogues for the nonstatistical cases.

6. The null hypothesis is an assertion about a parameter in the corresponding null model.

7. One way to state the corresponding argument from error would be: "It is learned that an error is present when (and only to the extent that) a procedure of inquiry with a very high probability of rejecting the hypothesized error if the error is absent nevertheless does not reject the hypothesized error."

8. Their null hypothesis might be expressed as: "It is an error to suppose that the effect of the earth's motion through the ether is detectable." In this case, there was a failure to reject the error hypothesis: the null hypothesis was accepted.

9. Error statistical tools include, but are not limited to, significance tests as developed by R. A. Fisher and testing and estimation methods as developed by Neyman and Pearson. The points on which I differ from the uses and interpretations advocated by these statisticians are discussed in Mayo (1996).

10. Howson concludes a recent article with this declaration: "Why it is taking the statistics community so long to recognize the essentially fallacious nature of NP [Neyman and Pearson] logic is difficult to say, but I am reasonably confident in predicting that it will not last much longer" (1997, 289). I respond to Howson in Mayo (1997).

11. One need not deny that an alternative theory might give one clues as to *how* one may err in taking e as evidence for the theory in question. But assessing the presence or absence of this error need not call for assessing P(e, given the alternative theory)—an assessment about which we may have no clue.

12. With even more limited information, they might instead appeal to what is known as Student's distribution and the corresponding *t*-test.

13. Taking advantage of work in a different area—the biochemistry of the cholesterol-carrying protein called ApoE—it turns out that the gene for ApoE is

located in the very place where Roses had found the suspect E4 gene in families with AD. Having the E4 version of the gene (rather than the E3) caused nerve cells to die sooner, which in turn caused the plaques.

14. I respond to Laudan in greater detail in Mayo (forthcoming).

15. All references to Galison will be to Galison (1987).

16. Accordingly this procedure is also very commonly called "bootstrapping," but it must not be confused with Glymour's (1980) notion.

17. See, for example, Mayo (1996, Chapter 1).

18. Their doing so follows formally from the acceptance of the likelihood principle. This is discussed in detail in Mayo (1996, Chapter 10).

REFERENCES

Earman, J. 1992. *Bayes or Bust? A Critical Examination of Bayesian Confirmation Theory.* Cambridge, Mass.: MIT Press.

Efron, B. 1979. "Bootstrap Methods: Another Look at the Jackknife." *Annals of Statistics* 7: 1–26.

Fisher, R. A. 1947. *The Design of Experiments.* 4th ed. Edinburgh: Oliver and Boyd.

Galison, P. 1987. *How Experiments End.* Chicago: University of Chicago Press.

Glymour, C. 1980. *Theory and Evidence.* Princeton, N.J.: Princeton University Press.

Howson, C. 1997. "A Logic of Induction." *Philosophy of Science* 64: 268–90.

Howson, C., and P. Urbach. 1989. *Scientific Reasoning: The Bayesian Approach.* La Salle, Ill.: Open Court.

Hubbard, A., and N. Gilinsky. 1992. "Mass Extinctions as Statistical Phenomena: An Examination of the Evidence Using Chi-Square Tests and Bootstrapping." *Paleobiology* 18(2): 148–60.

Jaynes, E. T. 1976. "Common Sense as an Interface." In W. L. Harper and C. A. Hooker, eds., *Foundations of Probability Theory, Statistical Inference and Statistical Theories of Science,* vol. 2. Dordrecht: Reidel, 218–57.

Kuhn, T. 1962. *The Structure of Scientific Revolutions.* Chicago: University of Chicago Press.

Laudan, L. Forthcoming. "Epistemology, Realism, and Rational Theory Evaluation."

Lehmann, E. L. 1990. "Model Specification: The Views of Fisher and Neyman, and Later Developments." *Statistical Science* 5(2): 160–68.

Mayo, D. 1996. *Error and the Growth of Experimental Knowledge.* Chicago: University of Chicago Press.

———. 1997. "Response to Howson and Landan." *Philosophy of Science* 64:323–33.

———. Forthcoming. "Making Progress with Laudan's Problems: 1977–1997."

———. 1997b. "Response to Howson and Laudan." *Philosophy of Science* 64: 323–33.

Mayo, D., and M. Kruse. Forthcoming. "Principles of Evidence and Their Consequences."

Signor, P., and N. Gilinsky. 1991. "Introduction to Analytical Paleobiology." In N. Gilinsky and P. Signor, eds., *Short Courses in Paleontology*, no. 4. Lawrence, Kans.: The Paleontological Society, 1–3.

Tanzi, R. E., D. M. Kovacs, T. Kim, R. D. Moir, S. Guenette, and W. Waco. 1996. "The Pre-senilin Genes and Their Role in Early-Onset Familial Alzheimer's Disease." *Alzheimer's Disease Review* 1: 91–98.

20

Inductive Logic and the Growth of Methodological Knowledge

Comment on Mayo

Bernd Buldt

Department of Philosophy, University of Konstanz

> Within the methodology of each science there are at times monists and dualists. In an epoch with a rationalistic frame of mind thinkers usually aim at overcoming dualism. But it is all too easy for this purification of methods to proceed by a preferential treatment of one particular side, and often one doesn't find the kind of appreciation of the other side that is needed for a reasonable refutation. This way there is a relapse into dogmatism.
>
> —Peter (1934, 260)

A good deal of Deborah Mayo's chapter consists of descriptions of impressive examples of scientific research, and concerning these I find nothing to comment on. But in addition to these case studies, she also presents us with a concise introduction to her project of providing an error-statistical reading of Neyman-Pearson test theory. There is much to comment on here—perhaps even too much for a short note—since by doing this we become caught between the fronts in an "ideological war," as Howson calls it. In this dispute between different schools of scientific inference and different views of probability, I will—at least most of the time—attempt to steer a noncollision course. Obviously the law of the equal distribution of vices and virtues holds on both sides, and as long as this is so, there is no good reason to commit ourselves to one side or the other in this dispute.

Putting Mayo's Project into Perspective

The basic scheme of western epistemology requires two ingredients: data and rules are what provide knowledge. Within this scheme there is plenty room for variety: data could be sense impressions, representations of pure intuition, or simply propositions. Rules could be Humean laws of association, Kantian principles of pure reason, or simply laws of logic. What is also distinctive is the precise mechanism by which the rules operate on data: what exactly is the mode in which specific rules have to operate on what kind of data in order to arrive at knowledge proper? Within this kind of epistemology the focus is usually on knowledge, with at best a marginal concern for its counterpart, namely, error. It is hard to think of a philosopher who made error the topic of his inquiry. But the main problem is that the data we have are obviously insufficient to act as a warrant for all the knowledge we claim to have. This is at bottom what philosophers of analytical training have come to call Hume's problem of induction. But of course this challenge of explaining knowledge in the face of insufficient data has been with us since antiquity.

Mayo's interest in this traditional scheme is the problem of scientific induction, but she deviates in two important respects from the standard approach to this enigma. First, she focuses on the possibility of getting things wrong, being mistaken, being led astray—in short, on error. But unlike a skeptic, whom we expect to focus on error, she proceeds in a constructive manner: she inquires whether it is possible to squeeze knowledge out of controlling error. It is important to see the difference: whereas received wisdom regards error as a negligible factor (e.g., as something that merely results from being inattentive to the proper way of doing things), Mayo makes error an unavoidable factor and hence something of systematic interest to the whole epistemological enterprise.

Second, Mayo draws our attention to the fact that we cannot take data for granted. By making light of the data, she claims, we not only run the risk of distorting our views of the mechanism by which the rules (should) operate on the data, we also create such self-inflicted problems as Duhem's problem, the underdetermination of hypotheses, and the Bayesian catchall problem.[1] To stress the difference again: unreliable data and more or less unsystematic solutions have been discussed since antiquity, but Mayo makes it an epistemological duty to scrutinize data in a systematic way. Both points run contrary to the usual way of treating epistemology and philosophy of science, respectively, but I think Mayo (1996) gives ample reason to take them both seriously.

Where Mayo's Project Scores

First, the demand for honesty. Within the traditional philosophy of science (roughly: before Carnap) the Baconian strand showed a great awareness of the complexities involved in framing hypotheses from alleged data. These philosophers rejected simpleminded approaches, such as enumerative induction, as childish, and I really cannot imagine them, for example, as serious investigators of Goodman's "grue." Instead they struggled to achieve a canon embracing a wide variety of methods, all which would act together in a mutually interwoven way to set up a sensitive apparatus that would enable us to unravel truthfully what Mill called the "tissue of nature." The honesty to acknowledge that the difficulties involved in experimental research do not allow for a simple model, it seems to me, was lost in our post-Carnapian times in what Mayo calls logics of evidential relationship or E-R logics. But I find this honesty again in Mayo's approach.

Second, the need for real-life input. Whereas working scientists were formerly often fond of the philosophy of their times—especially if they saw the honesty just mentioned realized—present-day scientists tend to ridicule the lofty business that Mayo characterizes as a "white glove analysis" of science. Simplification and idealization do certainly have their proper place and function in any theoretical investigation, but they should not distract a philosophy of science from also "saving the phenomena" on the non-glorious side of science, from the laborious task of producing sound data and making data and hypothesis finally fit. Concerning this tension between neat models and what one finds in real life, I agree with John von Neumann (1961 [1947], 9) that there is just one cure for a science that runs the risk of becoming what he calls "baroque," that is, entangled solely with self-created problems. This cure, he said, is a return to the source of empirical problems. Now I do not expect the working scientists to applaud us, but I do think it is a warning sign if they do not even get the point. Nor am I

opposed to E-R logics; on the contrary, I greatly appreciate these for-
malistic approaches, not least because they help us to clarify our intuitions. But I think Mayo's insistence on the mismatch between the
research work actually done in real life and the models offered by E-R
philosophers could serve as an antidote for a philosophy of science in
danger of becoming "baroque" and to counteract the propensity of
each theoretical science to develop along the line of least resistance.

Third, the advantage of virtual variation. The possibility of varying
and modifying an experimental setup is at the heart of the experimental method and crucial for its success. Without such variation no reliable information could be obtained to study the correlation and/or
independence of parameters, to distinguish background noise from
real effects, and so on. So far, philosophy of science has carried in its
methodological toolbox only variations of experimental setups that
can actually be performed; I call them "real" variations. What Mayo
adds to this traditional toolkit from statistics are "virtual" variations,
that is, variations to which no real-world experiments correspond.
Instead of modifying a real experimental setup, simulate a virtual one
by postulating an appropriate null hypothesis H_0 and pretend that H_0
is supported by the data one has. Considering the importance of variation for research work, Mayo's virtual variation—or, as she calls it,
learning from error—has the following virtues: first, it is a major
addition to methodology, not as it is practiced by scientists but as it has
entered discussions in philosophy of science; second, it unites real
experiments and thought experiments, which have previously lived
separate lives.

Putting the Opponents into Perspective

Giere (1997) has suggested that we regard Bayesianism and error statistics as the successors to the inductive logics put forward by Carnap
and Reichenbach, respectively. I agree—in a way—although I think
this is only part of the truth. Clearly, Bayesianism took over when
Carnap's inductive project came to an end. But even more, Bayesianism also inherited the basic orientation underlying Carnap's project,
and nowhere else than here is it more apparent not only that Carnap
started out as a neo-Kantian, but also that he kept to his commitment
and that the current Bayesianism still has some strong neo-Kantian
leanings. There is no room here for a lengthy argument based on

textual evidence to support this claim.[2] I point instead to a common blueprint underlying both projects. Kant answered Hume with a new logic, a transcendental logic, which was modeled upon and molded out of traditional logic, and his transcendentalism stems from the accompanying approach to focus solely on functional aspects in epistemology and to justify these functions as epistemological norms.[3] Likewise Carnap and the Bayesians answer Hume with a new logic that shares the same essential features: the intentional closeness to classical logic, the focus on function (no concern for the "hardware"), and the stress on normativity (see the arguments from coherence and Dutch Book).

Turning our attention to error statistics, we find it indebted to the tradition in which Neyman and especially Pearson stood (see Mayo 1996, especially 377–411). Neyman and Pearson were both deeply influenced by the heritage of what one might call the "heroic age of inductivism" (i.e., the nineteenth century), represented by such figures as Herschel, Whewell, Mill, Jevons, Apelt, Sigwart, and Wundt, and whose last product in the British line was Karl Pearson's *The Grammar of Science* (1991 [1892]). Egon S. Pearson was Karl's son (and there is no sign of an Oedipal complex or any other kind of rebellion that might have made him hostile to his father's views), while Neyman acknowledged a deep and lasting influence of Pearson's *Grammar* on his research. He writes in an autobiographical note: "All my research papers are a reflection of many divergent influences I experienced over the years." And the only philosophical source he mentions, besides all the mathematical ones, is Pearson senior: "I must acknowledge the lasting influence of Karl Pearson, particularly of his *Grammar of Science,* which I read at the suggestion of [Serge] Bernstein and which proved to be a revelation" (Neyman 1967, ix). In order to assess the meaning of this heritage, one should recall that in this "heroic age" one of the core projects of classical British empiricism, namely, to analyze the mechanism that frames knowledge from sense data, was transformed into the inductive project of analyzing the mechanisms that frame scientific knowledge from experimental data and hypotheses. I regard error statistics as the rightful heir to this particular empirical strand, and hence it comes as no surprise to see Mayo side with the "new experimentalists" (see Mayo 1996, Chapter 3).

The moral of this is that we do not see current versions of Carnap and Reichenbach at work, as Giere has suggested, but that the dispute between error statistics and Bayesianism is a recent version of the dispute between core elements of (modern developments of) empiricism and rationalism.[4]

An Attempt to Reconcile Hostile Siblings

One of the disputed points—possibly the most crucial one—between statistics and Bayesianism is the status of probability. This involves not only the question of whether probabilities are to be understood as subjective or objective, but also the question of whether it makes sense at all to ascribe probabilities to hypotheses. This last allegation is simply incomprehensible for someone raised as a statistician, and it was instrumental in getting the whole theory of statistical inference off the ground, in opposition to the hegemony of applied probability theory, the forerunner of current Bayesianism. Now the point in dispute indeed does not seem open to a congenial settlement, as shown by the hostility exhibited by both sides up to now. But I think it ain't necessarily so. In order to see why, I start with the thesis that, roughly speaking, Bayesianism is (inductive) *logic* of science, while (error) statistics belong to the *methodology* of science.

Bayesianism deals with propositions, and just like logic it does not care how the truth values or the degrees of belief were established. This was and is incontrovertibly clear from the very beginnings of nineteenth-century inductive logic throughout recent expositions of Bayesianism. Thus Bishop Whately states: "But universally, the degree of evidence for any proposition we set out as a Premiss [in an inductive argument] . . . is not to be learned from mere Logic, nor indeed from any one distinct Science; but is the province of whatever Science furnishes the subject-matter of your argument" (Whately 1848 [1826], 155; emphasis suppressed). Likewise Howson writes: "Inductive logic—which is how we regard the subjective Bayesian theory—is the theory of inference from some exogenously given data and prior distribution of belief to a posterior distribution" (Howson and Urbach 1989, 290; 1993, 419; with a little addition, suppressed here). Consequently, this means that "the frailty of much scientific data is therefore, however valuable in its own right, beside the point of evaluating the adequacy of the Bayesian theory of inference" (Howson and Urbach 1989, 272; 1993, 407).

On the other hand, it is just this "frailty of data" with which statistical theories grapple. In the words of Pearson: "To the best of my ability I was searching for a way of expressing in mathematical terms what appeared to me to be the requirements of the scientist in applying statistical tests to his data. . . . Indeed, from the start we shared . . . [the] view that in scientific enquiry, a statistical test is 'a means of learning' [from data], for we remark the tests themselves give no final verdict, but as both help the worker . . . to form . . . his final or provisional decision" (1955, 204–6). And Mayo also subscribes to this. First, since data in themselves can be unreliable, we need a method for assessing such data. Second, even if data are reliable or have been made reliable to a certain extent, we need an account to settle how good specific data are as evidence for a hypothesis in question. But doing both of these things—that is, "scrutinizing evidence," as Mayo dubs it—is impossible for a Bayesian, since it is not merely "a matter of logical or probabilistic relationships between statements, but turns on *empirical* informations about . . . the data" (emphasis hers).

Clearly, scrutinizing evidence does not belong to logic, which takes evidence for granted, but belongs to the methodology of science, which takes on as one of its concerns how to establish evidence reliably. Hence what we get is the (admittedly rough) picture of two projects operating on opposite sides of a spectrum (certainly a multidimensional one, but one that is reduced here to just one dimension), which extends from mere data on the one hand over interpreted data, tentative hypotheses, and successfully confirmed hypotheses to propositions on the other hand. Statisticians legitimately operate toward the side of data, whereas the Bayesians' place is toward the side of propositions.

This schema offers the advantageous possibility of having one's cake and eating it too. First, the dispute about assigning probabilities turns out to be idle. This is so because it seems very natural not to assign probabilities to hypotheses near the data end. Priors, insofar as they are unavoidable, essential for statistical work, or both, play their rightful role in the process of selecting which hypotheses are to be tested and with which data samples, but neither is there, at this point, any need to assign priors to hypotheses, nor does it even seem advisable. On the other hand it is clear either that evidence, once it has reached the stage of a definite or tentative scientific assertion, should already come with a degree of belief, or that one could reasonably demand that a degree of belief be assigned. (This should not be taken to imply that there are no mixed cases on the sides of propositions or of data. Without doubt, testing hypotheses involves some inductive reasoning, but conditionalizing belief must also include some testing of evidence.)

Second, this proposal does not settle the question of objective versus subjective probabilities. But it leaves room for both—just as Carnap or Lewis did before—simply by locating them at different ends. (A basic rationale behind this reconciliation proposal is then that crucial notions, such as "scientific inference" or "evidence," are ambiguous; their exact meanings depend on where on the continuum between data and propositions they are being applied.)

Third, the suggestion might result in some relocation of familiar problems. As mentioned previously, Mayo makes some promising suggestions to the effect that Duhem's problem, the problem of the underdetermination of hypotheses, and the problem of the catchall hypothesis might be better taken care of if one locates them toward the side of the data on our continuum and treats them error-statistically, rather than in a Bayesian fashion.

Mayo's Third Way

Mayo proposes in her chapter that there is a "third way" besides "Bayes or Bust," namely error statistics. Throughout her book and articles Mayo is—contrary to her favorite target Colin Howson—very well aware of the fact that Bayesians and statisticians are talking at cross-purposes, because of their fundamental difference in aims. But sometimes Mayo leaves her balanced attitude behind, as when she charges Bayesians with not scrutinizing evidence or with the inability to do so within their framework. The first reproach blames an inductive logic for being logic; the second neglects (semi-)Bayesian efforts like Jeffrey conditionalization. She also makes some claims that are unnecessarily bold. For example, she states without qualification that "data are inexact, noisy, and incomplete": an experimental nuclear physicist, to draw on her HWPF example, is more likely to explain systematic errors in the results than to calculate significance levels, not to speak of clearly sound data. She also alleges, by referring to *Nature* and *Science,* that statistical methods are ubiquitously applied in all sciences: a chemist is more likely to calculate the kinetics of a reaction than confidence intervals, not to speak of classificatory activities and the like. I mention these minor points only because they seem to be intimately connected to another overly bold claim, namely that "experimental learning . . . is learning from tests that satisfy the severity criterion" (Mayo 1996, 184)—as if there were no experiments without statistical evaluation. (Even if one admits that, as experiments become more and more involved, there is a growing need for statistical tools, one expects an argument here that statistics-free experiments will finally become extinct.) Maybe a "false diet of examples" and her successes in tackling notorious conundra (e.g., Duhem's problem; see above) has led Mayo to such claims, of which the boldest is to offer error statistics as a third way. First of all, it does not seem possible to square this claim with her own conception of statistical methods as simply being scientific instruments (see Mayo 1996, 409). But even if one went as far as saying that all scientific activity can be understood as applying "instruments" in some proper way, this leaves the problem of objectivity or normativity. There is nothing objective or normative in an instrument. But since both problems are dealt with in theories of rationality, of which Bayesianism is an example, statistical tools will never be a full substitute. Hence error statistics cannot be a third way besides "Bayes or Bust." The alternative I see to either Bayes or bust is to place statistics and Bayesianism at opposite ends of the scientific enterprise as sketched previously—an alternative, by the way, that conforms very well with her conception of statistics as an instrument.

Finally, a quick comment concerning the topic of this volume. Mayo's implicit thesis seems to be that, insofar as limits of science are connected to limited data, there is an error-statistical way out. I disagree. Certainly probabilistic reasoning has made impressive progress from the days when Mill (1974 [1843], 8:1142) could charge it with being alchemy, trying to turn the lead of ignorance into the gold of knowledge. But it is precisely one of Mayo's own examples, namely high-energy physics, that seems to prove her wrong, because, to the best of our knowledge, we can never be absolutely sure whether or not we suffer from limits in physical science as long as we are unable to conduct experiments in the region beyond 10^{25} GeV.

NOTES

I extend thanks to Luc Bovens and Peter McLaughlin, who were kind enough to improve my English.

1. Re Duhem's problem, see Mayo (1996, 106–111; 1997). Re the underdetermination of hypotheses, see Mayo (1996, 174–76). Re the Bayesian catchall versus her piecemeal approach, see Mayo (1996) Chapters 4.3, 6.3, and 10.1; 116–18, and Chapter 5.

2. Concerning Carnap, see Coffa (1991, especially Part II) and Carnap (1963, 978–79).

3. A more scholary note for the curious: Kant's transcendental analytic, with its pure concepts and pure principles, is derived from the traditional doctrine of

(concept and) judgment. His dialectic likewise draws on the traditional doctrine of inference, that is, syllogistics, while the question of application was dealt with in a (transcendental) doctrine of method, which has been an additional chapter in logic texts since the Renaissance. His focus on function is evident not only from his restricting the discussion to functional aspects, but also from his constant refusal to make any assertion about the underlying machinery that performs these functions; on the contrary, he even argues that it is impossible for us human beings ever to do so.

4. This perspective might not only help to explain the heat in the debate about the status of probabilities but also make room for a spark of hope, since it means that we are now able to study these classical rivals, empiricism and rationalism, within a rather formalized field.

REFERENCES

Carnap, R. 1963. "Replies and Systematic Expositions." In P. A. Schilpp, ed., *The Philosophy of Rudolf Carnap*. La Salle, Ill.: Open Court, 859–1013.

Coffa, J. A. 1991. *The Semantic Tradition from Kant to Carnap to the Vienna Station*. L. Wessels, ed. Cambridge: Cambridge University Press.

Giere, R. N. 1997. "Scientific Inference: Two Points of View." *Philosophy of Science* 64: S180–S184.

Howson, C., and P. Urbach. 1989. *Scientific Reasoning: The Bayesian Approach*. La Salle, Ill.: Open Court.

——. 1993. *Scientific Reasoning: The Bayesian Approach*, 2nd ed. La Salle, Ill.: Open Court.

Mayo, D. 1996. *Error and the Growth of Experimental Knowledge*. Chicago: University of Chicago Press.

——. 1997. "Duhem's Problem, the Bayesian Way, and Error Statistics, Or 'What's Belief Got to Do with It?'" *Philosophy of Science* 64: 222–44.

Mill, J. S. 1974 [1843]. "A System of Logic Ratiocinative and Inductive." In J. M. Robson, ed., *Collected Works*, vol. 7–8. London: Routledge and Kegan Paul.

Neyman, J. 1967. "Author's Note." In *A Selection of Early Statistical Papers of J. Neyman*. Cambridge: Cambridge University Press, ix.

Pearson, E. S. 1955. "Statistical Concepts in Their Relation to Reality." *Journal of the Royal Statistical Society Series B* 17: 204–7.

Pearson, K. 1991 [1892]. *The Grammar of Science*. Bristol: Thoemmes.

Peter, H. 1934. "Statistische Methoden und Induktion." *Allgemeines Statistisches Archiv* 24: 260–67.

Von Neumann, J. 1961 [1947]. "The Mathematician." In A. H. Taub, ed., *Collected Works*, vol. 1: *Logic, Theory of Sets and Quantum Mechanics*. Oxford: Pergamon Press, 1–9.

Whately, R. 1848 [1826]. *Elements of Logic*, 9th ed. London: Parker.

21

On Modeling and Simulations as Instruments for the Study of Complex Systems

Manfred Stöckler

Department of Philosophy, University of Bremen

Research on complex systems is a central topic in contemporary science. The analysis of complex systems exhibits features that normally cannot be found in early textbooks of philosophy of science. The dynamics of the system is generally not known in detail. In many cases, we do not even know the quantities that are decisive for the dynamics of the system. When the basic equations describing a system are known, very often the solutions cannot be calculated: analytical methods are usually not applicable, if the interactions are nonlinear.

Complex systems expose the limits of our capacities to deal with objects by means of calculations, using simple deductive approaches. Digital computer simulations, with their ability to compensate for our shortcomings, have spread throughout all areas of science. In this chapter I investigate the scope of these new methods and consider whether or not the triumphant advance of simulations is a revolution from the point of view of philosophy of science.

Paul Humphreys argued that computer simulations require "a new conception of the relation between theoretical models and their applications" (Humphreys 1991, 497). Fritz Rohrlich believes that computer simulations provide a qualitatively new and different methodology for the physical sciences, lying somewhere between theoretical physics and methods of experimentation and observation: "Scientific activity has thus reached a new milestone, somewhat comparable to the milestones that started the empirical approach (Galileo) and the deter-

ministic mathematical approach to dynamics (the old syntax of Newton and Laplace)" (1991, 507). In contrast to such interpretations, I
argue that models and simulations are very similar to traditional tools
of science. Computer simulations enable tremendous progress on a
pragmatic level, as a matter of degree in terms of speed and quantity,
but not a revolution in the principles of methodology.

A Traffic Jam in the Computer: A Case Study

I begin by sketching a simple model of a complex system: *traffic on a
highway.* (You may consider highways as a typical German topic; in
any case, we have the advantage of being quite familiar with these
particular phenomena as objects of simulation!)

The basic model (figure 21.1) is a rather simplified description of
traffic. The traffic flows in only one direction and the cars drive in a
single lane. The highway forms a closed ring (an "Indianapolis situation"). There are no passing, no places to park (that feature might be
quite realistic), and no gas stations. The system is characterized by
discrete quantities. The highway is divided into L segments of length
7.5 m. The segments are either empty or occupied by one car. The
number of cars, N, is constant. The vehicles have only one internal
parameter, velocity. The value of the velocity can be expressed as inte-
gers from 1 to 5. The maximal velocity, $v_{max} = 5$, is equivalent to
85 mph.

The dynamics of the model is based on the assumption that the
behavior of the driver is influenced only by the car ahead of him. In
order to employ computers we need a precise formulation of the basic
assumptions of our model[1]:

*On a ring with L boxes every box can either be empty or occupied by one
vehicle with velocity* v = 0,1, . . . , v$_{max}$.

At each discrete time step $t \rightarrow t + 1$ an arbitrary arrangement of N cars
is updated according to the following rules:

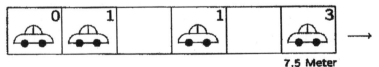

7,5 Meter

Figure 21.1 The model of Schreckenberg et al. (1995). Reprinted by
permission.

1. *Acceleration:* If the velocity v of a vehicle is lower than $v_{\max}$, the speed is advanced by 1 ($v_1 = v + 1$).

2. *Deceleration (due to other cars):* If the distance d to the next car ahead is not larger than v_1 ($d \le v_1$), the speed is reduced to $d - 1$ ($v_2 = d - 1$).

3. *Randomization:* With probability p, the velocity of a vehicle (if greater than zero) is reduced by one ($v_3 = v_2 - 1$).

4. *Car motion:* Each vehicle is advanced $v = v_3$ sites.

Rule (2) takes care of the safe distance and leads to a coupling of the motion of the cars. Rule (3) imitates dawdling (retarded acceleration) or overreactions at braking and leads to fluctuations in the dynamics. Only a few characteristics of traffic are taken into consideration. All the technical details of the cars are omitted, as well as most of the psychology of the drivers (e.g., whether a given driver is thinking of Kant or Wittgenstein).

A typical result of a single run of the simulation is shown in the space-time diagram in figure 21.2. The rows represent the state of the system, with time increasing downward. The dots represent empty sites, whereas numbers indicate that a car is present. The specific number gives the velocity v of the car.

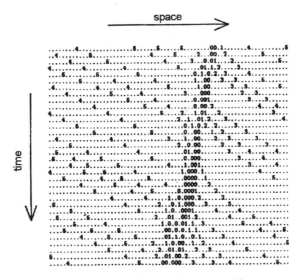

Figure 21.2 A typical result of a single run (showing traffic congestion). (From Schreckenberg et al. 1995. Reprinted by permission.)

We are mainly interested in the calculation of the so-called *funda-mental diagram* (flow f against density $c = N/L$, i.e., the average num-ber of cars in a segment). These results can be directly compared with measurements of real traffic.

There exists an analytical solution for the special case $v_{\max} = 1$:

$$f(c,p) = \tfrac{1}{2}[1 - (1 - 4\,(1 - p)c(1 - c)\,)^{1/2}]$$

where f is the flow (i.e., the number of cars passing during an hour), c is the density of cars, and p is the probability of dawdling.

The diagrams in figure 21.3 show the results of simulations and measured values. They demonstrate that the flow increases with the number N of cars on the highway, as long as the cars do not obstruct each other. It is an astonishing fact that even so simple a model repro-duces interesting features of reality. In this way, macroscopic phenom-ena can be explained by the "microscopic" assumptions of the model. The simulations show how such surprising effects as traffic jams arise "out of nowhere" (see figure 21.2). We have learned from such simula-tions that the manner of acceleration is more important for avoiding traffic jams than the manner of braking. Surprisingly, the optimal state, with the greatest throughput, is a critical state with traffic jams of all sizes (see Paczuski and Nagel 1996, 83).

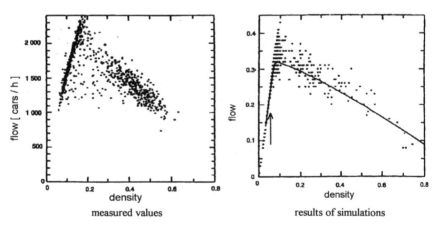

Figure 21.3 Fundamental diagrams (measured and simulated). (From Schreckenberg et al. 1995. Reprinted by permission.)

On Computer Simulations

Nowadays, simulations are used in nearly all fields of the natural and social sciences, and even in philosophy.[2] However, there is no generally accepted definition of "simulation," even if we focus on computer simulations. Sometimes the concept is used in a broad sense, including nearly all forms of approximate solutions of equations with the help of computers. Others propose definitions of simulations that rely on more restrictive criteria.[3] But no attempts to reach a standard definition have met with success.

Paul Humphreys states that the main objective of many computer simulations is to turn analytically intractable problems into ones that are computationally tractable. He proposes the following working definition: "A computer simulation is any computer-implemented method for exploring the properties of mathematical models where analytic methods are unavailable" (Humphreys 1991, 501). Considering simulations as a part of computational science, he indicates that there should be an additional criterion, as not every numerical method for solving equations with the help of computers should qualify as a computer simulation.

Stephan Hartmann is probably on the right track when he emphasizes that simulations rely on a model and aim to calculate the time development of the state of a system. In this approach simulations are closely related to dynamical models. A simulation results when the equations describing the time evolution of the underlying model are solved in a special way. Keeping this in mind, we can follow his definition: "A simulation is an imitation of a process by another process. A simulation imitates one process by another. If the simulation is run on a computer it is called a computer simulation" (Hartmann 1996, 83). But this definition is also in need of clarification. For example, if some computer program determines the analytical solution of the equations of motion for a simple system, we would not say that this procedure is a simulation. But we do call it a simulation when the solutions are constructed by special approximative methods or visualized as processes in space and time.

It should be added that, in most cases, the use of the word *simulation* seems to imply that the description given by the simulation is idealized and abstracts from many details of the phenomena (Simon

1969, 16). This feature is included in a definition offered by Francis Neelamkavil: "A . . . simulation . . . is the process of imitating . . . important aspects of the behavior of the systems (or plans or policies) in real time, compressed time, or expanded time by constructing and experimenting with the model of the system" (1986, 6).

Collecting all these ideas, I propose the following definition: a computer simulation is a tool to deduce the time evolution of the states of a system in cases for which there is no easy analytical solution at hand. The basis of the simulation is a model of the system including the (normally idealized) rules of time evolution. Then a computer is used for calculating approximative solutions for the equations of motion. Very often the solutions are visualized as processes in space and time.[4]

In any case, analyzing the functions of simulations is more fruitful than discussing problems of definition. Again I follow the classifications and ideas of Stephan Hartmann (1995a, 1995b). The main advantage of computer simulations is that they allow us *to deduce consequences* of the basic assumptions in the model *very quickly*. In this way, it is possible to screen the range of possible consequences of the model for a whole field of values of the parameters. Regularities of behavior may be discovered merely by varying the model parameters. The computer model can be extrapolated into regions that are experimentally inaccessible. Some call such methods "numerical experiments." Other simulations are part of "experimental mathematics": by modeling special processes, regularities and candidates for general theorems are found, which can afterward be proven by analytical methods.

Computers are also useful for studying preliminary hypotheses on the dynamics of the system. By comparing the results of the simulations with given data, one can check assumptions and initiate a new simulation using a corrected model. The high speed of a computer allows one to run many trials of feedback circles in a short time (Humphreys 1991, 503).

Another important advantage of computer simulations is that the results of the calculations (as well as the different states of the dynamical process) can easily be presented by powerful methods of *visualization*. People normally understand results presented in this way much better and much more quickly than the formulas of analytical solutions.

Models as Elements of Simulations

A model as a part of a simulation is a set of assumptions or propositions about an object. Of course, not all assumptions about an object are models. In order to understand the role of modeling in the sciences of complex systems we should distinguish between two sorts of models (see also Humphreys 1991, 503).

Models in Fundamental Theories. Models in fundamental theories (especially models in theoretical physics) are often discussed under the heading of *theoretical models* (see Achinstein 1968, 212–18). A theoretical model is a combination of a theory and a model object. *Model objects* are descriptions of an object and assumptions about a system that are necessary for applying a general theory such as Maxwell's equations of electrodynamics to concrete fields and given distributions of charges. Such fundamental theories are universal and they do not contain descriptions of any objects in the world. In most cases the model objects are very simplified and idealized descriptions of the real objects. For example, a planet can be represented by a point particle. Mercury and a pointlike particle are similar in some respects, and this common structure might be sufficient for performing calculations in a given context. In fundamental theories the form of the models is strongly influenced by the universal equations. The universal theory determines the types of the individuals and many properties of the model objects.

Models in Theories of Complex Systems (Phenomenological Models). Working with complex systems is different. Unlike modeling simple applications of fundamental equations in theoretical physics, complex systems do not allow the construction of a model object in terms of the fundamental theory, for practical reasons. We cannot write down the Schrödinger equation for highway traffic and we cannot analyze the cars in terms of quantum mechanics. Here we face pragmatic limits on applying and working with fundamental theories. Therefore descriptions of complex systems must be found that are simple enough to permit handling but allow us nevertheless to capture the main features of the process under investigation. Those who construct *phenomenological models* are guided much more by contexts and special purposes than by general laws and theories (see generally Rohrlich 1991).

Our case study is a typical example of a phenomenological model. There are many ways of modeling traffic. Fluid models were fashionable for many years. Choosing cars and cells as basic entities seems to be quite natural, but no one could tell whether the rules of car motion we studied earlier were sophisticated enough without testing the model in various simulations. The methods for constructing phenomenological models differ in some respects from those for constructing fundamental models. Computer simulations are especially useful for developing phenomenological models; we will return to this point later.

Models serve a variety of purposes in the sciences (see Morrison 1998). For some philosophers phenomenological models suggest an instrumentalist interpretation of science as a whole. For this reason I consider here whether modeling and simulation can also be understood from a realist point of view.

Both theoretical and phenomenological models are *incomplete* and *idealized* descriptions. The idealizations of the models limit the precision of the simulations. But we should not overstate the consequences of this fact for the practice of science. For most purposes, it turns out to be quite satisfying to represent the object of Sol Steinberg (figure 21.4) as a cube. In the same way, phenomenological models are very often quite successful.

Even believing that, in principle, an exact description of a complex system could be given, no realist would reject pragmatic reasons for using idealizations. In contrast to Nancy Cartwright (1983, Chapters 6 and 7; 1989, Chapter 5; see also Stöckler 1998), I believe that the practice of modeling and idealization can be understood by realists as well as by instrumentalists. People like me, who believe in the truth of fundamental theories, do of course accept the practical limits of analytical methods and the limitations on our ability to handle many parameters quickly and simultaneously. This requires new methods (e.g., simulations), which are different from calculating the consequences of the fundamental equations. Realists and reductionists must distinguish between a pragmatic and an idealized point of view: what can be done in practice indicates the limits of our capacities, and it is different from what is possible in principle. Giving a pragmatic account of phe-

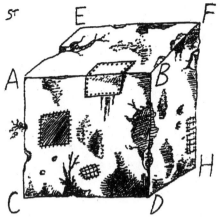

Figure 21.4 Steinberg's cube. (From Kenny 1994, by permission of Oxford University Press.)

nomenological models does not imply the abandonment of realistic intuitions.

Computer simulations are relatively new tools for science. But many features of phenomenological models that are used in simulations are well known within the methodology of dealing with complex systems. There are many traditional techniques for analyzing such systems, which have been developed, for example, in biology and the social sciences. For that reason philosophers who focus solely on physics often overlook these tools. Descriptions of complex objects are idealized and omit many details of the objects. We analyze such systems as hierarchically organized, meaning that one has to find the right level of description for the system. In these cases model building involves additional problems: How is the field of research to be organized? What are the relevant individuals (things)? What are the decisive properties?

Simulations of Complex Systems

I am not able and do not want to give a precise definition of *complexity*. Instead I rely on a short and appropriate characterization by Herbert Simon:

Roughly, by a complex system I mean one made up of a large number of parts that interact in a nonsimple way. In such systems, the whole is more than the sum of the parts, not in an ultimate, metaphysical sense, but in the important pragmatic sense that, given the properties of the parts and the laws of their interaction, it is not a trivial matter to infer the properties of the whole. In the face of complexity, an in-principle reductionist may be at the same time a pragmatic holist. (1969, 86)

What we classify as complex depends on the specific purpose of the investigation. Protons are complex for elementary particle physicists; on the other hand even whole galaxies might be considered as simple components of a complex system (Simon 1981, Chapter 7). These examples show that there are complex systems not only in biology but also in physics, even if we may find only a few examples in introductory *textbooks* of physics and even fewer in the textbooks of *philosophy of science*. The behavior of complex systems seems to be surprising for us and often counterintuitive. John L. Casti (1997, 86) argues in a popular book that it is this feature that most clearly distinguishes complex systems from those that are simple. He gives many examples and further characterizations of complex systems (e.g., instability, uncomputability, connectivity, emergence).

As we have seen, fundamental equations are not useful for analyzing complex systems; therefore, for example in biology and in the social sciences, new levels of description are introduced. This process includes new individuals and new kinds of regularities. In our traffic example, we considered the velocity of the cars, but not the motion of their pistons and valves. The search for fruitful levels of description and relevant parameters requires creativity and ingenuity, and this is why the science of complexity would remain an attractive business even after the construction of the "final theory." Simulations help to make traditional methods of investigating complex systems more precise.

There are at least three reasons why simulations are extremely helpful for analyzing complex systems. First, the regularities of interaction are very often given as simple local rules acting in many places. Our cellular automaton model of highway traffic is also quite typical in this respect. But as we have seen in this case study, knowledge of the local rules does not help in surveying the global features. In most cases, we cannot "see" the pattern evolving from these interactions, even if the rules are simple. That is the typical case, especially when nonlinear couplings are involved. Computers, however, are able to do the calculations in tidy order and in a short time, and they can exhibit the results in ways that take advantage of our efficient capacities to read visual information.

Second, simulations help to develop a "feeling" for the decisive features of higher-level description. The high speed of computers helps in testing many tentative hypotheses within a short time and in finding the appropriate values of the parameters by trial and error. We can try to develop an understanding of the relevant processes by calculating "macroscopic" properties (such as the flow of cars depending on the density in our example) from microscopic dynamics (see also Hartmann 1996). Such "hypothetical" simulations must also be defined in a precise way. Hartmut Kliemt correctly called this application of simulations "disciplined speculations" (Kliemt 1996, 20). From such back-of-the-envelope simulations Kliemt distinguishes simulations using more settled models of systems that are quite well understood (e.g., a flight simulator). In this case, the simulation could serve for solving practical problems ("thick simulations").

Third, the characterization of complex processes requires many parameters and produces a huge amount of data. The mechanisms for processing data in a computer simulation allow one to document the individual steps of the computation and to visualize the development of the system over time by means of often very impressive pictures. Sophisticated methods of visualization allow us to understand the result of the computations by employing our remarkable faculties for interpreting images. Our visual system has evolved to permit us to perceive a huge amount of information at a glance.

We have seen that the main advantages of computers are their speed of computation and the format of their information processing, which enables easy visualization. The examples in this section have shown that modeling and simulations are useful heuristic tools to compensate for our pragmatic shortcomings in studying complex systems. A reductionist would never claim that fundamental theories are the best way to handle complex systems. We must construct intermediate levels and phenomenological models. In addition to such pragmatic reasoning, a more fundamental metaphysical question remains: do we have any reason to hope that higher levels of descriptions and phenomenological models can be reduced to, or at least made understandable by, invoking theories and models from deeper levels? As we shall see in the following section, simulations can help us to answer these questions.

In order to address that topic, however, we will have to go beyond heuristic considerations.

Simulations and Explanations

In this section, I investigate whether simulations serve not merely heuristic but also explanatory and inferential functions. This brings us to the question that Herbert Simon (1969, 15; see also Latané 1996, 290) correctly considers to be the crucial one in the context of simulations: *How can simulations ever tell us anything that we do not already know?*

Simon agrees with the common assertion that "a simulation is no better than the assumptions built into it." He argues that simulations might nevertheless provide new knowledge: "even when we have correct premises, it may be very difficult to discover what they imply. All correct reasoning is a grand system of tautologies, but only God can make direct use of that fact. The rest of us must painstakingly and fallibly tease out the consequences of our assumptions" (Simon 1969, 15). Simon gives the example of weather prediction by means of local atmospheric equations, which can be considered as a sufficient basis for the dynamics. A computer is necessary to work out the implications of the interactions for a vast number of variables starting from complicated initial conditions. Such deductions are a central element of any explanation.

Even when we do not have fundamental equations that govern all the detailed behaviors of the system, however, simulations are useful for *understanding* such systems. Simon gives two reasons. The first is that we are seldom interested in explaining or predicting phenomena in all their particularity. Usually we are interested in only a few properties that are important for pragmatic reasons. For that reason even simulations employing extremely simplified models can provide useful explanations. Second, the description of the behavior of a system on each level depends only on a very approximate, simplified, abstracted characterization of this system at the next lower level. "This is lucky, else the safety of bridges and airplanes might depend on the correctness of the 'Eightfold Way' of looking at elementary particles" (Simon 1969, 15).

Again, simulations relying on very simplified approximations provide useful explanations (Simon 1969, 17). Many complex systems can be described in such a way that the main aspects of their behavior are explained by the structure of the interaction of a few properties of the components. A detailed knowledge of, for example, the actual physical mechanisms or the material properties of the components is not necessary. In these cases the respective structure is the most important aspect of the interaction, and computers are especially qualified to imitate and represent such formal properties and structures. In the model of highway traffic that we have been studying, the motion of the cars depends primarily on the distance to the next car ahead.

Given the model and the alleged initial state of the system, a simulation has the form of an argument and is an explanation. In many respects simulations are explanations, as reconstructed by C. G. Hempel and P. Oppenheim. Very often the premises are highly hypothetical or bluntly false, but we know many other cases of counterfactual reasoning in science. Most explanations, even in physics, rely on idealized models.

Explanations by simulation resemble typical explanations in biology or the social sciences. The explanatory force of our traffic model does not rely on a fundamental law of nature. Nevertheless, in this case too we find general rules. The rules for updating the velocities of the cars are the same for all arrangements, that is, for all initial conditions. This traffic simulation, however, develops persuasive power because it provides a *mechanism*. This mechanism shows how the interaction of plausible behaviors of the components leads to the observed behavior of the system. By showing how surprising phenomena on the macro level follow from the interaction of components on a deeper level, simulations make our body of knowledge more coherent.

As far as explanations of complex systems are concerned, simulations provide new instruments for the reduction and understanding of emergent properties. Here I use the concept of emergence in the same way as it is understood by most scientists (also in the social sciences) and by some philosophers who believe that emergent properties could be reduced, at least in principle. In any case, simulations help to show how special phenomena on the macro level result from the dynamics on the micro level (e.g., the traffic jam as the result of the behavior of individual cars in our case study). We would say that an emergent property is explained when we can show by means of a computer simulation that, for a special set of boundary conditions, the emergent feature results from the combined interactions of the components.

That is an important step, especially for individualistic approaches in the social sciences.[5]

Very often, the mechanisms for producing emergent properties are self-organizing processes. As a rule, the behavior of a system cannot be predicted from the knowledge of what each component does in isolation (i.e., in more homogeneous surroundings, without the other components of the system). This conception of emergence emphasizes the pragmatic aspect of generating surprise (Casti 1997, 82, 92), whereas in many philosophical contexts emergent properties are defined from a more ontological point of view.

In addition, simulations can help to *identify* emergent structures. Working with computer simulations, we can see regular patterns on the screen, like a zoologist facing an exemplar of a previously unknown species. (Think of the crowded cells representing the traffic jam in our traffic simulation.) In many cases, there is a natural way of defining emergent properties with the help of visualized results of simulations of complex systems (see Nowak and Lewenstein 1996, 257). People believe in the existence of emergent properties, especially when they find these structures in many different runs using different values of the parameters.

Here again the advantage of the computer does not depend on new principles, but on the ability to compute quickly the consequences of a set of assumptions. The easy variation of assumptions allows us to produce a huge number of explanations in a short time, and the best explanation can then be selected. In this way the speed of computers allows the extension of the basis for inference to the best explanation.

In principle, there is nothing in a simulation that could not be worked out without computers. It is only because of their natural limitations that human beings are not able to carry out these tasks within a finite period of time. Of course, computers also have advantages in the pragmatics of science. The sort of insight and understanding provided by computer simulations is similar to that provided by simulations that do not rely on computers or by other ways of analyzing complex systems. Herbert Simon reminds us of model basins and wind tunnels, where small models are used to study large objects, and he shows that simulations predate the digital computer (Simon 1969, 14). Here again, the advantage of computers is their high speed. Most progress in understanding complex systems has been made possible by the great number of explanations that can be computed within a comparatively short time, for many "would-be worlds" and in the end for our real world.

Philosophical Conclusions

Some have argued that simulations are a new type of activity, somewhere between theories and experiments. Fritz Rohrlich, for instance, claims that computer simulations provide a qualitatively new and different methodology for the physical sciences and that this methodology lies somewhere between traditional theoretical physical science and its empirical methods of experimentation and observation (Rohrlich 1991; see also Fredkin 1990). Indeed, in some respects, simulations are similar to experiments. One can repeat a simulation (an experiment) with different parameters. And in both areas one can find the psychological aspect of surprising results. As Peter Galison (1996, 137, 138, 142) emphasizes, we can find similarities with the activities of researchers, especially in their more pragmatic aspects.[6] For example, "computational errors" are introduced by finite mesh size and other artifacts (see Hegselmann 1996), and the search for such errors is similar to the search for experimental errors. Simulations, however, cannot supplant the semantic and epistemological role of experiments.[7] Simulations are arguments, not experiences. (Sometimes we are surprised by the conclusions of our arguments!)

When I argue that we should not overstate the methodological revolution sparked by simulations, I must emphasize the substantial progress they have made possible, especially in sciences dealing with complex systems. Even though no one probably thought about simulations in writing the old philosophy of science textbooks, I believe that working with simulations can be reconstructed in traditional philosophy of science terms.

The speed of computers makes a quantitative difference in our way of doing science, which has always been restricted by our practical abilities and pragmatic limits. Today, computer simulations might have the same value as logarithmic tables had for Johannes Kepler. Each tool is particularly useful for particular purposes. Most of us are more readily capable of analyzing pictures than interpreting equations. Simulations and visualizations help to transcend the special "hardware limits" inherent in our scientific practice. As Paul Humphreys indicates, the development of tractable mathematics is one of the main features that drives scientific progress. Computers made a new class of mathematics tractable, and that is an additional source of progress in simulations (Humphreys 1991, 498).

There are some reasons why I do not believe that modeling and simulations introduce a new methodology that could be compared to the methodological revolutions of Galileo or Darwin. For example, we do not find a rejection of received aspects of methodology or an establishment of new goals for the scientific enterprise. We would have engaged in all the tasks now carried out by computers if we had had the appropriate tools before. We cannot find a special revolutionary point on the smooth transition from the slide rule to the workstation. The progress made by simulations concerns pragmatic, not principal, aspects of methodology.

I agree, computers are in general an extremely powerful new tool. Simulations as special applications of computers allow us to tackle many new problems. All the nonlinear processes and most classes of complex systems were simply out of the range that was accessible to science in the time before computers were available. In addition, the programming of computers itself stimulated new mathematical approaches, such as the development of cellular automata and further areas of discrete mathematics. Certainly new objects of research are brought into focus by scientists. But these new areas previously remained outside their focus because no one saw a chance to solve such problems without powerful computers.

I have learned much from computer scientists (see Simon et al. 1997; Valdés-Pérez 1994a, 1994b) about the progress of research in computational scientific discovery. For example, computer systems can discover patterns of sequences in chemical formulas of complex systems and even scientific laws. Again, these new heuristic approaches would also have been possible before, *if* we had been able to spend more time and resources on such approaches.

What marks the boundary between pragmatic progress and revolutionary methodological change? In order to find a substantiated answer to this question, we must develop a precise concept of methodological change. A survey of the history of theories of scientific method (see Laudan 1968) shows the richness and complexity of the historical development of methodological practice and thought. Considering this complexity, I restrict myself to the argument advanced here, that modeling and simulations show so many similarities to traditional elements of methodology that computer simulations should not be considered as a revolution of modern science.

In any case, even if the progress made possible by simulations is mainly in the pragmatic aspects of science, they are still an important new part of science. Modeling and simulations remind us of our special human limitations, which are revealed when we work with fundamental theories in order to understand complex systems. At the same time, they are new instruments enabling us to move beyond these boundaries.

NOTES

I gratefully acknowledge instructive discussions with and invaluable help received from Bjoern Haferkamp, Stephan Hartmann, Ulrich Kuehne, Meinard Kuhlmann, and Martin Carrier.

1. This cellular automaton model was introduced by Nagel and Schreckenberg (1992). See Schadschneider and Schreckenberg (1993), Schreckenberg et al. (1995), and their contributions to Wolf et al. (1996).

2. See Kaufmann and Smarr (1993) and Casti (1997) for popular accounts. Many examples from the social sciences can be found in Hegselmann et al. (1996).

3. See Rohrlich (1991, 516): "One could define computer simulation in a narrow sense as a method devoted entirely to the modeling of complex systems by means of the CA [cellular automata] syntax."

4. I thank Martin Carrier for clarification.

5. Simon (1981, Chapter 7): In general, many states are explained by reconstructing the process that produced the state. For an explanation of emergent properties in the social sciences, see Nowak and Lewenstein (1996, 256).

6. Galison also sketches some selected chapters of the history of simulations.

7. Like any argument, simulations can be extremely helpful for experimentalists (Hartmann 1995a, 139). They inspire experiments by showing the best choice of parameters and the most interesting places to look for new effects. In high-energy physics simulations imitate background radiation and help to pinpoint optimal detector locations.

REFERENCES

Achinstein, P. 1968. *Concepts of Science*. Baltimore: Johns Hopkins University Press.
Cartwright, N. 1983. *How the Laws of Physics Lie*. Oxford: Oxford University Press.
———. 1989. *Nature's Capacities and Their Measurement*. Oxford: Oxford University Press.

Casti, J. L. 1997. *Would-Be Worlds*. New York: John Wiley & Sons.

Fine, A., M. Forbes, and L. Wessels, eds. 1991. *PSA 1990*, vol. 2. East Lansing, Mich.: Philosophy of Science Association.

Fredkin, E. 1990. "Digital Mechanics." *Physics D* 45: 254–70.

Galison, P. 1996. "Computer Simulations and the Trading Zone." In P. Galison and D. Stump, eds., *The Disunity of Science*. Stanford, Calif.: Stanford University Press, 118–57.

Hartmann, S. 1995a. *Metaphysik und Methode*. Konstanz: Hartung-Gorre Verlag.

——. 1995b. "Simulation." In J. Mittelstrass, ed., *Enzyklopädie Philosophie und Wissenschaftstheorie*, vol. 3. Stuttgart: Metzler, 807–9.

——. 1996. "The World as a Process: Simulations in the Natural and Social Sciences." In Hegselmann et al. 1996, 77–100.

Hegselmann, R. 1996. "Cellular Automata in the Social Sciences: Perspectives, Restrictions, and Artefacts." In Hegselmann et al. 1996, 209–34.

Hegselmann, R., U. Müller, and K. G. Troitzsch, eds. 1996. *Modelling and Simulation in the Social Sciences from the Philosophy of Science Point of View*. Dordrecht: Kluwer.

Humphreys, P. 1991. "Computer Simulations." In Fine et al. 1991, 497–506.

Kaufmann, W., and L. Smarr. 1993. *Supercomputing and the Transformation of Science*. New York: Scientific American Library.

Kenny, A., ed. 1994. *The Oxford Illustrated History of Western Philosophy*. Oxford: Oxford University Press.

Kliemt, H. 1996. "Simulation and Rational Praxis." In Hegselmann et al. 1996, 13–26.

Latané, B. 1996. "Dynamic Social Impact: Robust Predictions from Simple Theory." In Hegselmann et al. 1996, 287–310.

Laudan, L. 1968. "Theories of Scientific Method from Plato to Mach." *History of Science* 7: 1–63.

Morrison, M. C. 1998. "Modelling Nature: Between Physics and the Physical World." *Philosophia Naturalis* 35: 65–85.

Nagel, K., and M. Schreckenberg. 1992. "A Cellular Automaton Model for Freeway Traffic." *Journal de Physique I* (France) 2: 2221.

Neelamkavil, F. 1986. *Computer Simulation and Modelling*, New York: John Wiley & Sons.

Nowak, A., and M. Lewenstein. 1996. "Modelling Social Change with Cellular Automata." In Hegselmann et al. 1996, 249–86.

Paczuski, M., and K. Nagel. 1996. "Self-Organized Criticality and $1/f$ Noise in Traffic." In Wolf et al. 1996, 73–85.

Rohrlich, F. 1991. "Computer Simulation in the Physical Sciences." In Fine et al. 1991, 507–18.

Schadschneider, A., and M. Schreckenberg. 1993. "Cellular Automaton Models and Traffic Flow." *Journal of Physics A: Mathematical and General* 26: L679–L683.

Schreckenberg, M., A. Schadschneider, and K. Nagel. 1995. "Zellularautomaten simulieren Straßenverkehr." *Physikalische Blätter* 52: 460–62.

Simon, H. 1969. *The Sciences of the Artificial.* Cambridge, Mass.: MIT Press.
——. 1981. *The Sciences of the Artificial.* 2nd ed. Cambridge, Mass.: MIT Press.
Simon, H. A., R. E. Valdés-Pérez, and D. H. Sleeman. 1997. "Scientific Discovery and Simplicity of Method (Editorial)." *Artificial Intelligence* 91: 177–81.
Stöckler, M. 1998. "On the Unity of Physics in a Dappled World: Comment on Nancy Cartwright." *Philosophia Naturalis* 35: 35–39.
Valdés-Pérez, R. E. 1994a. "Algebraic Reasoning about Reactions: Discovery of Conserved Properties in Particle Physics." *Machine Learning* 17: 47–67.
——. 1994b. "Conjecturing Hidden Entities by Means of Simplicity and Conservation Laws: Machine Discovery in Chemistry." *Artificial Intelligence* 65: 247–80.
Wolf, D. E., M. Schreckenberg, and A. Bachem, eds. 1996. *Traffic and Granular Flow.* Singapore: World Scientific.

Contributors

Robert Almeder
Department of Philosophy
Georgia State University

Thomas Breuer
Department of Mathematics
Fachhochschule Vorarlberg

Bernd Buldt
Department of Philosophy
University of Konstanz

Martin Carrier
Department of Philosophy
University of Bielefeld

Gordon N. Fleming
Department of Physics
Pennsylvania State University

Giora Hon
Department of Philosophy
University of Haifa

Paul Humphreys
Department of Philosophy
University of Virginia

Andreas Hüttemann
Department of Philosophy
University of Bielefeld

Peter McLaughlin
Department of Philosophy
University of Konstanz

Gerald J. Massey
Department of Philosophy
University of Pittsburgh

Deborah G. Mayo
Department of Philosophy
Virginia Polytechnic Institute and
    State University

Jürgen Mittelstrass
Department of Philosophy
University of Konstanz

Margaret Morrison
Department of Philosophy
University of Toronto

Alfred Nordmann
Department of Philosophy
University of South Carolina

Richard Raatzsch
Department of Philosophy
University of Leipzig

Nicholas Rescher
Department of Philosophy
University of Pittsburgh

Alex Rosenberg
Department of Philosophy
University of Georgia

Laura Ruetsche
Department of Philosophy
University of Pittsburgh

Hans Julius Schneider
Department of Philosophy
University of Potsdam

Manfred Stöckler
Department of Philosophy
University of Bremen

Michael Stöltzner
Vienna Circle Institute
Vienna, Austria

Index